Key Issues in Environmental Change

Preface to the series

The study of environmental change is a major growth area of interdisciplinary science. Indeed, the intensity of current scientific activity in the field of environmental change may be viewed as the emergence of a new area of 'big science' alongside such recognized fields as nuclear physics, astronomy and biotechnology. The science of environmental change is fundamental science on a grand scale: rather different from nuclear physics but nevertheless no less important as a field of knowledge, and probably of more significance in terms of the continuing success of human societies in their occupation of the Earth's surface.

The need to establish the pattern and causes of recent climatic changes, to which human activities have contributed, is the main force behind the increasing scientific interest in environmental change. Only during the past few decades have the scale, intensity and permanence of human impacts on the environment been recognized and begun to be understood. A mere 5000 years ago, in the mid-Holocene, non-local human impacts were more or less negligible even on vegetation and soils. Today, however, pollutants have been detected in the Earth's most remote regions, and environmental processes, including those of the atmosphere and oceans, are being affected at a global scale.

Natural environmental change has, however, occurred throughout Earth's history. Large-scale natural events as abrupt as those associated with human environmental impacts are known to have occurred in the past. The future course of natural environmental change may in some cases exacerbate human-induced change; in other cases, such changes may neutralize the human effects. It is essential, therefore, to view current and future environmental changes, like global warming, in the context of the broader perspective of the past. This linking theme provides the distinctive focus of the series and is mentioned explicitly in many of the titles listed overleaf.

It is intended that each book in the series will be an authoritative, scholarly and accessible synthesis that will become known for advancing the conceptual framework of studies in environmental change. In particular we hope that each book will inform advanced undergraduates and be an inspiration to young research workers. To this end, all the invited authors are experts in their respective fields and are active at the research frontier. They are, moreover, broadly representative of the interdisciplinary and international nature of environmental change research today. Thus, the series as a whole aims to cover all the themes normally considered as key issues in environmental change even though individual books may take a particular viewpoint or approach.

John A. Matthews (Co-ordinating Editor)

Titles in the series

Already published:
Cultural Landscapes and Environmental Change (Lesley Head, University of Wollongong, Australia)
Environmental Change in Mountains and Uplands (Martin Beniston, University of Fribourg, Switzerland)
Glaciers and Environmental Change (Atle Nesje and Svein Olaf Dahl, Bergen University, Norway)

Forthcoming:
The Oceans and Environmental Change (Alastair Dawson, Department of Geography, Coventry University, UK)
Pollution of Lakes and Rivers: a Palaeoecological Perspective (John Smol, Queen's University, Canada)
Atmospheric Pollution: an Environmental Change Perspective (Sarah Metcalfe, Edinburgh University, UK)
Biodiversity: an Environmental Change Perspective (Peter Gell, Adelaide University, Australia)
Climatic Change: a Palaeoenvironmental Perspective (Cary Mock, University of South Carolina, USA)

Natural Hazards and Environmental Change

Bill McGuire
Benfield Greig Hazard Research Centre,
University College London

Ian Mason
Department of Space and Climate Physics,
University College London

Christopher Kilburn
Benfield Greig Hazard Research Centre,
University College London

A member of the Hodder Headline Group
LONDON
Co-published in the United States of America by
Oxford University Press Inc., New York

First published in Great Britain in 2002 by
Arnold, a member of the Hodder Headline Group,
338 Euston Road, London NW1 3BH

http://www.arnoldpublishers.com

Co-published in the United States of America by
Oxford University Press Inc.,
198 Madison Avenue, New York, NY10016

British Library Cataloguing in Publication Data
A catalogue record for this book is available from the British Library

Library of Congress Cataloging-in-Publication Data
A catalog record for this book is available from the Library of Congress

ISBN 0 340 74219 4 (hb)
ISBN 0 340 74220 8 (pb)

1 2 3 4 5 6 7 8 9 10

Production Editor: Wendy Rooke
Production Controller: Martin Kerans
Cover Design: Mousemat Design

Typeset in 10/11½pt Palatino by Phoenix Photosetting, Chatham, Kent
Printed and bound in Malta by Gutenberg Press Ltd

Contents

Preface

Natural hazards are exacting an increasing toll across the planet, particularly in the developing world but also in the industrialised countries, and last year they affected one in thirty people. As we show later, hazards and a changing environment are intimately linked, and as the world enters a period of unprecedented environmental change that we ourselves are driving, prospects for a reduction in natural hazard impacts look bleak. A combination of growing vulnerability and global warming seems set to ensure that few will escape – in years to come – the direct or indirect consequences of flood, windstorm, landslide, wildfire and other geophysical hazards. Since the following chapters were written, two significant events have conspired to make the picture even more depressing. Early in 2001, the Intergovernmental Panel on Climate Change (IPCC) published its third assessment report. This revealed that the speed and impact of anthropogenic climate change is likely to be substantially greater than previously thought, with temperatures forecast to rise by up to 8 degrees Celsius by the end of the century. Hard on the heels of this gloomy assessment, the United States pulled out of the Kyoto Protocol for the reduction of Greenhouse gases, and despite attempts to push ahead with the proposed reductions at an international conference in Bonn the initiative currently appears to be dead in the water. With greenhouse gas emissions rising faster than the IPCC's 'business as usual' scenario, it is clear that it is now too late to avoid many of the more hazardous implications of global warming. Indeed, it is difficult to imagine certain elements of the international community taking the threat of contemporary environmental change seriously, without being shocked into action by events of a dramatic and unambiguous nature. Perhaps natural hazards will have a pivotal role to play here, and unless and until a hurricane obliterates Miami or wildfires rage through downtown Sydney, the dramatic consequences of our impact on the global environment will continue to be played down or disregarded.

Bill McGuire
Hampton, September 2001

Acknowledgements

The following are gratefully acknowledged: Dr Simon Day, of the BGHRC at UCL, for permission to include his most recent thoughts on the link between volcanic ocean island collapses and climate, and for permission to elucidate these in Figures 6.10 and 6.11; Dr John Grattan, of Aberystwyth University, for his Laki weather graphic (Figure 5.7); Dr Bill Murphy, of the University of Leeds, for permission to use his photograph of crown cracking at the Black Ven landslide (Figure 4.6); Dr Maria Carmen Solana, of the BGHRC at UCL, for permission to use her photograph of mudflow-damaged buildings in Honduras (Figure 4.11); Dr Steven Ward of the University of California, Santa Cruz for his graphic of tsunami generated by asteroid impact (Figure 7.8).

Abbreviations

AU	astronomical unit
CEI	Climate Extremes Index
CFB	continental flood basalt
CFC	chlorofluorocarbon
CRE	catastrophic rise event
DVI	Dust Veil Index
EAIS	East Antarctic Ice Sheet
ELA	equilibrium-level altitude
ENSO	El Niño–Southern Oscillation
GCM	general circulation model
GHG	greenhouse gas
IDNDR	International Decade for Natural Disaster Reduction
IPCC	Intergovernmental Panel on Climate Change
ISDR	International Strategy for Disaster Reduction
ITCZ	intertropical convergence zone
IVI	Ice Core Volcanic Index
K/T	Cretaceous/Tertiary
MCC	meso-scale convective complex
MCS	meso-scale convective system
MJO	Madden–Julian Oscillation
MOID	minimum orbit intersection distance
MPI	maximum potential intensity
NAO	North Atlantic Oscillation
NEA	near-Earth asteroid
NEC	near-Earth comet
NEO	near-Earth object
PDO	Pacific Decadal Oscillation
PDSI	Palmer Drought Severity Index
PFJ	polar front jet stream
PHA	potentially hazardous asteroid
PNA	Pacific/North American
QBO	Quasi-biennial Oscillation
SLP	sea-level pressure
SOI	Southern Oscillation Index
SST	sea surface temperature
TC	tropical cyclone

TCLV	tropical cyclone-like vortex
TOMS	Total Ozone Mapping Spectrometer
VEI	Volcanic Explosivity Index
WAIS	Western Antarctic Ice Sheet
WASA	Waves and Storms in the North Atlantic (group of researchers)

1

Natural hazards: an introduction

1.0 Chapter summary

Here we set the scene by introducing the most important issues relating to natural hazards and environmental change, which will be addressed in greater depth in later chapters. We define and describe natural hazards and summarise their impact on human society throughout history. The dynamic nature of our planet, and the fortuitous coincidence of the development of a global technological civilisation and a period of climatic and geological calm, are stressed, with the proviso that such benign conditions cannot prevail. We describe and explain the rising trend in natural hazard impacts in terms of a combination of changing climate and increasing vulnerability, the latter particularly in developing countries. Exacerbatory influences, such as increasing urbanisation, deforestation and the occupation of marginal land, are introduced in the context of raising the sensitivity of many developing-world communities to natural hazards. We examine environmental change in the nineteenth and twentieth centuries, primarily in the context of anthropogenic global warming, and address the consequences in terms of increasing natural hazard frequency and intensity.

1.1 What are natural hazards?

Although the nomenclature is sometimes ambiguous, natural hazards are usually defined as extreme natural events that pose a threat to people, their property and their possessions. Natural hazards become natural disasters if and when this threat is realised. Rapid-onset natural hazards, which form the focus of this book, can be distinguished from the often disastrous consequences of environmental degradation, such as desertification and drought, not only by their sudden occurrence but also by their relatively short duration. Natural hazards that are geophysical in nature, rather than biological, such as insect infestations or epidemics, arise from the normal physical processes operating in the Earth's interior, at its surface, or within its enclosing atmospheric envelope. Most geophysical hazards can be conveniently allocated to one or other of three categories: *geological* (earthquakes, volcanoes and landslides), *atmospheric* (windstorms, severe precipitation, temperature extremes, and lightning) and *hydrological* (floods and debris flows). Others, however, are less easy to pigeonhole. Tsunami, for example, can be regarded as hydrological hazards in the sense that they form and are transported within the hydrosphere. Their origin, however, is almost invariably geological, and usually the result of a large submarine earthquake. Similarly, wildfires and snow avalanches are not so easily compartmentalised. In the context of environmental change, *extraterrestrial hazards* must also be considered, with a large and growing body of evidence pointing to asteroid and comet impactors as major initiators of environmental change and associated extinctions throughout the geological record.

Increasingly, during the last half of the twentieth century, the impact of natural hazards on communities was exacerbated by human action, particularly urbanisation, changes in land use and agricultural practice, and deforestation. In 1998, for example, large-scale logging was a major contributory factor to the devastating flooding in the Yangtze basin of China. During the previous year, huge forest fires raged out of control across Sumatra and Borneo, when the seasonal rains that put them out in other years failed to materialise, generating an enormous pall of smoke over South-East Asia that lasted for many weeks. The term *na-tech* (for natural-technological) is increasingly being used to describe these complex disasters for which there is a recognisable and significant human contribution.

In this book we focus on current knowledge of the relationships between natural hazards and environmental change, gleaned both from the geological record and from studies of contemporary geophysical phenomena that are linked to significant short-term environmental changes, such as ENSO (El Niño–Southern Oscillation). We restrict ourselves to rapid-onset geophysical phenomena (Table 1.1), rather than epidemics or environmental degradation, although drought and desertification are briefly touched upon. Furthermore, although social and economic issues, particularly where they relate to increasing vulnerability, are addressed, we concentrate primarily on the science of natural hazards and environmental change. The idea that natural hazards can also trigger environmental change as well as occurring in response to it is particularly emphasised, and in this context low-frequency but globally significant hazards (McGuire, 1999) such as volcanic 'super-eruptions' and asteroid and comet impacts are included. Inevitably, the human dimension is also explored, both through the exacerbation of natural hazards as a result of human activities and in terms of increasing vulnerability due to multiplying populations and increasing urbanisation.

1.2 Natural hazards and environmental change

It is almost certainly no coincidence that modern human society has developed against a background of generally equable climate and geological calm. Evidence exists for catastrophic events, unprecedented today, occurring as recently as 7000 years BP – most associated with rapid post-glacial sea-level rise and its consequences. These include major, tsunami-generating landslides from continental margins, the emptying of gigantic glacial lakes, the breaching of land bridges and the filling of new marine basins such as the Black Sea. Furthermore, some support exists for episodic encounters between the Earth and cometary debris swarms as recently as 4000 years ago. For the past two millennia, however, the Earth's environment has been characterised by a general level of climatic and geological stability sufficient to ensure the growth of a technology-based global society. There has certainly been some climate variability due to changes in the Sun's irradiance; for example, the *Maunder Minimum* in sunspot activity in the late 1600s may have helped to produce the coldest phase of the 'Little Ice Age' at that time. There have also been short-term climatic perturbations due to large volcanic eruptions (e.g. El Chichón,

TABLE 1.1 The principal rapid-onset geophysical hazards

Geological	Atmospheric	Hydrological	Other
Earthquake	Windstorm	Flood	Tsunami
Volcanic eruption	Severe precipitation	Debris flow	Snow avalanche
Landslide	Lightning		Wildfire
	Temperature extremes		Asteroid/comet impact

Mexico, in 1982 and Pinatubo, the Philippines, in 1991), although no geological or climatological hazard has had a sufficiently global impact – over this period – to hinder or reverse the trend of growing human influence across the planet.

Notwithstanding contemporary planetary warming – whether anthropogenically attributable or not – the current period of relative environmental stability cannot be expected to last. Not only can a return to glacial conditions not be ruled out in the longer term, but also a global natural catastrophe – due to either a volcanic *super-eruption* (e.g. Rampino and Self, 1992) or a collision with a kilometre-scale impactor (e.g. Chapman and Morrison, 1994) – is not impossible within the next few tens of millennia, and is likely within the next 100,000 years. Furthermore, evidence from the Quaternary (the past 1.6 million years; Adams *et al.*, 1999) reveals that large-scale global and regional climatic change over the last two and a half million years occurred rapidly rather than incrementally. Over the past 150,000 years in particular (Taylor *et al.*, 1993), most significant climatic variations – involving regional changes in mean annual temperature of several degrees Celsius – appear to have occurred over periods ranging from centuries to as little as a few years. For example, the step-like switch from the cold *Younger Dryas* to the warmer Holocene 11,500 years ago took place over only a few decades (Dansgaard *et al.*, 1989; Taylor *et al.*, 1997). Similarly, the coldest phase of the Little Ice Age began with an abrupt (though much smaller) drop in temperature during the mid-1500s (IPCC 1996). The current period of global warming and changes to the global hydrological cycle (e.g. UK Meteorological Office, 1999; Harvey, 2000) is in fact only the latest of a series of major climatic events that have occurred during the Quaternary and late Tertiary (Fig. 1.1) (Maslin, 1998). A picture of a benign planet characterised by an unchanging or slowly changing environment has always, therefore, been a wildly inaccurate one. Viewed on a sufficiently long timescale, the Earth's natural environment is constantly changing, either in response to long-term astronomical, climatic or geological forcing, or as a consequence of discrete, catastrophic geological or astronomical events. Environmental change is the norm rather than the exception. Climate-related environmental modifications on a

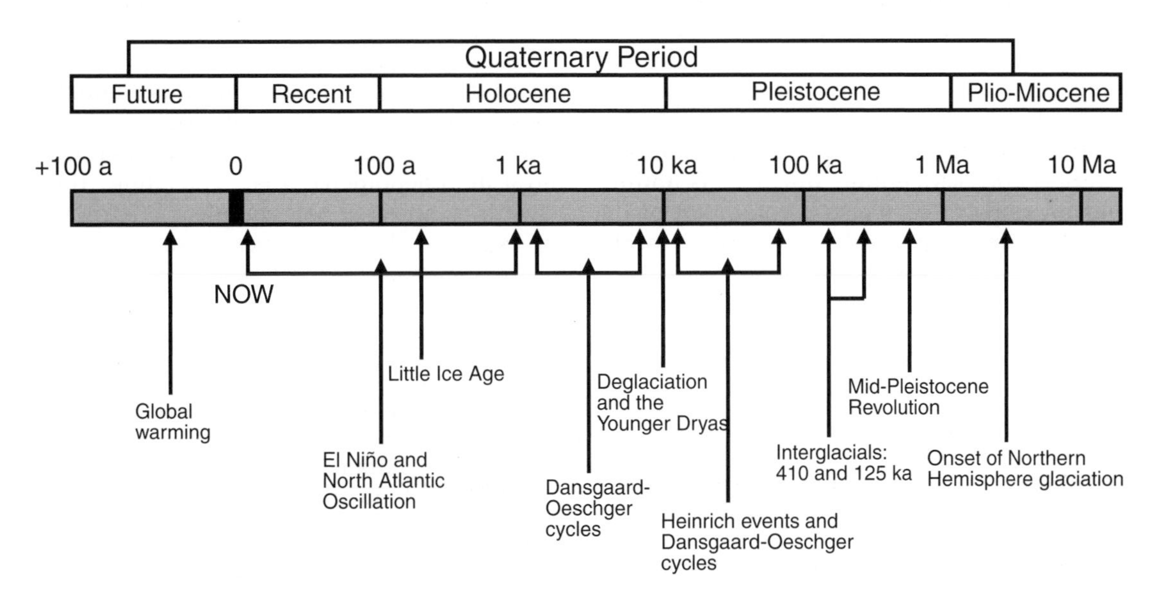

FIGURE 1.1 Major climate events of the Quaternary and late Tertiary. Sudden and dramatic environmental change has occurred throughout Earth's history and particularly during the Quaternary period. (Figure modified from Maslin, 1998, and Adams *et al.*, 1999)

decadal scale have been held responsible for various regional cultural changes in the past, including the onset of agriculture in the Middle East (Wright, 1993) and the collapse of the Classic period of Mayan civilisation (Hodell *et al.*, 1995). From the perspective of modern human society the important issue is how well it can adapt to the current period of environmental change characterised by planetary warming and its consequences, and whether society can change its ways so as to reduce the anthropogenic impact on the environment. In a sense, how well society copes will demonstrate just how well the human race is suited to a long and fruitful existence on a dynamic planet that is full of surprises.

One of the most interesting aspects of natural hazards and environmental change is their two-way relationship. On the one hand, major changes to the environment, such as the large and rapid sea-level rises that characterised the early Holocene, invariably result in an increase in geophysical phenomena that today would clearly be hazardous to human society. On the other, certain individual geological or extra-terrestrial natural phenomena are capable, provided they are of sufficient magnitude, of promoting – both drastically and rapidly – global environmental change. In some cases the connection between natural hazards and environmental change is complex and the establishment of a cause-and-effect relationship problematical. In recent decades an additional complication has become apparent as anthropogenic impacts have also become part of the equation from the local to the global scale. Overpopulation and overexploitation have led increasingly to degradation of the natural environment, which in turn has contributed to an increase in potentially destructive geophysical phenomena. For example, deforestation in many parts of the world has led to increased frequencies of flooding, debris flow formation and landsliding. Of most concern is the impact of rapidly rising concentrations of carbon dioxide (Fig. 1.2) and other 'greenhouse' gases in the atmosphere on surface temperatures and sea levels, and in turn on the frequency and severity of accompanying hazardous geophysical phenomena (e.g. Downing *et al.*, 1999a; Saunders, 1999). To some extent the implications for natural hazards of the current period of environmental change can be forecast. Clearly, for example, if sea levels are elevated, then the impacts of storm surges and tsunami are liable to be correspondingly more severe. Some consequences are, however, less certain: will we, for example, experience an increase in the frequency and severity of windstorms and other extreme meteorological events (see Chapters 2 and 3)? Yet others are speculative: could a warmer, wetter, climate promote the large-scale collapse of ocean island volcanoes and the formation of ocean-wide giant tsunami (see Chapter 6)?

Detailed figures are now available that record the impact of natural hazards on human society (see Section 1.3). Information on the absolute numbers of the rapid-onset geophysical phenomena that constitute these hazards is, however, less complete. It is known, for example, that every year – on average –

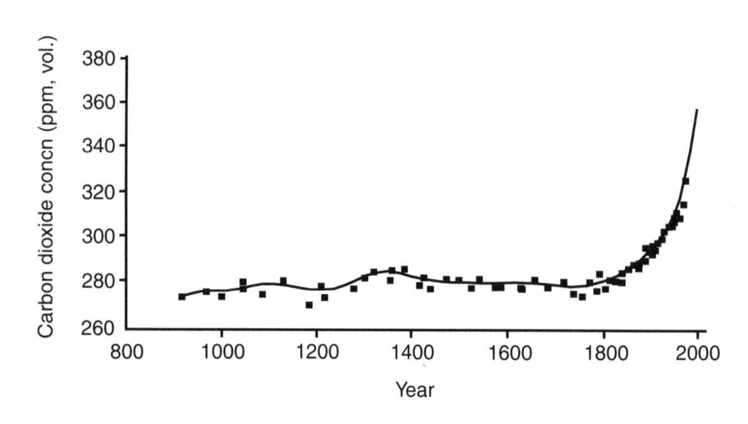

FIGURE 1.2 Atmospheric CO_2 concentrations since AD 1000 show a dramatic increase. Since the Industrial Revolution, CO_2 levels in the atmosphere have risen by around a third. This is broadly comparable to the increase in CO_2 emissions arising from the burning of fossil fuels. (Figure modified from IPCC, 1994)

approximately 50 volcanoes will be in eruption, around three thousand earthquakes will exceed Richter magnitude 6, and between 30 and 40 tropical cyclones will be generated. The absence of monitoring in poorly inhabited or uninhabited regions ensures, however, that the numbers of rapid-onset geophysical phenomena that are less easy to detect remotely, such as floods, severe precipitation and landslides, remain unknown. Consequently, in considering whether or not the incidence of natural hazards is increasing during the current warming episode, it is important to be aware that the picture we obtain from monitoring the numbers and severity of natural catastrophes – in other words, the *impact* of rapid-onset geophysical phenomena on human society – is unlikely to be the complete one. This is partly because many potentially damaging phenomena occur away from concentrations of human habitation, but also because increasing population and vulnerability in particularly hazard-prone regions are skewing the true picture.

If we look back in time, the situation is even more problematical. Prior to a few millennia or so ago we have little or no cultural record of natural-hazard impacts to provide even a rough guide to the incidence of geophysical phenomena, and in fact, only for the last half of the twentieth century is this catalogue anything like representative of the true situation. Similarly, the scientific record is not comprehensive enough to determine whether or not there has been a real change, over the same period, in the frequency, scale and type of potentially hazardous geophysical phenomena. Going back further in time, the Quaternary represents one of the most dramatic periods of fluctuating environmental conditions, certainly during the recent geological record. Even here, however, establishing a link between environmental change itself and rapid-onset geophysical phenomena remains speculative. Clearly, the incidence of flooding would have been orders of magnitude higher during the early Holocene due to the rapid melting of the continental ice sheets and the emptying of giant meltwater-filled lakes (e.g. Hillaire-Marcel *et al.*, 1981). Similarly, the loading and unloading of the continental margins due to the redistribution of planetary water accompanying glaciation–deglaciation cycles may have increased global seismicity (e.g. Nakada and Yokose, 1992), triggered the formation of large landslides along the margins (e.g. Bugge *et al.*, 1988), or even raised the level of volcanic activity (McGuire *et al.*, 1997). Such hypotheses, although intellectually reasonable, remain speculative, and few hard data exist for changes over time in the numbers and nature of rapid-onset geophysical phenomena, even during periods of dramatic environmental change. The corollary of this is that we have no precedent for determining whether and how the current warming phase will be reflected by a concomitant change in the absolute numbers and other parameters of natural hazards. This can be accomplished only by a combination of careful monitoring of contemporary geophysical hazards and by modelling based on the predicted consequences of the current warming, such as sea-level rise and changing meteorological conditions. The paucity of hard data relating to the hazard implications of global warming is one of the reasons why many scientists, environmentalists and others warn that human society is currently engaged in a planetary experiment, the consequences of which remain largely unknown.

1.3 Natural hazard impacts: a historical perspective

Rapid-onset geophysical hazards, in particular windstorms, floods and earthquakes, have always taken a devastating toll on human society, in terms of both damage to property and loss of life. During the past millennium it is estimated (Munich Re, 1999) that over 12 million lives were lost as a result of more than 100,000 natural catastrophes generated by rapid-onset hazards. Although there is no way of validating the figures, the reported death tolls for the most destructive events are shockingly large (Table 1.2). Between the thirteenth and fifteenth centuries, for example, severe storm surges in the North Sea are held

TABLE 1.2 Selected major natural catastrophes from the eleventh to the nineteenth century (Modified after Munich Re, 1999)

Year	Country	Event	Fatalities
1042	Syria	Earthquake	50,000
1202	Israel, Lebanon, Jordan, Syria	Earthquake	30,000
1268	Turkey	Earthquake	60,000
1281	Netherlands	Storm surge	80,000
1287	Germany (North Sea)	Storm surge	50,000
1290	China	Earthquake	100,000
1303	China	Earthquake	200,000
1362	Germany (North Sea)	Storm surge	100,000
1421	Netherlands	Storm surge	100,000
1498	Japan	Earthquake	41,000
1531	Portugal	Earthquake	30,000
1556	China	Earthquake	830,000
1622	China	Earthquake	150,000
1668	Azerbaijan	Earthquake	10,000
1668	China	Earthquake	50,000
1693	Italy	Earthquake	60,000
1721	Iran	Earthquake	40,000
1731	China	Earthquake	100,000
1737	India	Tropical cyclone	300,000
1739	China	Earthquake	50,000
1755	Portugal	Earthquake	30,000
1780	Iran	Earthquake	50,000
1780	Barbados, Guadeloupe, Martinique	Tropical cyclone	24,000
1815	Indonesia	Volcanic eruption and following famine	90,000
1822	Bangladesh	Tropical cyclone	50,000
1850	China	Earthquake	300,000
1852	China	Flood	100,000
1864	India	Tropical cyclone	50,000
1876	Bangladesh	Tropical cyclone	215,000
1881	Vietnam	Tropical cyclone	300,000
1882	India	Tropical cyclone	100,000
1883	Indonesia	Volcanic eruption and tsunami	36,000
1887	China	Flood	900,000

responsible for the loss through drowning of hundreds of thousands of lives. Tropical cyclones in the eighteenth and nineteenth centuries are reported to have killed millions of people in India and what is now Bangladesh, while the Great Henan Flood of 1887 in China may have taken 900,000 lives, making it the most lethal natural catastrophe on record. In the twentieth century alone the death toll may have been as high as 3.5 million (Table 1.3), and there is no sign, at present, of the situation improving. In fact, the past three decades have seen a considerable and worrying rise, both in the numbers of natural catastrophes caused by rapid-onset hazards (Fig. 1.3, p. 8), and in the accompanying economic and insured losses. This is partly due to the increasing concentration of both people and values in regions of high vulnerability to such hazards (e.g. Changnon *et al.*, 2000), as indicated by the accompanying rise in technological hazards. It may, however, also reflect an increase in

TABLE 1.3 Selected major natural catastrophes of the twentieth century (Information from Munich Re, 1999 and Swiss Re, 2000)

Year	Country (and region/city)	Event	Fatalities	Economic losses (US$ million)	Insured losses (US$ million)
1900	USA (Texas, Galveston)	Tropical cyclone	6000	30	
1906	USA (California, San Francisco)	Earthquake	3000	524	180
1908	Italy (Messina)	Earthquake	25,926	116	
1915	Italy (Avezzano)	Earthquake	32,610	25	
1920	China (Ganzu)	Earthquake	235,000	25	
1923	Japan (Tokyo–Yokohama)	Earthquake	142,800	2800	590
1931	China (Yangtzekiang)	Flood	140,000		
1936	Pakistan (Quelta)	Earthquake	35,000	25	
1938	USA (New England states)	Tropical cyclone	600	300	
1939	Chile (Concepción)	Earthquake	28,000	100	
1939	Turkey (Erzincan)	Earthquake	32,740	20	
1942	Bangladesh and India	Tropical cyclone	61,000		
1953	Netherlands and SE UK	Storm surge	1932	3000	
1954	China (Dongting)	Floods	40,000		
1959	Japan (Honshu)	Tropical cyclone	5100	600	
1960	Morocco (Agadir)	Earthquake	12,000	120	
1965	USA (Florida and the Gulf)	Tropical cyclone	75	1420	715
1970	Bangladesh (Chittagong)	Tropical cyclone	300,000	63	
1970	Peru (Chimbote)	Earthquake and landslide	67,000	550	14
1976	Western and central Europe	Windstorm	82	1300	508
1976	Guatemala (Guatemala City)	Earthquake	22,084	1100	55
1976	China (Tangshan)	Earthquake	290,000	5600	
1985	Mexico (Mexico City)	Earthquake	10,000	4000	275
1985	Colombia (Nevado del Ruiz)	Volcanic eruption	24,700	230	
1988	USA, Caribbean, Central America	Tropical cyclone	355	3000	800
1988	Armenia (Spitak)	Earthquake	25,000	14,000	
1989	USA (California, San Francisco)	Earthquake	68	6,000	950
1990	Western Europe	Windstorm	230	14,800	10,200
1991	Bangladesh	Tropical cyclone	139,000	3000	100
1991	Japan (Kyushu, Hokkaido)	Tropical cyclone	62	6,000	5200
1992	USA (Florida and the Gulf)	Tropical cyclone	62	30,000	17,000
1993	India (Maharashtra)	Earthquake	9475	280	
1994	USA (California, Los Angeles)	Earthquake	61	44,000	15,300
1995	Japan (Kobe)	Earthquake	6348	>100,000	3000
1998	China (Yangtze, Songhua)	Flood	3650	30,000	1000
1998	Honduras, Nicaragua	Tropical cyclone	9200	5500	150
1999	Turkey (Kocaeli)	Earthquake	19,118	20,000	1000
1999	Taiwan (Nantou)	Earthquake	3400	14,000	
1999	India (Orissa)	Tropical cyclone	15,000	2500	
1999	Venezuela	Landslides	~50,000	10,000	

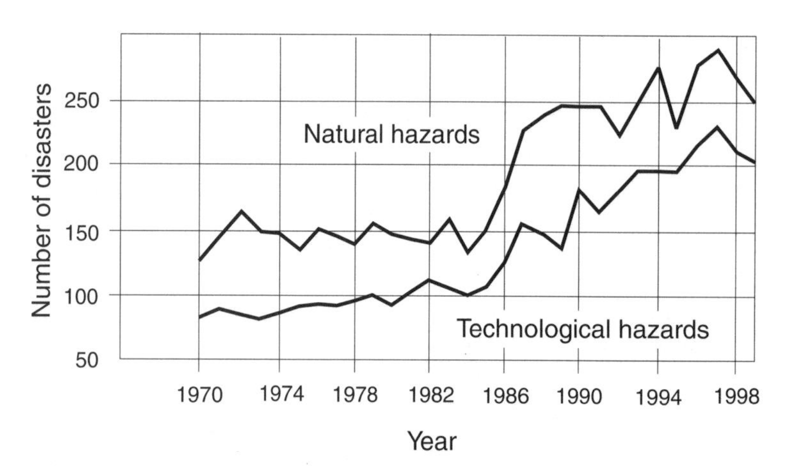

FIGURE 1.3 Trends in natural catastrophes, 1970–99. The numbers of both natural and technological catastrophes have increased dramatically over the past two decades. In the former case this is primarily due to increased vulnerability of populations in hazard-prone regions, but it may also reflect a real increase in the number and intensity of hazardous geophysical phenomena. (Data from Swiss Re, 2000)

Table 1.4a The ten most lethal catastrophes due to natural hazards in 1999 (Source: Munich Re, 1999)

Event	Dead or missing	Total loss (billion US$)	Percentage of GDP
Venezuela landslides	50,000	10	10.2
Turkey (Izmit) earthquake	19,118	20	10.1
India (Orissa) tropical cyclone	15,000	2.5	0.6
Taiwan (Nantou) earthquake	3,400	14	4.8
Mexico floods & landslides	1,300	0.2	0.04
Columbia (Armenia) earthquake	1,185	1.5	2
Turkey (Duzce) earthquake	834		
Pakistan and India tropical cyclone	751		
China (Yangtze) floods	725	3.4	0.3
Vietnam floods	662	0–1	0.3

TABLE 1.4b The ten most costly financial losses due to natural hazards in 1999 (Source: Swiss Re, 2000)

Event	Insured loss (US$ billion)	Total loss (US$ billion)	Percentage of GDP
Europe (Winter storm Lothar)	4.5	9	
Japan (Typhoon Bart)	3	3.3	
USA and Bahamas (Hurricane Floyd)	2.4	7	
Europe (Winter storm Martin)	2.2	4.5	
Turkey (Izmit earthquake)	2	20	10.1
USA (tornadoes)	1.5	—	
Taiwan (earthquake)	1	14	4.9
Australia (Sydney hailstorm)	1	—	
USA (snowstorm)	0.8	1	
Europe (Winter storm Anatol)	0.5	2	

extreme meteorological events that are arguably linked to the current period of global climate change (e.g. Karl *et al.*, 1995b; Alexandersson *et al.*, 1998).

Since the 1950s the numbers of major natural catastrophes have increased fourfold, while resulting economic losses have risen by a factor of 14. Owing to increasing insurance density in vulnerable areas, insured losses have risen at twice this rate and the US$100 billion single insured loss due to a rapid-onset hazard cannot now be far off. The year 1999 was a particularly bad one, and it provides a snapshot of the type, scale and impact of natural hazards with which modern society is faced. Inevitably, vulnerable developing countries once again bore the brunt in terms of injury, loss of life and damage to infrastructure (Table 1.4a; International Federation of Red Cross and Red Crescent Societies, 2000). Natural hazards took close to 100,000 lives, with heavy-rain induced debris flows and landslides in Venezuela accounting for perhaps half. In addition, almost 20,000 lives were lost in a severe earthquake in the Kocaeli region of north-west Turkey, while a further 15,000 died in tropical cyclone 05B, which devastated Orissa in north-east India. The most costly disasters, as might be expected given the higher wealth concentration, occurred mostly in developed countries (Table 1.4b), and resulted largely from tropical cyclones hitting Japan and the US east coast, severe winter storms pounding northern Europe, and earthquakes in Turkey and Taiwan. Economic losses of around US$80 billion and insured losses totalling approximately US$18 billion made 1999 the third (after 1995 and 1998) most costly year ever in terms of damage due to natural hazards.

Examination of the geographical breakdown of major natural catastrophes during 1999 (Fig. 1.4) reveals at a glance which regions of the planet are most impacted by natural hazards, although this is a function as much of vulnerability as it is of hazard distribution. For example, North America scores highly mainly because of its large geographical area and the extensive range of hazards to which it is exposed. The high numbers of natural catastrophes that affected South Asia, South-East Asia and South and Central America over the past century are largely, however, a function of the exacerbation of exposure to hazards by particularly high vulnerability. The picture presented in Fig. 1.4 is also skewed because it represents a geographical breakdown of natural catastrophes as defined by a major financial institution. To be included in *Sigma*, the Swiss Reinsurance company's annual catalogue of natural catastrophes, economic losses of US$66 million (or insured losses of US$33 million)

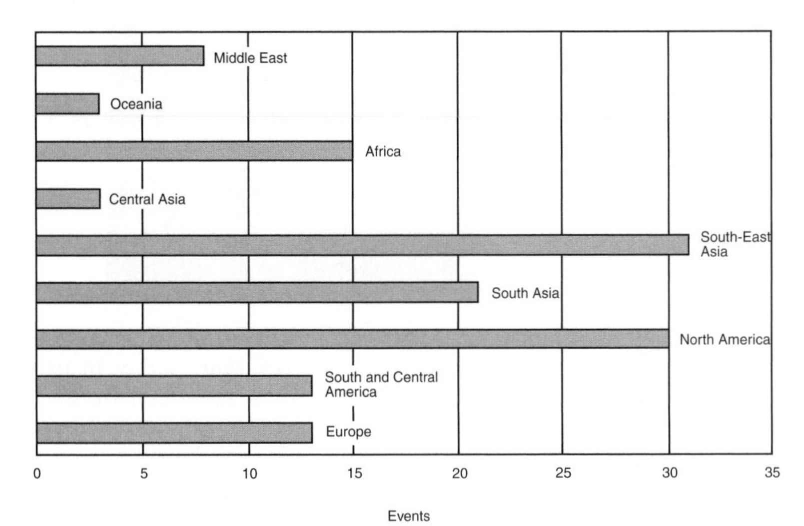

FIGURE 1.4 The geographical breakdown of natural catastrophes in 1999 shows that both developed and developing countries are affected. In the former case, high insured and economic losses are a characteristic feature, while in the latter a high death rate is the parameter which most often defines a natural 'catastrophe'. (Data from Swiss Re, 2000)

must be incurred, at least 20 people have to die, or 2000 become homeless. This ensures that many non-lethal but costly natural 'disasters' in Europe and North America can dramatically swell their share of a year's disastrous events. In contrast, the great majority of natural disasters in developing countries in South Asia, South-East Asia and elsewhere are characterised by high death rates and relatively small economic and insured losses.

1.4 A natural hazards primer

Some hazardous phenomena, such as flooding, are widespread, but many are more or less constrained by weather patterns or geology. With some notable exceptions, geological hazards are confined to the margins of the plates that comprise the crust and uppermost mantle. Active volcanoes are associated with both constructive plate margins, such as the Mid-Atlantic Ridge and East Pacific Rise, where new oceanic lithosphere is being created, and destructive margins (subduction zones), where oceanic lithosphere is consumed within deep submarine trenches. Major volcanic chains parallel such trench systems along the length of western South America, through southern Alaska, the Aleutian Islands and the Kamchatka Peninsula (eastern Russia), and down the island archipelagos – such as Japan and the Philippines – that border the western Pacific. Volcanoes associated with constructive margins tend to be submarine and pose little hazard. In some places, however, sub-aerial edifices have been constructed, for example at Tristan da Cunha and the Azores, and have hosted destructive eruptions, at Tristan da Cunha as recently as 1961. Most of the world's hazardous eruptions occur at volcanoes associated with subduction zones, where high magma viscosities contribute to high gas overpressures and violent explosive activity. The principal destructive hazards here tend to be pyroclastic flows and debris flows rather than the effusions of lava that characterise most constructive-margin volcanoes. Recent damaging eruptions at subduction-zone volcanoes have occurred in 1982 (El Chichón, Mexico), 1985 (Nevado del Ruiz, Colombia), 1991 (Pinatubo, the Philippines, and Unzen, Japan), 1994 (Rabaul, Papua New Guinea), and from 1995 to the present (Soufrière Hills, Montserrat, Caribbean) (Fig. 1.5). Some active volcanic provinces are remote from plate margins and owe their origin to mantle plumes within which decompressional melting leads to often voluminous eruptions. In oceanic environments, such plumes underlie both Iceland and Hawaii, where damaging eruptions of fluid basaltic lavas occur periodically. Plumes located beneath continental crust feed active volcanic systems in the East African Rift and the Yellowstone region of the north-western United States. In the latter case, remelting of granitic crust by plume-sourced basaltic magmas has resulted – over the past two million years – in some of the largest and most destructive explosive eruptions ever recorded. Current estimates of the numbers of

FIGURE 1.5 Eruption of the Soufrière Hills volcano (Montserrat, Caribbean) continues today. In 1996 and 1997, pyroclastic flows generated by lava-dome collapse and explosive eruptions destroyed the capital, Plymouth, and led to the evacuation of 70 per cent of the island's population. (Photo © Bill McGuire)

active volcanoes are not well constrained. Although just over 1500 volcanoes have erupted during the Holocene (Simkin and Siebert, 1994), many pre-Holocene volcanoes could still have the potential to erupt again. Consequently, a figure of around 3000 active volcanoes is probably more accurate. Of these, however, many have been long dormant, and evidence of historic activity exists for fewer than 600. In any one year, around 50 volcanoes are active. Some, such as Etna and Stromboli (Italy) or Kilauea (Hawaii), are almost continually in eruption, while others may have become reactivated following decades or centuries of dormancy.

Over 500,000 earthquakes are recorded every year, many of which are too weak to be detected other than by seismometers. Low-energy tremors can occur on fault systems anywhere on the planet, and may even be associated with deep mine workings in largely aseismic countries such as South Africa. Large and potentially destructive earthquakes are, however, broadly – although not solely – confined to either destructive plate margins or to conservative margins (such as that represented by the San Andreas Fault) where two plates slide past one another. Around 100 earthquakes a year exceed a Richter magnitude of 6.5 and are therefore capable of causing significant damage if they impact upon urbanised centres that are poorly prepared. On average, however, only one or two major quakes of the order of Richter magnitude 8 occur annually, often remote from centres of high population. Because strain accumulation and release rates in rock masses vary from one region to another, the frequency of movement along seismogenic faults is also variable. In northern and southern California, earthquake-related displacements along the San Andreas Fault and related structures occur several times a year, and major earthquakes of the order of Richter magnitude 7 affected inhabited areas (San Francisco and Los Angeles respectively) in both 1989 (Loma Prieta) and 1994 (Northridge). Similarly, segments of the North Anatolian Fault in Turkey shifted every few decades during the twentieth century, generating serious earthquakes, the latest of which caused severe destruction to the Izmit region in August 1999. Strain accumulation on deep faults remote from plate margins may also eventually be released, triggering large earthquakes. Here, however, the accumulation process typically takes centuries, resulting in long recurrence intervals between seismic events. Major *intraplate* earthquakes of this type occurred in the New Madrid region of Missouri in 1811 and again the following year, and similar events in the future pose a serious threat to the city of Memphis and adjacent urban centres. In fact, more than 20 notable intraplate earthquakes have occurred in North America over the past 200 years, some – such as those at Charleston (South Carolina) in 1886 and Hebgen Lake (Wyoming) in 1959 – causing loss of life and significant damage. From a hazard impact viewpoint, such infrequent intraplate quakes may be disproportionately damaging and disruptive, owing to less effective and less stringently enforced seismic building codes and to low public awareness. Such a situation is currently being addressed in the north-west United States, where only over the past decade has historical evidence for infrequent major earthquakes associated with the offshore Cascadia Subduction Zone come to light. Around 300 years ago a quake estimated to have been in excess of Richter magnitude 8 generated a series of Pacific-wide tsunami that seriously impacted on the Seattle region. A similar-sized earthquake today would cause severe damage and disruption to the Seattle–Tacoma urban centre and to the coastlines of Washington State, British Columbia and Oregon.

While tsunami are not, strictly speaking, geological hazards, their close link with earthquakes makes it appropriate to address them here. Most tsunami are generated by large submarine earthquakes in which enormous amounts of energy are impulsively transmitted to the sea floor. The most effective *tsunamigenic* earthquakes are not necessarily those with the highest Richter magnitude, although they tend to equal or exceed magnitude 6.5. More importantly, the earthquake needs to be shallow (generally less than 30 km) – so that most of the energy penetrates upwards into the water and is not absorbed by overlying rock – and should involve the vertical displacement of a large area of the sea floor. In the largest earthquakes, such as Alaska 1964, in excess of tens of thousands of square kilometres may be displaced. The size and destructive capacity of a seismogenic tsunami may be enhanced if the source earthquake triggers submarine landslides that may substantially add to the displacement of seawater. Such an event has been recognised as contributing to the formation of the larger than expected tsunami that devastated the Aitape area of Papua New Guinea in July 1998 taking around 3000 lives (Tappin *et al.*, 1999). Tsunami may be formed in a number of other ways, including non-seismogenic submarine landslides, sub-aerial landslides into the ocean or other body of water, and volcanic eruptions. In the latter context, the 1883 explosive eruption of Krakatoa (Indonesia) was responsible for one of the most lethal tsunami, with waves in excess of 15 m high taking around 36,000 lives along the shores of Java and Sumatra. The destructive capacity of tsunami is a consequence of a number of parameters. First, they can travel at velocities in excess of 500 km/h across entire ocean basins with little loss of energy. Second, they are barely detectable in deep water, where wave heights are commonly less than a metre. Third, because typical wavelengths are of the order of hundreds of kilometres, inundation of a coastline by a tsunami may last for tens of minutes – sufficient time to cause massive destruction. As might be expected, their close link with earthquakes ensures that most destructive tsunami – something of the order of 85 per cent – are sourced within the subduction zones that circle the Pacific Basin. Here around 400 tsunami have taken over 50,000 lives during the past 100 years alone. The generally aseismic nature of the Atlantic Basin contributes to its hosting less than 2 per cent of recorded tsunami. Worthy of note, however, are those generated during the 1755 Lisbon earthquake and an earthquake-related submarine landslide off Newfoundland in 1929. The Atlantic does, however, have the potential to generate very large, if infrequent, non-seismogenic tsunami. These are formed primarily as a result of the large-scale collapse of masses of accumulated sediment around the continental margins (for example, the Storegga Slide off Norway ~ 7000 BP) (e.g. Bugge *et al.*, 1988; Dawson *et al.*, 1988), or to the formation of giant landslides on oceanic island volcanoes such as those making up the Canary and Cape Verde archipelagos (e.g. Carracedo *et al.*, 1999; Day *et al.*, 1999).

Although landslide hazards are ubiquitous, occurring in all geomorphological and geological environments across the planet, their association with unstable, elevated terrain and earthquakes ensures that many of the most damaging occur at destructive plate margins. In terms of impact, landslides often tend to be undervalued. This is largely because their formation is often consequent upon another hazardous event, such as an earthquake, volcanic eruption or period of excessive precipitation. In Japan, for example, it is estimated that over half of all earthquake-related deaths result from landslides triggered by severe ground-shaking. An earthquake also caused one of the largest single landslide disasters of recent times. In 1970 a strong earthquake in Peru initiated the collapse of the peak and associated glacier of the Nevados Huascaran mountain, generating a landslide that within minutes killed 18,000 people and buried several towns. Countless lethal and destructive landslides and associated debris flows have also been triggered by exceptionally heavy rainfall. In 1998, torrential precipitation from slow-moving Hurricane Mitch is estimated to have mobilised over 1,000,000 landslides in Nicaragua and Honduras, while during 1999, landslides and debris flows generated by a similar storm-related deluge may have taken in excess of 50,000 lives in Venezuela. Less frequently, landslides may also be associated with volcanoes. In 1980 a mass of fresh magma emplaced within the recently reactivated Mount St Helens (Washington State, USA) triggered the collapse of the north flank, generating a catastrophic landslide and a major explosive eruption that lasted for several hours. More recently, in 1998, heavy precipitation associated with the aforementioned Hurricane Mitch led to collapse of part of the retaining wall around the crater lake of the

Casitas volcano in Nicaragua, forming a landslide and debris flows that took over 3000 lives.

Atmospheric hazards are free of the geophysical constraints that confine volcanic eruptions, earthquakes, tsunami and landslides to restricted geological environments. Consequently, windstorms and severe precipitation are much more widespread, although particular types of event are often limited by weather patterns to certain regions of the globe. Of all the atmospheric hazards, most destruction and loss of life is wrought by tropical cyclones, which occur at low latitudes in all three major ocean basins. Because they impact on major concentrations of wealth in Japan, the Caribbean, and the Gulf and eastern states of the USA, tropical cyclones have caused some of the largest economic losses on record. Hurricane Andrew, which devastated southern Miami in 1992 (Fig. 1.6), for example, was – at a cost to the economy of US$32 billion – the second most expensive disaster in US history. In the developing world, Bangladesh and India, together with the poorer countries of South-East Asia, Central America and the Caribbean, are particularly badly affected by cyclones, with Hurricane Mitch (Honduras and Nicaragua, 1998) and cyclone 05B (Orissa, India, 1999) alone taking over 25,000 lives between them. Between 40 and 50 tropical cyclones are generated each year, around 60 per cent of which occurred in the Pacific Basin. Fortunately, most cyclones blow themselves out before making landfall, thereby posing a threat to shipping but little else.

FIGURE 1.6 Hurricane Andrew devastated southern Miami in 1992, costing the US economy US$32 billion, and making the event the second most expensive natural disaster in US history. (Photo: courtesy of the United States National Oceanic and Atmospheric Administration/Department of Commerce)

Rarely, cyclones may retain their destructive potential even when they travel beyond the tropics, and in 1938 a rapidly moving hurricane took over 2000 lives in New England. Wind velocities close to those encountered in tropical cyclones may also characterise intense low-pressure weather systems at higher latitudes (extratropical cyclones). Such storms caused severe damage in north-western Europe in October 1987 and January/February 1990, and more recently late in 1999 and October 2000.

Of far more limited extent than tropical or extratropical cyclones are tornadoes. Owing to wind velocities that may be up to 500 km/h, however, local destruction can be total. Most damaging tornadoes are confined to the United States, and particularly to the 'tornado alley' states of Oklahoma, Kansas and Texas, although damaging tornadoes are not uncommon in South Africa, Australia and other continental regions characterised by convection-driven thunderstorms.

In addition to wind, severe precipitation, in the form of rain, hail or snow, may also prove to be extremely destructive. Localised deluges and hailstorms with the potential to devastate crops and cause major damage to property can be generated anywhere during major convective thunderstorms, but the mid-western United States, South Africa and Australia are particularly affected. Most recently, a major hailstorm that hit Sydney in April 1999 caused damage costing over US$1 billion. Bouts of severe precipitation are often associated with lightning strikes that can damage electrical installations and initiate wildfires where particularly dry conditions exist. Blizzards and ice storms (involving freezing rain) have also been responsible for major disruption and large economic losses, particularly in North America. In January 1998, for example, over 20 people were killed and economic losses totalled over US$2.5 billion when large areas of eastern

Canada and the United States were impacted by a severe ice storm. Unexpected blizzards are also responsible for loss of life in developing countries characterised by severe winters or high altitudes. In December 1999, for example, nearly 300 people were killed by extreme cold in the Indian state of Bihar. In mountainous terrain, such as the Alps and Himalayas, blizzard conditions often lead to dangerous accumulations of snow that collapse to form avalanches. In January and February 1999, particularly heavy snowfall over the whole of the Alps triggered a number of avalanches that killed over 70 people and caused damage of the order of US$ 1billion.

Even without excessive snowfall, extreme low temperature events can still constitute a hazard, causing frost damage to crops and pipes, and deaths through hypothermia. Extremes of high temperature, especially multiday heatwaves accompanied by high humidity, can also be hazardous, causing deaths by heat stress.

Flooding is the most ubiquitous of all natural hazards, taking lives and causing damage in virtually every country on an annual basis. The most severe flooding events in recent times have occurred along the coasts of low-lying countries such as Bangladesh, and on the floodplains of major river systems including the Mississippi–Missouri (USA, 1993), the Rhine (Germany, France, Belgium and the Netherlands 1995), and the Yangtze and neighbouring rivers (China, 1996 and 1998). Smaller river systems also experience serious flooding, however, and the severe flooding that occurred on relatively minor rivers across England and Wales during the autumn of 2000 caused billions of pounds worth of damage and widespread misery and chaos. Floods are more disruptive than any other hazard, in recent years regularly affecting over 100 million people and killing over 20,000 annually. Notwithstanding serious flood events in Mexico (1999) and Mozambique (2000), South and South-East Asia regularly bear the brunt of the world's flooding. In South Asia, Bangladesh and India suffered seven major flood events during 1999, mainly due to storm surges accompanying tropical cyclones, making over 1.5 million people homeless and taking around 17,000 lives. In addition to coastal flooding accompanying tropical storms, Asia also suffers as a result of the monsoon phenomenon, which involves 90 per cent of annual precipitation falling in only 5 months. Torrential rainfall, exacerbated by drought and deforestation in the surrounding countryside, rapidly leads to overspilling of major river systems, such as the Ganges–Bramaputra (Bangladesh), the Mekong (Vietnam) and the Yangtze (China), and inundation of their floodplains. China, where half a billion people live on floodplains, is particularly vulnerable, with over 200 million people affected and 5 million homes destroyed by severe flooding of the Yangtze, Yellow and Xi Jiang rivers in 1996.

Despite being far less frequent, hazards from beyond the Earth's atmospheric envelope must also not be ignored. Collisions between the Earth and asteroids or comets pose a serious threat to human civilisation owing to their ability to cause global devastation and to dramatically modify the environment. Impacts from objects in the 50 m size range – large enough to destroy a city or small country – occur on a timescale of decades, the best known being the Tunguska (Siberia) event of 1908. Of greater concern are objects of 1 km or more, which have the potential to wipe out over a billion people and trigger an *impact winter*. Following observation of the impact of the fragments of Shoemaker–Levy on the planet Jupiter in 1994, the threat from space has been taken far more seriously, and considerable efforts are currently being made to locate and log the orbits of all potentially threatening objects that approach the Earth.

1.5 Recent environmental change

Studies of surface temperature change, utilising a range of proxy measurements, reveal that the planet has been significantly warmer during the past 70 years than at any time in the past millennium (Jones *et al.*, 1998). Furthermore, the warming has accelerated in the past three decades, and 14 of the 15 warmest years on record have occurred since 1980, with 1998, followed by 1997 and 1995, being the warmest ever. Mean global temperatures are now higher than at any time since AD 1400. Reliable global mean surface temperature records, extending back to 1860 (Fig. 1.7) reveal that the planet has experienced a temperature rise of 0.6–0.7 degrees Celsius since the end of the twentieth century, and a rise of 0.4 degrees Celsius over the past three decades or so. Variability of the solar output has been shown to have made only a small (around 15–20 per cent) contribution to global warming over the past century and a half (Kelly and Wigley, 1992; Schönwiese *et al.*, 1994). Recent planetary warming has occurred mainly during two periods – between 1910 and 1940, and again since 1970 – and scientific consensus currently holds that the latter warming is primarily anthropogenic in nature (e.g. Tett *et al.*, 1999; Fig. 1.8). One of the exceptional aspects of current global warming is the rate of temperature rise. By the end of this century, the global average temperature may be up to 4 degrees Celsius higher than at the end of the twentieth century. This compares with a temperature change of 4–5 degrees Celsius between the end of the last ice age and the present interglacial period, but this post-glacial warming occurred over a period of 10,000 years. The Earth is warmer now than it has been for over 90 per cent of its history, and by 2100 the planet may be warmer than at any time during the past 150,000 years (Saunders, 1999).

Accompanying the warming process are a number of additional effects, all of which have the potential to contribute towards an increase in the threat posed by natural hazards. Most significantly, global sea level rose during the twentieth century at a greater rate than at any time in the past millennium – by between 10 and 25 cm (Douglas, 1997). The rise is explained partly by melting of ice caps and mountain glaciers and partly by thermal expansion of the oceans (Warrick *et al.*, 1996). The increased risk of flooding resulting from rising sea level is compounded by increased global precipitation over the past hundred years, due to surface warming raising evaporation rates and increasing the amount of water vapour in the atmosphere. Dai *et al.* (1997) report a 2 per cent

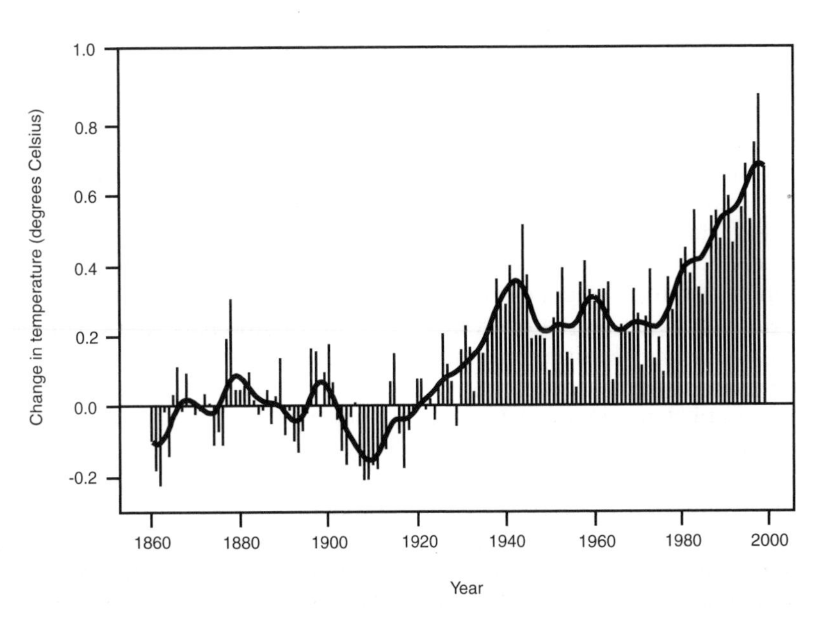

FIGURE 1.7 Global mean surface air temperature records for each year since 1860 are shown, relative to that at the end of the nineteenth century, together with a smoothed curve. The planet as a whole has experienced a temperature rise of 0.6–0.7 degrees Celsius since the beginning of the twentieth century, and a rise of 0.4 degrees Celsius over the past three decades or so. (UK Met Office, 1999)

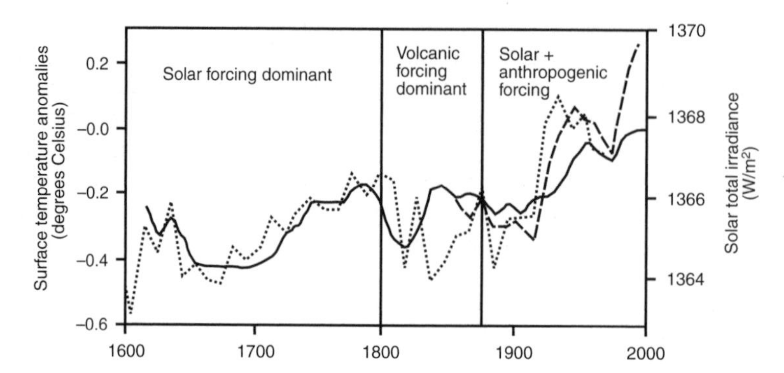

FIGURE 1.8 Comparison of the reconstructed solar total irradiance with the reconstructed Northern Hemisphere surface temperature record for the period 1600–1995. Surface temperature palaeo-reconstructions (continuous line) and instrumental records (dashed line) and reconstructed solar total irradiance (dotted line) are shown. Variations in solar radiation alone can explain most of the pre-Industrial Revolution, long-term temperature changes, but cannot account for the rapid temperature rise characteristic of the twentieth century. (After Lean, 1997)

rise in global precipitation during the twentieth century, while Karl *et al.* (1997) propose that in mid- to high latitudes the rise could be over 10 per cent. From the hazard point of view, however, it is not necessarily the total amount of annual precipitation that is important but its intensity. In this regard, Karl *et al.* (1995a) report a trend towards a greater proportion of rainfall taking the form of extreme precipitation events (greater than 50 mm per day) for the United States (1911–92), China (1952–89) and the former Soviet Union (1935–89). Reinforcing this observation, Karl and Knight (1998) showed that for the United States, the proportion of annual rainfall accounted for by the wettest 10 per cent of days increased from 36 per cent to 40 per cent, a rise that may have been a contributory factor in the disastrous Mississippi floods of 1993 (Karl and Haeberli, 1997).

Of considerable importance, from a hazard point of view, is establishment of a relationship between global warming and the number and intensity of windstorms, particularly tropical cyclones. As the formation and intensity of the latter are strongly related to sea surface temperature (SST) (e.g. Saunders and Harris, 1997; Henderson-Sellers *et al.*, 1998), it might be expected that a warmer ocean – resulting from general planetary warming – would cause more frequent and more intense tropical cyclones. Modelling, however, shows that such relationships are themselves likely to be affected by climate change, preventing any such simple analysis (Henderson-Sellers *et al.*, 1998). Certainly, evidence to support such a trend has not been forthcoming. In fact, for the Atlantic Basin, Landsea *et al.* (1996) report a weak *downward* trend in both the intensity and the frequency of the strongest hurricanes, while for the western North Pacific region Chan and Shi (1996) recognise a fall in tropical cyclone frequency from the 1960s to 1979, followed by a rise. The IPCC (1996) concluded, however, that on the basis of the limited evidence available, no obvious trends were discernible in either hurricane global numbers or strength. Most recently, a study by Mark Saunders and Frank Roberts (Saunders, 2001) of the Benfield Greig Hazard Research Centre at University College London has revealed a more confusing picture. They show that for the three main Northern Hemisphere ocean basins combined, the numbers of both hurricane strength and intense hurricane strength tropical cyclones show a broad linear upward trend over the period 1969–99. They also reveal, however, that a statistical best fit to the data indicates that tropical cyclone numbers have been falling since the early 1990s. Any potential link between global warming and tropical cyclones remains, therefore, unconfirmed, and observed trends may or may not result simply from natural multidecadal variability. The situation is equally confusing for extratropical windstorms, with Lambert (1996) proposing a sharp rise in Northern Hemisphere intense winter extratropical cyclones after 1970 in the Pacific, but Palutikof and Downing (1994) reporting no long-term trend in windstorm frequency and severity for Europe. One cause of the discrepancies may be inhomogeneities in the data. More recently, Alexandersson *et al.* (1998), using a more homogeneous data set, find an increase in winter wind speeds over the northern United Kingdom and Norway during the 1980s and 1990s, but the jury remains out on whether this can be unambiguously linked to global warming.

On the basis that economic losses caused by extreme weather attributable to the severe ENSO of 1997–98 exceeded US$20 billion, any link with the frequency and duration of this phenomenon is of crucial importance in terms of the impact of global warming. El Niño (and its cold-episode sister, La Niña) events have been recognised as far back as 15,000 years ago (Rodbell *et al.*, 1999), and they represent the strongest interannual climate signals on the planet. An El Niño is heralded by a significant rise in sea surface temperatures in the eastern tropical Pacific, and La Niña by a corresponding cooling. Both phenomena involve a perturbation of the ocean–atmosphere system that has knock-on effects across the planet in the form of changes in temperature, precipitation patterns and intensity, and cyclone numbers and strength. Worryingly, the frequency of El Niño events has increased since 1976, at the expense of La Niña. Trenberth and Hoar (1997) suggest that this increase is unlikely to result solely from natural variability, although this conclusion is disputed (Harrison and Larkin, 1997). Most of the 1990s were characterised by El Niño conditions, with the longest El Niño on record lasting from 1990 to 1995 followed by one of the strongest ever in 1997–98.

Climate change in the twentieth century was accompanied by many other anthropogenic changes to the environment that have dramatically increased the impact of natural hazards on society. A population increase from well under 2 billion to 6 billion since 1900 and accompanying global industrialisation have ensured accelerated exploitation of resources that has modified the environment of the planet at least as much as changing climatic conditions over the period, and has contributed to a more hazardous world. Deforestation, in particular, continues largely unabated as demand for agricultural land and living space increase alongside an expanding logging industry. In upland regions, such as the Himalayan foothills of north-west India or the slopes of the Yangtze River basin, this has led to a dramatic reduction in the ability of the soil to retain water and a consequent increase in surface run-off and severe flooding. In the Yangtze Basin, over 80 per cent of the forests have been destroyed, contributing to major floods in 1996 and 1998 that affected hundreds of millions of people. The flood problem is compounded by silt carried down from stripped uplands, which raises the level of river beds and increases the frequency of overspill of the banks. Deforestation also greatly exacerbated the impact of the heavy rains associated with Hurricane Mitch in 1998, generating hundreds of thousands of landslides and debris flows in Honduras and Nicaragua. Even in developed countries such as the UK, there can be little doubt that recent serious flood events are at least as much due to increasing urbanisation and changing land use causing more rapid surface run-off as to higher levels of precipitation. At lower altitudes, the loss of coastal wetlands – currently running at around 1 per cent a year – is increasing the vulnerability of coastal communities to erosion and flooding associated with storm surges, sea-level rise and tsunami. As discussed in more detail in the final chapter of this book, one of the most dramatic environmental changes launched by society over the past century is the increasing concentration of the planet's population in substantial urban communities. Perhaps more than climate change itself, it is the trend towards the growth of giant, poorly constructed and poorly managed *megacities* in highly vulnerable areas that is undoubtedly going to lead to an increase in the numbers and severity of natural disasters in the present century.

2

Windstorms in a warmer world

2.0 Chapter summary

In a world that is increasingly vulnerable to natural hazards, it is of great importance to assess how weather-related hazards such as windstorms and floods are likely to change in frequency or severity as a result of global warming, and whether such changes are already occurring. In this chapter we focus on windstorms (tropical and extratropical cyclones, tornadoes, derechos and dust storms), describing each of these hazards in the context of a changing climate, and reviewing the latest research into observations and predictions of change. First, however, we set the scene for such investigations by providing background material both for this chapter and for Chapter 3, which concentrates on floods and other weather-related hazards. Section 2.1 introduces these hazards in relation to anthropogenic climate change, and outlines three important natural patterns of climate variability: El Niño–Southern Oscillation (ENSO), the North Atlantic Oscillation (NAO) and the Pacific/North American (PNA) teleconnection pattern. Section 2.2 then describes the methodologies used for investigating weather-related hazard trends. We discuss how the hazards are characterised and quantified, and note the possibilities for large changes in extreme weather for relatively small changes in the general climate. We then summarise the difficulties involved in observing and predicting trends in frequency and severity, which are mainly related to the events' rarity, severity and small spatial scale. We note how these problems are being tackled in the latest observational, theoretical and numerical modelling studies.

2.1 Weather-related natural hazards and climate change

Recent weather-related disasters such as the examples in Chapter 1 have demonstrated the vulnerability of many communities to natural hazards caused by extreme weather: windstorms, floods, hailstorms, snow and ice storms, droughts, wildfires, heatwaves and cold waves. Therefore any increase in the frequency or severity of such events would probably be the most noticeable and damaging aspect of anthropogenic climate change, particularly where vulnerability to these hazards is also increasing (Houghton, 1997, chapters 1 and 6). Moreover, it has been known for some time that because such events are related to extreme statistical fluctuations in the weather about its average values, changes in *average* weather conditions (e.g. global warming) can be accompanied by significant changes in the frequency of *extreme* weather events too (Wigley, 1985). It certainly seems that extreme events such as the great storm of 1987 in the UK and the 1988 heatwave in the USA helped to stimulate public concern about climate change in the late 1980s (Ungar, 1999).

Nevertheless, no single weather-related hazard event, however severe, can necessarily be ascribed to climate change – it may just be associated with a large statistical fluctuation within an unchanged climate, or with natural variability of the climate (Box 2.1). Furthermore, any resulting disaster may just reflect the fact that the event happened to affect a large or vulnerable population centre. Similarly, although there have been dramatically rising trends in weather-related disasters and insurance losses over the past few decades (Downing *et al.*, 1999a; Kunkel *et al.*, 1999), these may not necessarily be the result of climatic effects. For example, Changnon *et al.* (2000) summarise recent research showing that in the USA, such trends can largely be explained by societal and insurance factors, with most studies showing an overall increase in vulnerability to weather-related extremes. To help with such assessments, and to help determine what the future holds, two important goals for climate researchers are to quantify any observed changes in the frequency or severity of these exceptional events, and to predict accurately any future trends in a world where global warming is a reality. This chapter and the next explore these issues, Chapter 2 concentrating on windstorms and Chapter 3 on floods and other weather-related hazards. We build on reviews of observed trends and model predictions by

Box 2.1 Climatic oscillations

The atmosphere exhibits irregular but apparently non-random oscillations between different patterns of circulation, often spending most of the time in one pattern for a number of years or even decades. The oscillations involve correlated changes in climatic variables at widely separated places; these linkages are known as teleconnections. Three particularly important oscillations (Fig. 2.1) are the El Niño–Southern Oscillation (ENSO) phenomenon, the North Atlantic Oscillation (NAO) and the Pacific/North American (PNA) teleconnection pattern, which is related to the Pacific Decadal Oscillation (PDO). These form part of the climate's natural variability against which any anthropogenic trends need to be distinguished, and they all have a significant effect on the occurrence of weather-related hazards.

The ENSO phenomenon is described in detail in many meteorology textbooks. It occurs in the region of the equatorial Pacific, being a quasi-periodic (2- to 10-year) oscillation of the climate system involving both ocean and atmosphere. It was first observed in terms of a seesaw in sea-level pressure (SLP) across the western tropical Pacific, known as the Southern Oscillation, with the Southern Oscillation Index (SOI) being related to the Tahiti minus Darwin pressure difference (Fig. 2.1a). Later it was also linked to a failure in the cold northward Peru Current (the weakening of which every year around December is known locally as *el Niño*, the Christ Child). The two extreme phases of ENSO are now often referred to as El Niño events and La Niña events. La Niña events are associated with a positive SOI, strong easterly trades in the tropical Pacific, convective activity and rainfall in the western Pacific, and a strong Peru Current extending a tongue of cool surface water westwards across the Pacific. El Niño events are associated with a negative SOI, surface winds in the western Pacific becoming westerly, and a tongue of warmer water moving eastwards along with the region of maximum convection. Events of both kinds vary in strength and longevity, as illustrated by the El Niño events in 1990–95 (exceptionally long) and 1997–98 (exceptionally strong). As well as these effects in the tropical Pacific, there are many teleconnections linked to ENSO (e.g. precipitation patterns), as the atmospheric circulation is affected worldwide.

The NAO (Fig. 2.1b) is a variation in the monthly average strength of the subpolar Icelandic low and the subtropical Bermuda–Azores high at sea level in the North Atlantic (van Loon and Rogers, 1978; Rogers and van

Loon, 1979), and is related to the West Atlantic teleconnection pattern with respect to mid-tropospheric airflow (Wallace and Gutzler, 1981). Positive winter NAO index values are associated with a strong mean Icelandic low, a strong mean Bermuda–Azores high, a lower Greenland surface air temperature, a higher northern European surface air temperature, strong surface westerlies across the North Atlantic on to Europe, and a strong mean jet stream (Wallace and Gutzler, 1981). In this phase the Atlantic storm track usually has a north-easterly orientation, and extends into the far north-east Atlantic (Rogers, 1990), leading to high (low) winter precipitation levels in northern (southern) Europe (Hurrell, 1995). Negative winter NAO index values are associated with the opposite tendencies for SLP, temperature, precipitation, westerlies and jet stream strength. In this phase, the storm track is shorter and further south, with an eastward orientation along latitudes 40°–45°. There is also a westward extension of the low-level Siberian high into eastern Europe (Rogers, 1997). Interestingly, however, this location of the storm track also occurs in a minority of positive NAO index years (Rogers, 1997).

The PNA teleconnection pattern (Wallace and Gutzler, 1981) refers to a variation in the monthly average winter mid-tropospheric airflow over the North Pacific and North America between two basic configurations (Fig. 2.2). With a negative PNA index there is a high-pressure ridge over the central North Pacific, and the flow across North America is fairly zonal (following latitude lines). With a positive PNA index the flow over North America is more generally meridional (following longitude lines), with a warm ridge in the west and a cold low-pressure trough in the east. In this phase there is also a broad trough over the central North Pacific and strong zonal flow south of it. This oscillation, which has also been linked to ENSO, has significant effects on the regional winter climate in North America (Leathers *et al.*, 1991). The main surface signatures, for a positive PNA index, are a lowering of winter SLP in the North Pacific and a cooling of the winter sea surface temperature (SST) there. These parameters therefore exhibit related oscillations, characterised by indices based on North Pacific SLP values (Fig. 2.1c) (Trenberth, 1990; Mantua *et al.*, 1997), and by the PDO index, based on North Pacific SST values (Mantua *et al.*, 1997).

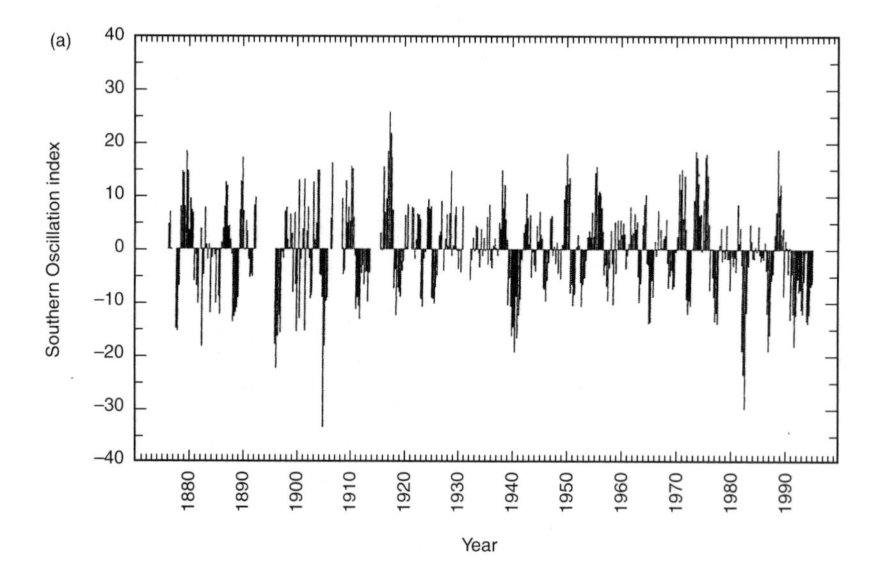

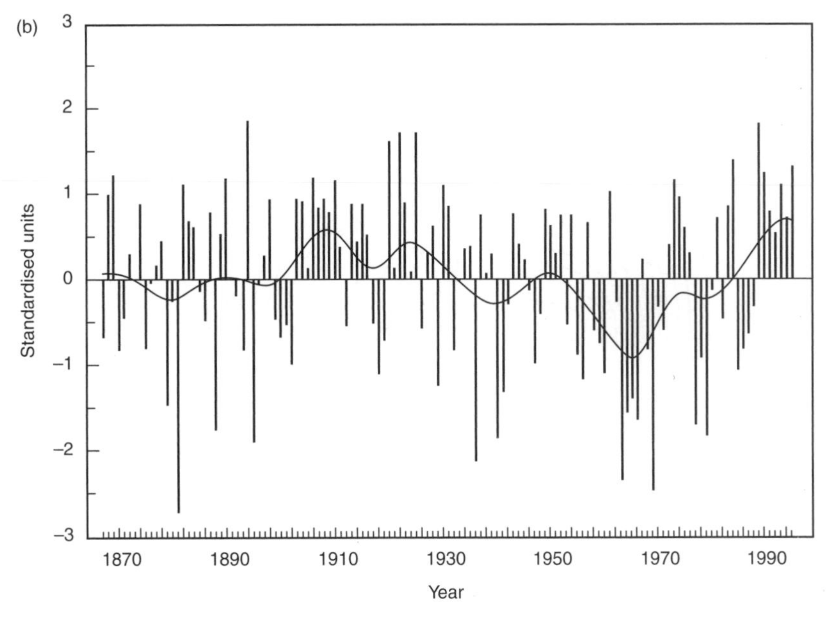

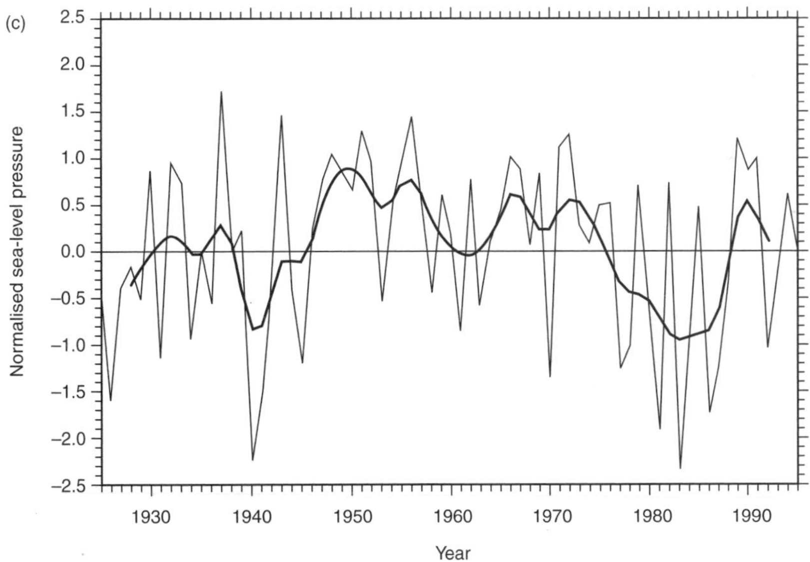

FIGURE 2.1 (opposite and above) Time series of indices of El Niño–Southern Oscillation (ENSO), the North Atlantic Oscillation (NAO), and the Pacific/North American (PNA) teleconnection pattern. (a) Seasonal values of the Southern Oscillation Index (SOI), March–May 1876 to March–May 1995, based on the Tahiti minus Darwin sea-level pressure (SLP) difference. SOI is plotted only when both Darwin and Tahiti data are available. (b) One of several winter NAO indices, being the standardised difference of December–February SLP between Ponta Delgada, Azores, and Stykkisholmur, Iceland, 1867 to 1995. The smooth curve was created using a 21-point binomial filter. (c) Time series of normalised mean North Pacific SLP averaged over 30–65° N, 160° E to 140° W for the months November–March from 1925 to 1995 and smoothed with a low-pass filter (thick line). This index is closely related to, but inverted with respect to, the PNA index. (All reproduced from Nicholls *et al.*, 1996 with permission from the Intergovernmental Panel on Climate Change)

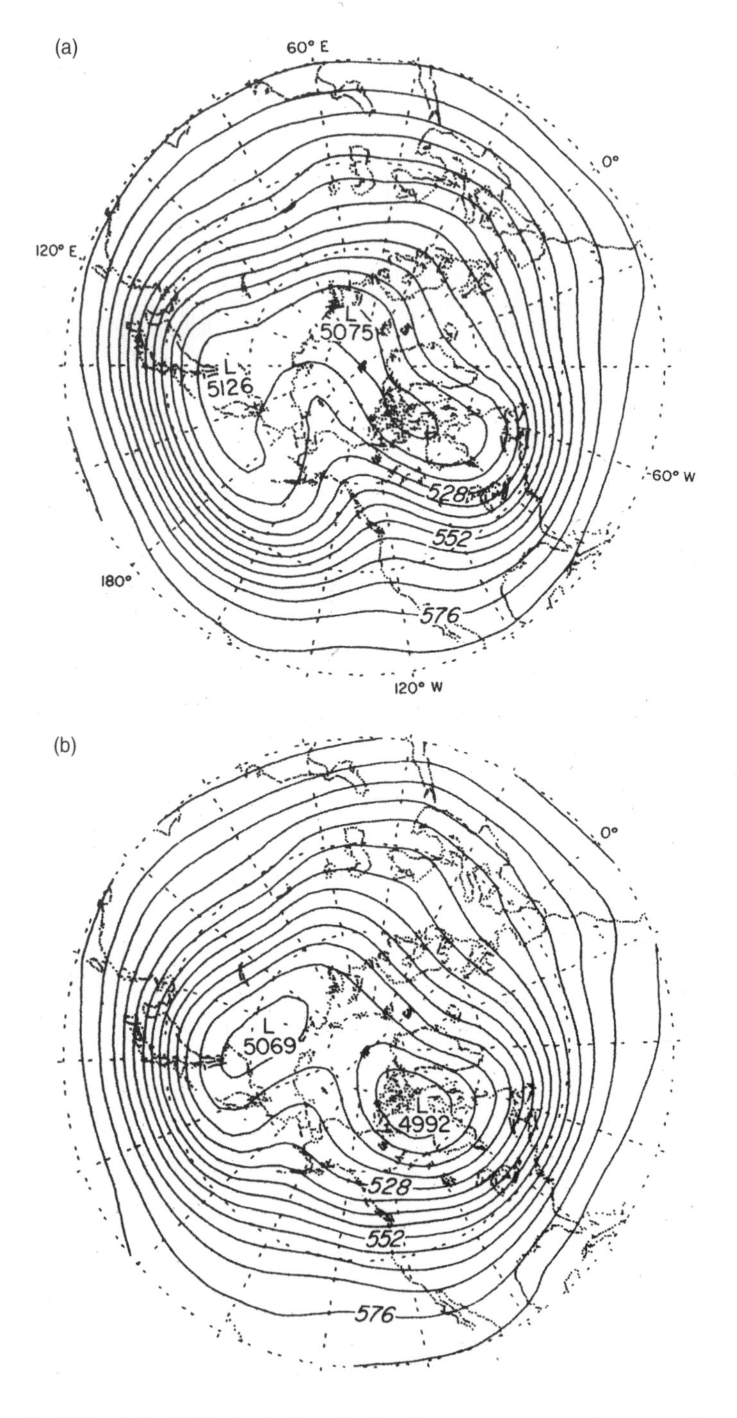

FIGURE 2.2 Composite 500-mb height charts (equivalent to pressure contours at about 5.5 km altitude, with high altitudes representing high pressures), based on the 10 winter months out of a 45 month data set (December, January and February from 1962–63 to 1976–77) with the strongest (a) positive and (b) negative values of the Pacific/North-American (PNA) index. (Reproduced from Wallace and Gutzler, 1981 with permission from the American Meteorological Society)

the Intergovernmental Panel on Climate Change (IPCC) (Nicholls *et al.*, 1996; Kattenberg *et al.*, 1996), and more recent reviews by Henderson-Sellers *et al.* (1998), Walsh and Pittock (1998), Saunders (1999), Karl and Easterling (1999), Harvey (2000), Meehl *et al.* (2000b) and Easterling *et al.* (2000a, b). We also provide brief overviews of the hazard events themselves. It is important to note that extreme weather phenomena can be fully understood only in the overall context of the Earth's weather and climate. Descriptions of these, including the phenomenon of global warming, can be found in many textbooks, so need not be repeated here. We do, however, briefly discuss climatic oscillations (Box 2.1, p. 19), because of their particular importance for the investigation of weather-related hazard trends.

2.2 Observing and predicting trends in weather-related hazards

Before we look at the methodologies used for investigating trends in weather-related hazards, it is important to clarify the scope of such investigations. A weather-related hazard may be defined as an event which is caused, at least in part, by unusual and extreme weather, and which threatens damage (loss of life, or economic or cultural loss) to a vulnerable community. Sometimes, however, rare and extreme events that pose no direct threat may also be included in investigations of hazard trends. Some of these form part of the same population of events, but are remote from people (e.g. a remote storm at sea). Others are not particularly severe events in an absolute sense, but may be susceptible to climate change in a similar way (e.g. moderate precipitation in an area that normally has light precipitation). Still others are hazard-related weather events such as heavy rainfall (which may cause a flood but is not usually *directly* hazardous). In addition, the inclusion of less extreme events, in larger numbers, may also help to show up trends.

Note that the extreme weather which causes the hazard event (e.g. high wind speed, heavy rainfall) is generally associated with particular regional and seasonal weather patterns, and often occurs within localised weather systems (e.g. tropical cyclones, thunderstorms). In some cases the extreme weather events themselves constitute the hazard (e.g. high wind events), whereas other types of hazard involve secondary non-weather processes (e.g. river floods, which involve drainage basin hydrology as well as heavy rainfall). For convenience, moreover, the extreme localised weather systems that cause these hazards are themselves often treated as hazards as well. Thus, for example, we can speak of a tropical cyclone hazard having high wind speeds and heavy rainfall, and hence providing both a wind hazard and a flood hazard. Finally, some hazard events are discrete occurrences (e.g. tornadoes, hailstorms), whereas others just represent the occurrence of extreme values in an essentially continuous record (e.g. high values of river discharge or wind speed in a daily record).

2.2.1 Quantifying the frequency and severity of the hazards

To observe or predict changes in the frequency and/or severity of a hazard (or of a related event such as heavy rainfall), it is necessary to quantify these properties. Parameters for measuring frequency include the interarrival time (the time between successive events), the return period (the mean interarrival time) and the mean annual frequency (the reciprocal of the return period in years). The severity of a weather-related hazard (or related parameter) is usually quantified mainly in terms of its magnitude or intensity (e.g. river discharge, wind speed, rainfall rate, temperature). Sometimes an index of severity is used, reflecting a number of intensity measures, or incorporating the duration and spatial extent of the event.

For any particular climatic conditions, the distribution of events in magnitude or intensity (often extended to include non-hazardous magnitudes) largely summarises the statistical properties of that particular type of event, and forms the basis for most other indices. For weather-related parameters, these frequency–magnitude distributions have a wide variety of shapes (for example, see Brooks and Carruthers, 1953, chapter 8, and von Storch and Zwiers, 1999, chapters 2 and 3). They are sometimes normalised to form probability distributions. For natural hazards, however, the most extreme events must also be rare, so the distributions will always have a low and/or high extreme value tail. The above frequency parameters generally refer to the cumulative (exceedance) frequency – that is, including all events that are equal to or less than (greater than) some intensity level in the low (high) value tail. In some cases, only the most extreme event within a certain time period (e.g. the maximum value of river discharge each year) is considered. The resulting distribution can then be modelled using *extreme value analysis* (e.g. von Storch and Zwiers, 1999, chapter 2; Bedient and Huber, 1992).

In using frequency–magnitude distributions to investigate extreme events, however, it is important to be aware of a number of potential problems (Smith and Ward, 1998, pp. 188–191; Essenwanger, 1976, pp. 143–189). First, it is possible for the data to be undersampled (missing some events) or oversampled (including events that are not independent). Second, the distributions themselves may be changing over the period of the observations (e.g. owing to climatic variations). Finally, there may be more than one population of events (i.e. events with different source mechanisms), leading to a mixed frequency distribution.

2.2.2 Observing and predicting trends in extreme weather-related events

In our discussion of methodologies for investigating trends, we first consider the case where more frequent, less hazardous events are *not* included. When it comes to observing or predicting trends in either the frequency or severity of rare and extreme events, the main method employed is a *peaks-over-threshold* approach. For frequency–magnitude distributions with a high value tail, for example, this involves considering changes with time in the frequency of events falling above a particular threshold in magnitude. The threshold is often chosen to provide events with a particular mean annual exceedance frequency or return period, calculated using data for a 'baseline' time period. In this case a more accurate threshold value may sometimes be obtainable by fitting a theoretical distribution to the data first (e.g. Jones *et al.*, 1999; Nicholls and Murray, 1999). For sufficiently rare events, the number of events per time interval follows the Poisson distribution (for example, see Brooks and Carruthers, 1953, pp. 76–9; von Storch and Zwiers, 1999, pp. 25–6), and the interarrival times will then follow an exponential distribution. However, the smaller the numbers of events per time interval, the larger, relatively, are their statistical fluctuations. This means that it can be difficult to observe any climate-related changes with time above the statistical spread of the data, especially if we confine our attentions to a particular vulnerable location. Nevertheless, Kiem and Cruise (1998) showed that accurate trend detection is possible for very rare events by considering interarrival times as a function of time, particularly if events are analysed in groups of two or more. In addition, they noted that if Poisson statistics apply, the data set can be partitioned so as to assess independently the trends of different populations within the data.

In an alternative approach, so far used mainly for prediction studies, the tails of frequency–magnitude distributions are compared for different periods or time. Generally, theoretical distributions are fitted to past or predicted data, and return periods are then calculated from the curves. For example, Zwiers and Kharin (1998) employed extreme-value analysis in a global predictive modelling study, using theoretical distributions to characterise annual maxima in temperature, precipitation and wind speed, and so calculate, for CO_2 doubling, the changes in magnitude corresponding to a particular return period.

As well as these statistical difficulties associated with the events' rarity, there are other problems associated with monitoring extreme weather-related events, largely stemming from their localised nature and their severity. They were reviewed by Nicholls (1995). In particular, it is often more difficult to obtain *homogeneous* data sets than is the case with means of variables. In this context, Karl and Easterling (1999) defined *inhomogeneities* as

> changes and variations in the record that are non-climatic or are not representative of the time and space scales of interest (e.g. urban heat island effects are climate-related, but are not the scales of interest for global temperature change analysis).

Extreme events are localised, so direct measurements are often unavailable. Thus their severity may have to be estimated indirectly from other information. Such estimates can lead to systematic biases if estimation or analysis methods are changed, or if the estimates are subjective. (In some cases data may even be missing altogether.) Even where instrumentation can be used, severe events can cause instrument malfunction, procedures can vary with time and location, and siting changes (e.g. sheltering by trees) can sometimes affect extreme events more than mean values. Again, therefore, indirect parameters are often used. For example, Alexandersson *et al.* (1998), in a study of north-west European wind extremes (see Section 2.3), estimated wind speeds indirectly using atmospheric pressure data, following Schmidt and von Storch (1993). This was to avoid inhomogeneities with direct wind measurements. (In fact, the method was even more indirect, in that their estimates were of the geostrophic wind rather than true wind.) Detailed recommendations regarding indices and indicators for climate extremes were given by Trenberth and Owen (1999), Nicholls and Murray (1999) and Folland *et al.* (1999). Finally, there are issues of data management that are causing difficulties in obtaining satisfactory global data records for climate-extreme monitoring (Karl and Easterling, 1999). These include cost recovery policies for historical climate records, and a lack of electronic digital data.

The concept of the indirect use of related parameters, noted above, has often been applied in a much more general way, to argue for changes in extreme events based either on known meteorological mechanisms or on statistical correlations with other parameters or weather patterns. Expected changes in various climatic parameters or processes are thus used to infer changes in the frequency or intensity of the hazards. The procedure is not always straightforward, however, as a number of different parameters are likely to be involved, and these may have opposing effects under global warming. For example, expected changes in the meridional temperature gradient and the ocean temperature have competing effects on extratropical cyclone intensity (see Section 2.3). In addition, any statistical correlations may themselves change as the climate warms. For example, some of the criteria for tropical cyclone formation appear to be tuned to the current climate (see Section 2.4). Generally, therefore, accurate prediction requires dynamic climate models (see Section 2.2.4). In some cases, however, a theoretical framework has been developed, which can help us understand and interpret the observational and dynamic modelling results. For example, Emanuel (1986, 1987) developed a simple thermodynamic model of tropical cyclones and used it to predict increases in maximum potential wind speed under global warming (see Section 2.4). Trenberth (1999) developed a conceptual framework for explaining how climate change is likely to affect hydrological extremes, and used it to predict an increase in heavy precipitation events under global warming, as well as an increase in global annual precipitation (see Section 2.3.4 and Chapter 3).

2.2.3 Observation and prediction using more frequent events

Other approaches to investigating changes in extreme events make use of the fact that extreme events represent the high (or low) value tail of the overall frequency–magnitude distribution for the parameter or phenomenon of interest. The basic idea is that it may be easier to observe or predict changes in the less extreme, non-hazardous events. They will certainly occur in larger numbers, and so allow smaller changes to be seen above the statistical 'noise' of the data. They also make up the bulk of the overall distribution. Therefore either changes in the frequency of less extreme events, or changes in the properties of the overall distribution, might be used to infer changes in the more extreme events. These two techniques are reviewed below.

In considering such techniques, however, it is first useful to distinguish various basic ways in which frequency–magnitude distributions might change so as to cause a change in numbers in the extreme event tail. Considering, for convenience, the case of increases involving high value tails, we indicate these distribution changes schematically for a normal distribution in Fig. 2.3, and categorise them as follows. (Similar arguments apply to decreases, or to low value tails.) A type A change (Fig. 2.3a), representing an overall increase in the frequency of all events, can apply only to discrete events such as precipitation events, tropical cyclones or tornadoes. However, for continuously varying or regularly sampled parameters (e.g. temperature, wind speed, river discharge, daily rainfall), such a change is meaningless: the total number of 'events' is just the number of observations which happens to be used in the distribution (e.g. one per day, etc.). The other two cases can occur for all types of event and have been widely noted (e.g. Mitchell *et al.*, 1990; Houghton, 1997; Downing *et al.*, 1999a; Wolter *et al.*, 1999; Meehl *et al.*, 2000a). A type B change (Fig. 2.3b) represents a shift in the mean, i.e. an overall shift of the whole distribution to higher magnitudes. A type C change (Fig. 2.3c) represents a broadening of the distribution and/or a change in its shape (e.g. a change in its skewness or peakiness); here an increase in the tail occurs at the expense of a decrease for less extreme values. In any given situation, a certain distribution may be experiencing a combination of these different types of change, each individually having either positive or negative influence on extreme numbers. For type A changes, whatever the width and shape of the distribution (provided it remains the same), the relative change in the frequency of extreme events will clearly just be the same as the

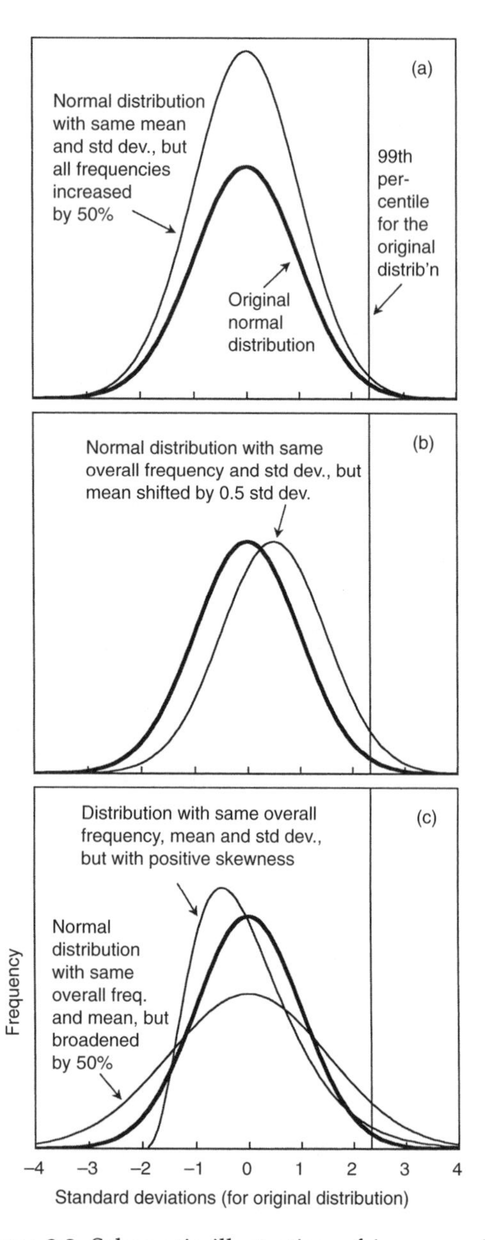

FIGURE 2.3 Schematic illustration of increases in the frequency of extreme events of high magnitude caused by changes in the overall frequency distribution of events, for an initially normal distribution. (a) Type A change: an increase in the frequency of all events (can only occur for discrete events, not for regularly sampled data). (b) Type B change: increase in the mean of the otherwise unchanged distribution. (c) Type C change: a broadening (increase in the standard deviation) or a change in shape (here, an increase in skewness).

relative change in the frequency of all events. For type B or type C changes, however, the relative change in the frequency of extreme events will depend on the amount of shift, and on the width and shape of the distribution. For the case of a normal distribution, the dramatic effects of a type B shift in the mean were noted by Parry (1978), Mearns *et al.* (1984) and Parry and Carter (1985), and further quantified and summarised by Wigley (1985, 1988) (Box 2.2).

Katz and Brown (1992) and Wagner (1996) showed that for a normal frequency distribution, the relative frequency of extreme events is more sensitive to type C changes in its width than to type B changes in its mean value, provided the changes are of the same magnitude. For other distributions, and especially for events in their extreme tails, the effects of such changes will be different. For events exhibiting an exponential tail, for example, a type B shift would provide *exponential* relative changes in probability with shift, but the relative change would be *independent* of initial probability – that is, the same however extreme the event (Katz and Brown, 1992).

As regards techniques for inferring extreme changes, let us first consider the use of overall distribution properties. One example of where this concept can be directly and accurately applied to hazard prediction is that of coastal flooding. In the absence of any other changes, an increase in mean sea level provides a simple type B shift to the flood-level distribution, allowing new probabilities and return periods to be calculated (see Chapter 3). Such predictions are subject to error, however, if climate change also gives rise to type C changes in the flood-level distri bution. Changes in the standard deviation of a distribution (a type C width change) can also be considered as a proxy for extreme event changes, as was done by Mitchell *et al.*

Box 2.2 The effect of a shift in the mean on extreme event exceedance frequency, for a normal distribution

When a normal distribution experiences a relatively small type B shift in its mean (see main text), the relative change in extreme event exceedance frequency (or exceedance probability) can be surprisingly large. There are three aspects to this effect:

The relative change in exceedance frequency is steep function of the shift in the mean
Relatively small changes in the mean value produce very large relative changes in the exceedance frequency (or exceedance probability) of extreme events. For example, with a shift in the mean equal to only 0.5 standard deviation, an event with an exceedance probability of 0.01 (i.e. above the 99th percentile) becomes 3.4 times more frequent, i.e. with a new probability of 0.034.

The relative change in exceedance frequency is non-linear with the shift in the mean
The relationship between relative change in exceedance frequency and shift in the mean is non-linear, with the slope of the function increasing as the shift increases, though the relationship is less than exponential. For example, in the above case, a shift in the mean of 1 standard deviation (twice the above shift) makes the event 9.2 times more frequent, i.e. with a new probability of 0.092. (For a linear relationship the figure would be $3.4 \times 2 = 6.8$; for an exponential relationship the figure would be $3.4^2 = 11.5$.)

The relative change in exceedance frequency is greater and more non-linear, the rarer (and more extreme) the event
For example, an event with an exceedance probability of 0.001 becomes 4.8 times and 18.3 times more frequent for shifts in the mean equal to 0.5 and 1 standard deviation respectively (i.e. with new probabilities of 0.0048 and 0.0183 respectively). This effect was quantified by Katz and Brown (1992).

(1990) for a range of parameters in a predictive modelling study.

In a more general application of this concept, the above analysis on type B shifts for normal distributions can be used to argue that any observed or predicted shift in the mean value of such distributions should be associated with a large change in the frequency of extreme values. For example, if Fig. 2.3b represented daily maximum temperature for a region in summer, a rise in average temperature due to an enhanced greenhouse effect should produce more days with extremely high temperatures. This example, in fact, has support from modelling and empirical studies (e.g. Mearns *et al.*, 1984). But this serves to remind us that such ideas should not be generalised or extended to any particular parameter or hazard, particularly in a quantitative way, without other observational or modelling support. First, many of the relevant distributions are not normal, so the sensitivity to type B shifts may be different. Second, associated with the type B shift there may be type C changes in the width/shape of the distribution, or even type A changes in total numbers of events, affecting any increases in the numbers of extreme events (for example, see Heino *et al.*, 1999).

The second technique for inferring changes in extreme events involves the direct use of changes in the numbers of less extreme events. One approach simply involves setting a lower magnitude threshold, further back up the tail, so as to provide larger numbers of high-value events. For example, in the study referred to earlier, Alexandersson *et al.* (1998) used a wind speed threshold that included about 10 high wind events per year. Though event numbers will now be dominated by less hazardous ones, observed changes can, with caution, be related to changes in the more extreme events, by making assumptions about related changes in the tail of the distribution. A mixed distribution, however, with several populations, could still cause difficulties for this approach. If the whole frequency–magnitude distribution is available, another approach is to consider possible changes with time in various high but not particularly extreme percentile values for the distribution, for example the 95th or the 99th percentile. This was also done by Alexandersson *et al.* (1998), who found good correlation between changes in these values and changes in the numbers of high wind events (as described above) whose threshold happened to be close to the 99th percentile.

2.2.4 Modelling approaches to the prediction of extreme event changes

Accurate prediction of changes in extreme weather-related hazards can be achieved only by modelling. The basic workhorses for climate change prediction are dynamic climate models, which have been used extensively to predict changes over the next century under a variety of human activity scenarios. They have evolved from general circulation models (GCMs) of the atmosphere, originally developed for short-term (up to a few days) weather forecasting. These numerical computer models divide the atmosphere into a number of vertical layers and a large number of grid points horizontally. The basic equations governing the behaviour of the atmosphere are then solved at each grid point for a long series of small time steps. Though the precise weather details are not predictable for more than a week or so, such models can be used to predict long-term statistical changes in the climate as the external constraints on the climate system (such as greenhouse gas concentrations) are changed.

The best climate models are now coupled atmosphere–ocean models. In some of these the ocean is represented in a simple way, but others simulations incorporate ocean general circulation models similar to the atmospheric ones, as well as modelling the growth, decay and motion of sea ice. A standard investigation using such models is the response of the climate to a doubling of CO_2. In 'equilibrium' simulations a constant CO_2 value is used and the model is run until an equilibrium (constant climate) state is reached. Thus a comparison of the $1 \times CO_2$ and $2 \times CO_2$ cases can be made. More realistic 'transient' experiments are able to simulate the fairly slow response of the climate system to any CO_2 change, and the fact that CO_2 is increasing gradually rather than stepwise. In these experiments, CO_2 is increased at an expected rate (typically 1 per cent per year), and the climate is observed until CO_2 has doubled (typically after about 70 years).

The problems with regard to hazard prediction with GCMs were discussed by Kattenberg *et al.* (1996). A major difficulty stems from the small scale of many weather-related hazard events. Typical current global models have a resolution of several degrees in longitude and latitude (i.e. several hundred kilometres), so they can only just resolve individual extratropical cyclones (Carnell and Senior, 1998). They cannot accurately model tropical cyclones, still less individual thunderstorms or tornadoes. In addition, the development of hazard events can depend on sub-grid-scale features (e.g. coastlines) and processes (e.g. cloud formation) which may be inadequately approximated by global models. This scale problem is being addressed, however, in two ways. The first method is *statistical downscaling,* whereby statistical relationships are first developed between local climate parameters and the large-scale variables that are provided by GCMs. These relationships are then applied to GCM outputs so as to predict local climatic change. The second method is by the use of *nested models,* where a higher-resolution regional model uses the output of a global model as its boundary conditions. (Typical regional models currently have a resolution of tens of kilometres.) Another difficulty is that of discerning trends above the natural variability of the system, as relevant processes such as ENSO are still not being modelled accurately enough. Finally, the rarity of extreme events means that very long model runs are necessary for trends not to be lost in the statistical noise. This is obviously easier to deal with in an equilibrum experiment than a transient one, though ensembles of transient runs can improve the signal-to-noise ratio (e.g. Carnell and Senior, 1998).

2.3 Mid-latitude storminess in a warmer world

In the rest of this chapter we concentrate on the windstorm hazard and the associated weather systems: extratropical cyclones, tropical cyclones, tornadoes, derechos and dust storms. Cyclones are also associated with floods, through both storm surges and precipitation, but that hazard will be dealt with in Chapter 3. We first examine extratropical cyclones.

2.3.1 Extratropical cyclones: an overview

Extratropical cyclones generally have less destructive power than do tropical cyclones or tornadoes, but are able to provide damaging winds over a wide area, and also wave damage in coastal areas. The wind speed tends to be higher in open country, on hills and over the sea, though topographic features can increase speeds locally. The wave hazard depends on the wind's fetch (the distance of water over which the wind blows) and duration, as well as its speed. Extratropical cyclones are described at length in most meteorology textbooks; we briefly summarise their main features here. As their name suggests, they mainly occur in mid-latitudes, and are intimately associated with the westerly airflow there. Sometimes known as mid-latitude storms or depressions, they involve regions of low sea-level pressure with closed isobars, whose scale is typically thousands of kilometres in diameter, and whose life span is usually four to seven days (Fig. 2.4). Deflection by the Coriolis force produces winds that blow cyclonically around the low, i.e. anticlockwise (clockwise) in the Northern (Southern) Hemisphere. However, near the surface, friction allows the wind to blow obliquely across the isobars towards the storm centre, tending to increase the pressure there. The storm can therefore intensify only if there is diverging air aloft, so that the converging surface air can rise and be removed aloft faster than it flows into the low at the surface. The central pressure then reduces, increasing the sea-level pressure gradient and hence the surface wind speed.

Extratropical cyclones generally form and develop along the polar front, where strong meridional temperature gradients give rise to strong pressure gradients in the upper troposphere, resulting in the westerly polar jet

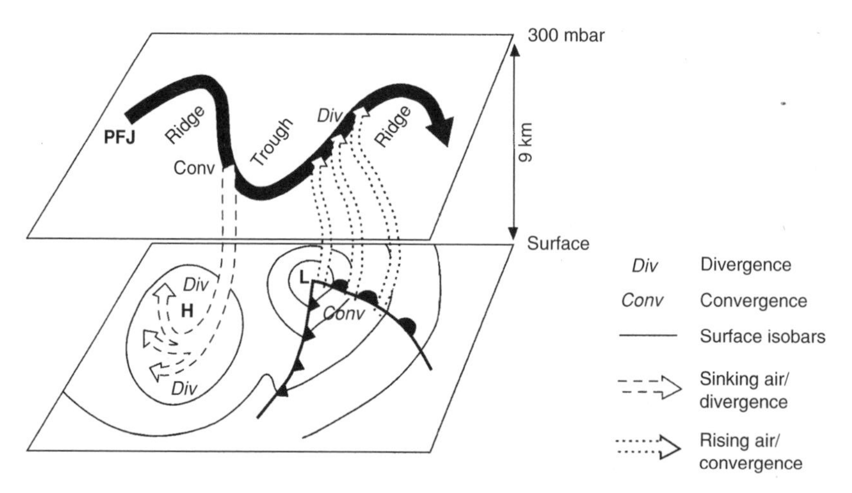

FIGURE 2.4 Schematic diagram of an extratropical cyclone in the Northern Hemisphere. At sea level the pressure chart shows a low central pressure with cold and warm fronts on either side of a 'warm sector'. A cool air mass behind the cold front is pushing under the warm-sector air mass. At the warm front, warm-sector air is moving up over the cooler air mass ahead of the front. In both cases the rising warm air produces cloud and rain. At upper levels the polar front jet stream (PFJ) is shown meandering round ridges of high pressure and a trough of low pressure, with a region of divergence (Div) helping remove rising air faster than it converges at the surface, thus intensifying the storm. Similarly, a region of convergence (Conv) in the PFJ helps to build a nearby anticyclone at the surface. (Reproduced from *Atmospheric Processes and Systems* by Russell D. Thompson, 1998 with permission from Routledge)

stream. In the Norwegian cyclone model, a wave in the polar front develops, producing, near the surface, a warm sector on the equatorward side, between a warm front and a cold front (Fig. 2.4). As the storm travels and intensifies, the cold front overtakes the warm front, producing an occluded front, after which the storm dissipates. Moist air from the warm sector rising along the fronts (particularly if the atmosphere is unstable to convection) condenses into cloud and precipitation, and gives the mature storm a comma-shaped appearance on satellite images. Intensification of the storm is helped by the associated latent heat release, which causes divergence aloft by expanding the air column and producing an outward pressure gradient. Thus storms will intensify more if the sea surface is warmer, providing the air with more moisture by evaporation.

The other causes of upper tropospheric divergence are changes in the direction and speed of the associated jet stream. A number of effects involving the Coriolis force create particularly strong divergence downstream of an equatorward meander around a trough, and at the exit from a jet streak (Fig. 2.4). In all cases, greater divergence occurs for a higher overall speed of the jet stream, which in turn requires a sharper temperature gradient across the polar front. Thus storms will intensify where this occurs, either due to an increase in large-scale latitudinal heating differences, or due to a regional effect such as the land cooling more than the ocean in winter. Sharp temperature gradients also favour initial genesis of the storms. The gradients can in turn be affected by advection of warm and cold air within the depressions themselves as they develop.

The jet stream also largely determines the paths of the cyclones, which therefore occur in fairly well-defined regions known as storm tracks. In the Northern Hemisphere, seasonal changes in the temperature gradient cause the polar jet stream to be stronger and generally further south in the winter, so this is usually when the most hazardous storms occur. The major winter storm tracks (Fig. 2.5) are across the Atlantic and Pacific. There is also a storm

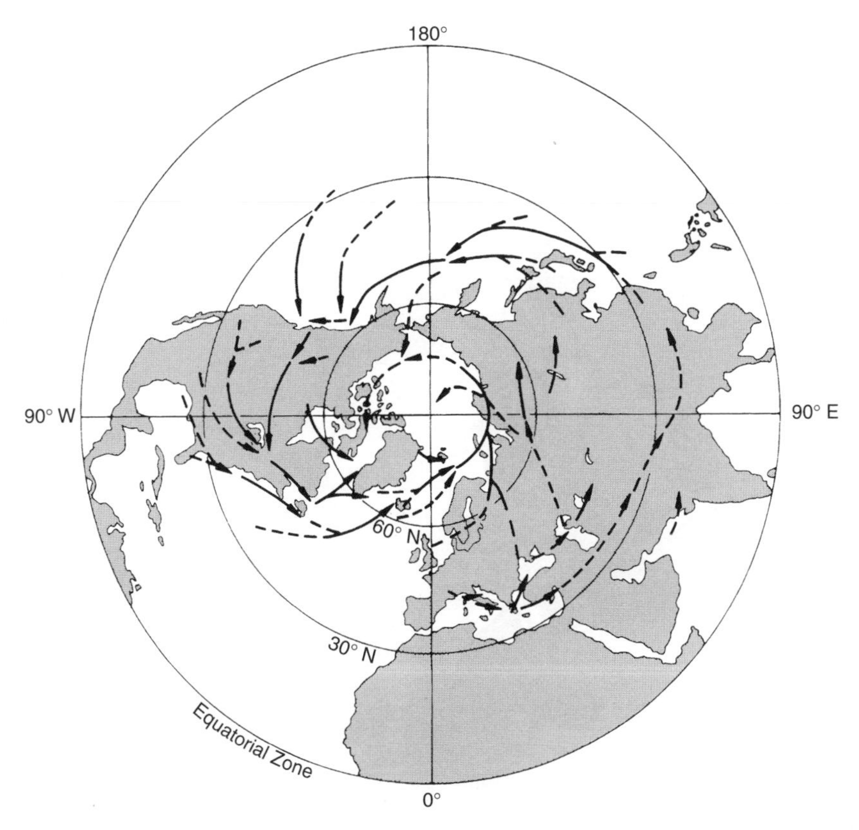

FIGURE 2.5 The main Northern Hemisphere winter extratropical cyclone tracks. The full lines show major tracks, the dashed lines are tracks that are less frequent and less well defined. The frequency of lows is a local maximum where the arrowheads end. (Reproduced from *Atmosphere, Weather and Climate* by Roger G. Barry and Richard J. Chorley, 1998 with permission from Routledge)

track in the Mediterranean area, and three major ones across North America. The track length and location are affected by any blocking highs in the upper air flow, and by the presence of shallow temperature-related surface high-pressure regions over the cold continents (principally Asia), which prevent cyclones from penetrating far inland. In the Southern Hemisphere the main storm tracks occur in the ocean. An important set of intense storms are occasional *east coast cyclones* in the middle and southern parts of Australia's east coast, which are a major cause of heavy rainfall and flooding.

Storm track locations are affected by oscillations in the patterns of atmospheric circulation (Box 2.1). The relationships between extratropical cyclones and the NAO, the PNA pattern and ENSO can be summarised by referring to a study by Sickmöller *et al.* (2000) using meteorological reanalysis data from 1979 to 1997. In the North Atlantic and Europe, their results are in general agreement with those of Rogers (1990) and Sereze *et al.* (1997). For positive (negative) values of the winter NAO index they found more (less) stationary cyclones (most of which occur near Iceland). For the same NAO phases the cyclone density (the percentage of time an area is occupied by lows) was also higher (lower) to the north, and lower (higher) to the south, of a boundary running mainly along the 60° latitude line but dipping south to include Great Britain (Fig. 2.6). However, they found no correlation between the NAO index and *total* numbers of travelling cyclones. In El Niño (La Niña)

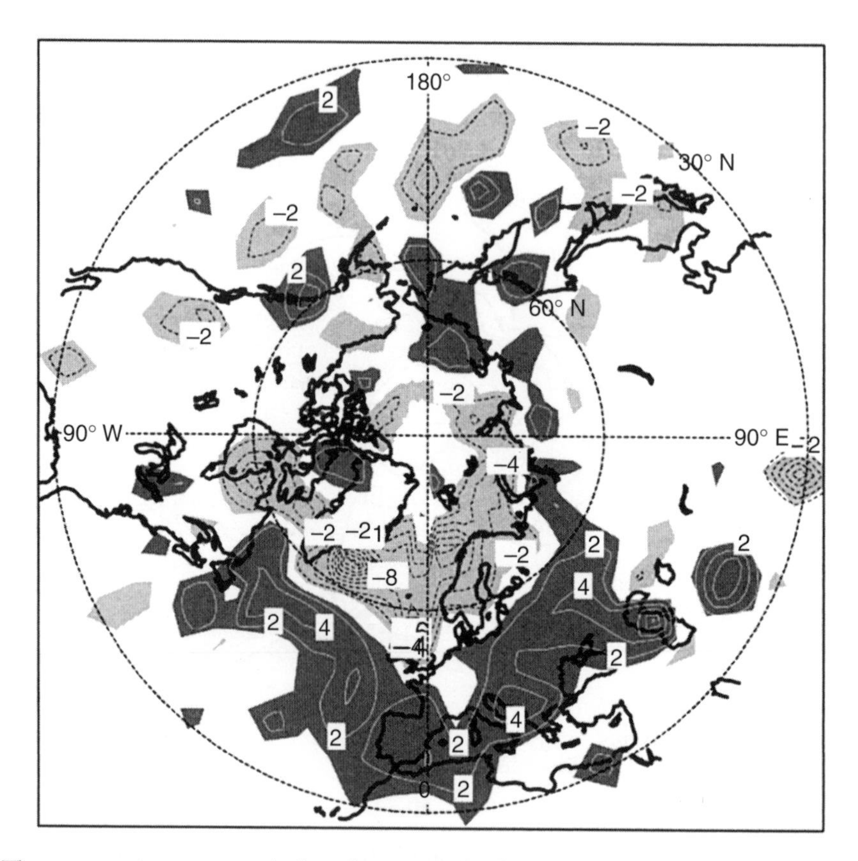

FIGURE 2.6 The seesaw in extratropical cyclone activity between northern and southern regions of Europe and the North Atlantic as the winter NAO index changes its sign. The data, for the period 1979–1997, show the difference in winter (December–February) cyclone density (the fraction of time an area is occupied by lows) between nine negative-NAO and nine positive-NAO winters. Positive values (*dark shade*) mean that there are more (fewer) cyclones for a negative (positive) NAO index; negative values (*light shade*) mean that there are more (fewer) cyclones for a positive (negative) NAO index. (The numerical values are scaled.) (Reproduced from Sickmöller *et al.*, 2000 with permission from the Royal Meteorological Society)

winters they found that cyclones were less (more) frequent in the north-east Atlantic and more (less) frequent in eastern Europe. In the Pacific, a positive (negative) PNA phase was dominated by winter cyclones that were further south (north), zonal (north-eastwards) and mid-ocean (reaching the coast of North America). El Niño (La Niña) winters were associated with the zonal (north-eastwards) cyclones.

2.3.2 Quantifying the mid-latitude windstorm hazard

Before looking at the observational record of mid-latitude windstorms, we need to consider how trends in the windstorm hazard can be measured. Individual hazardous storms are often fairly well documented, and their severity can be characterised in a number of ways (Box 2.3). However, characterising the

Box 2.3 Characterising the severity of mid-latitude windstorms

Though the earliest reports of mid-latitude windstorms are lacking in quantitative information, the use of the *Beaufort scale*, developed in 1806 as an aid for mariners, provided useful estimates of wind speed in the nineteenth century, prior to the widespread use of instruments (Lamb, 1991, p. 5). This scale, the standard for all wind observations till 1946, is based on observation of the effects of wind, and so essentially defines the wind hazard on land, as follows:

- Force 8 (18–21 m/s): Gale; small twigs break off; impedes all walking
- Force 9 (21–24 m/s): Strong gale; slight structural damage
- Force 10 (24–28 m/s): Storm; considerable structural damage; trees uprooted
- Force 11 (28–32 m/s): Violent storm; widespread damage
- Force 12 (>32 m/s): Hurricane; extensive damage

The equivalent anemometer wind speed values are 10-minute averages measured 10 m above the surface. A storm's maximum 10-minute or hourly mean wind speed in a particular region is sometimes used as a measure of its severity (Dorland *et al.*, 1999). However, the worst damage is caused by gusts (irregular variations in wind speed caused by turbulence, with durations of seconds to minutes), so the highest measured gust speed is also noted. For example, the extratropical cyclone Daria, which caused widespread destruction in the Netherlands and surrounding region (including southern England) in January 1990, generated a maximum gust speed in the Netherlands of 44 m/s, the second highest there during the previous 80 years. However, its maximum hourly mean of 29 m/s was not so rare, having been equalled or exceeded on 10 occasions during the same period (Dorland *et al.*, 1999).

In characterising the hazard or disaster potential of storms it has sometimes been considered helpful to create an index based on a variety of severity measures. For example, Lamb (1991) proposed a severity index based on the area affected and the duration, as well as wind speed. The situation is complicated because the hazard from extratropical cyclones is not limited to wind damage, but includes flooding from precipitation or storm surge (see Chapter 3), and damage from wave action. For *nor'easter* cyclones on the east coast of North America, the destructive potential is mainly due to coastal waves rather than wind, so the Dolan–Davis scale of severity was developed, using a wave power index based on their duration and the maximum significant wave height (Davis and Dolan, 1993).

severity of individual hazardous storms over a large region may not automatically show up trends with time. If we restrict our observations to windstorms that are hazardous, we find that even though such storms can be well monitored, their relative rarity, and the large variations between individual cyclones (e.g. in maximum wind and gust speeds, duration, affected area), make trends difficult to detect. The problem is made worse by the fact that different localities may have quite different frequency distributions for the storm-related parameters. For example, gust speeds measured in southern England during the great storm of October 1987 (some at more than 50 m/s) were estimated to have return periods of about 200 years for that region, agreeing with estimates that this was the worst storm there since that of 1703. However, such gust speeds are several times more common in the north of Scotland (Lamb, 1991, pp. 189–90). Therefore, it is necessary to consider the hazard trends in different locations, separately monitoring any regional and local-scale variations. This restricts the numbers of hazardous cyclones still further, making trend analysis (e.g. for the southern England example) almost impossible.

Therefore most researchers also include, in their analysis, extreme events that are not hazardous to communities (e.g. those over the ocean). Using the recent instrumental record it is also common to include events that are somewhat less extreme. It is also considered crucial to use homogeneous data. However, although these approaches are more likely to produce reliable trend estimates, they do make it difficult to relate any measured trends to changes in the most extreme hazardous events in any location (see Section 2.2).

How, then, are trends monitored in practice? Hazard events are by definition threatening (often causing disasters) and are relatively infrequent, and so historical records of the most severe storms can be fairly complete, especially in well-populated areas with extensive shipping and a tradition of record-keeping (Lamb, 1991, pp. ix, 22). Furthermore, the use of the Beaufort scale for wind speed (Box 2.3) means that nineteenth-century records are more accurate. However, there is probably a tendency for historical records to be biased towards disasters, and, as we see in the modern era, these are dependent on factors other than climate, such as vulnerability and whether the storm happened to affect major centres of population.

For observing trends in extreme wind events since the late nineteenth century using instrumental records, Trenberth and Owen (1999) reviewed the available data and their limitations. The most direct parameter to use is the frequency of obtaining wind speeds above a certain threshold, using station values that are noted regularly (e.g. every few hours) to ensure that all extreme wind events are sampled. However, the use of anemometer data presents difficulties of inhomogeneity with regard to instrumentation accuracy, measurement methodologies and siting changes (Trenberth and Owen, 1999). As an alternative, pressure gradients from weather charts have been used to derive wind speeds, but these data are subject to changes in sampling and analysis procedures (Schmith *et al.*, 1998). Therefore, following Schmidt and von Storch (1993), some investigators now recommend the use of geostrophic wind values derived from sea-level pressure measurements or analyses, as being a more homogeneous data source (Trenberth and Owen, 1999). Over the ocean, measurements of wave height have also been used to indicate wind speed, but they are less reliable, as they can also include swell caused by distant storm activity (Bouws *et al.*, 1996). Changes in the frequency of extreme wind speed will also be accompanied by changes in the overall frequency–magnitude distribution, so parameters such as a fairly high percentile, or even the standard deviation, may be considered as proxy indicators, provided that their relationship to the frequencies of extreme events can be determined (see Section 2.2.3). It may also be of interest to compare variations in the mean or median values of wind speed.

The other main set of parameters for observing trends is based on the fact that the wind hazard in any particular location will be strongly related to the intensity and tracks of the associated extratropical cyclones, as obtained from meteorological analyses. In the absence of intensity information, data on cyclone tracks will provide a record of the frequency and location of *all* cyclones (Trenberth and Owen, 1999). High-frequency fluctuations in the mid-tropospheric pressure are often used as a general indicator of the location and strength of the storm track, even though such data will also include anticyclonic variations. Ideally, however, investigations of extreme wind event trends require sea-level pressure data for individual cyclones. One common proxy measure is the central pressure. If this is used, an individual cyclone system can be associated either with a single data point (the minimum pressure reached by the cyclone) or with many data points (e.g. if all lows in a daily record are counted). Lambert (1996) reports that around 5 per cent of all extratropical Northern Hemisphere daily lows have central pressures below about 970 mb. (The deepest depressions, such as Daria, noted in Box 2.3, can reach minimum values below 950 mb.) However, Schmith *et al.* (1998) note that cyclone sea-level pressure data can be inhomogeneous; in particular, there is a tendency for recent meteorological analyses to capture deeper and deeper lows. There are also large differences in the accuracy and temporal coverage of analyses between different regions (Trenberth and Owen, 1999). In addition, the relationship between maximum pressure gradient (which determines the wind speed) and minimum pressure will depend on the mean climatological pressure of the region (WASA Group, 1998; Sickmöller *et al.*, 2000). Therefore, sea-level pressure gradient data for individual cyclones are expected to yield more accurate results (Sickmöller *et al.*, 2000). More indirectly, high-frequency sea-level fluctuations have also been used to indicate sea-level pressure variations due to the passage of cyclones (WASA Group, 1998).

2.3.3 Observed changes in the mid-latitude windstorm hazard

Most instrument-based observational studies of extreme wind events stretch back only as far as the late nineteenth century. For earlier periods, historical records have sometimes been used. Lamb (1988, chapters 8, 9 and 11; 1991) suggested that there were periods of increased storminess in the North Sea and surrounding areas during the Little Ice Age, a period of generally colder climatic conditions in the Northern Hemisphere, whose coldest phase lasted approximately from AD 1570 to 1730 (Nicholls *et al.*, 1996). Lamb (1977, pp. 35–6, 466–7) argued that reduced temperatures and increased ice cover at high latitudes were accompanied by a southward shift of the polar jet stream, which was frequently stronger (owing to the increased meridional temperature gradient). Such a jet stream would act both to steer extratropical cyclones further south and to increase their severity, a conclusion supported qualitatively by a historical survey of severe storms in the region (Lamb, 1991). In one case study, Lamb (1979) presented evidence that between 1675 and 1705 there was a particularly large meridional temperature gradient between 55° N and 65° N. He speculated (Lamb, 1988, p. 125) that this helped to produce at least four exceptionally severe windstorms in the North Sea area between 1690 and 1703, including the great storm of 1703 noted earlier (Lamb, 1991, pp. 51–72).

Using the modern instrumental record, a somewhat larger area of north-west Europe and the north-east Atlantic has been studied by the WASA (Waves and Storms in the North Atlantic) group of researchers. The WASA project was set up in the early 1990s to determine whether recent extreme waves and storms in the area formed part of a real increasing trend (WASA Group, 1998). In one of the studies, Alexandersson *et al.* (1998, 2000) used homogeneous station pressure data to calculate geostrophic wind speed three times a day for a series of triangular areas over the period 1880–1998. High percentile values of the resulting wind speed distributions (see Section 2.2.3) show significant decadal-scale variations, though with considerable local differences. For a typical area, the smoothed exceedance frequency (ø 25 m/s for the geostrophic wind) shows overall variations for the period of about a factor of 2.5. Combined plots for the west and east of the region both show a general declining trend from a high in the 1880s to a minimum in the 1960s, followed by a rise to 1995 (Fig. 2.7). Correlations with the NAO index (see Box 2.1 and Section 2.3.1) were found, the best being for the western region and the winter NAO index, though only from 1900 onwards. Variations in the smoothed curves for this period (Figs 2.1b and 2.7a) also appear qualitatively comparable.

Of particular interest with regard to climate change is the period of the past few decades, when the rapid rise in global temperatures is thought to have a major anthropogenic component (Fig. 1.8). The WASA Group (WASA Group, 1998; Schmith *et al.*, 1998) also found a general tendency for an increase in storminess over this period using less direct measures of wind speed, namely pressure tendency (the 24-hour change in pressure at a station) and high-frequency sea-level variations due to atmospheric pressure effects. By tracking cyclones over the period 1979–97, Sickmöller *et al.* (2000) found an increasing trend of typically a few percentage points per year in cyclone density (see Section 2.3.1) in the WASA area (mainly in the west of the region), accompanied by a decrease in central and eastern Europe.

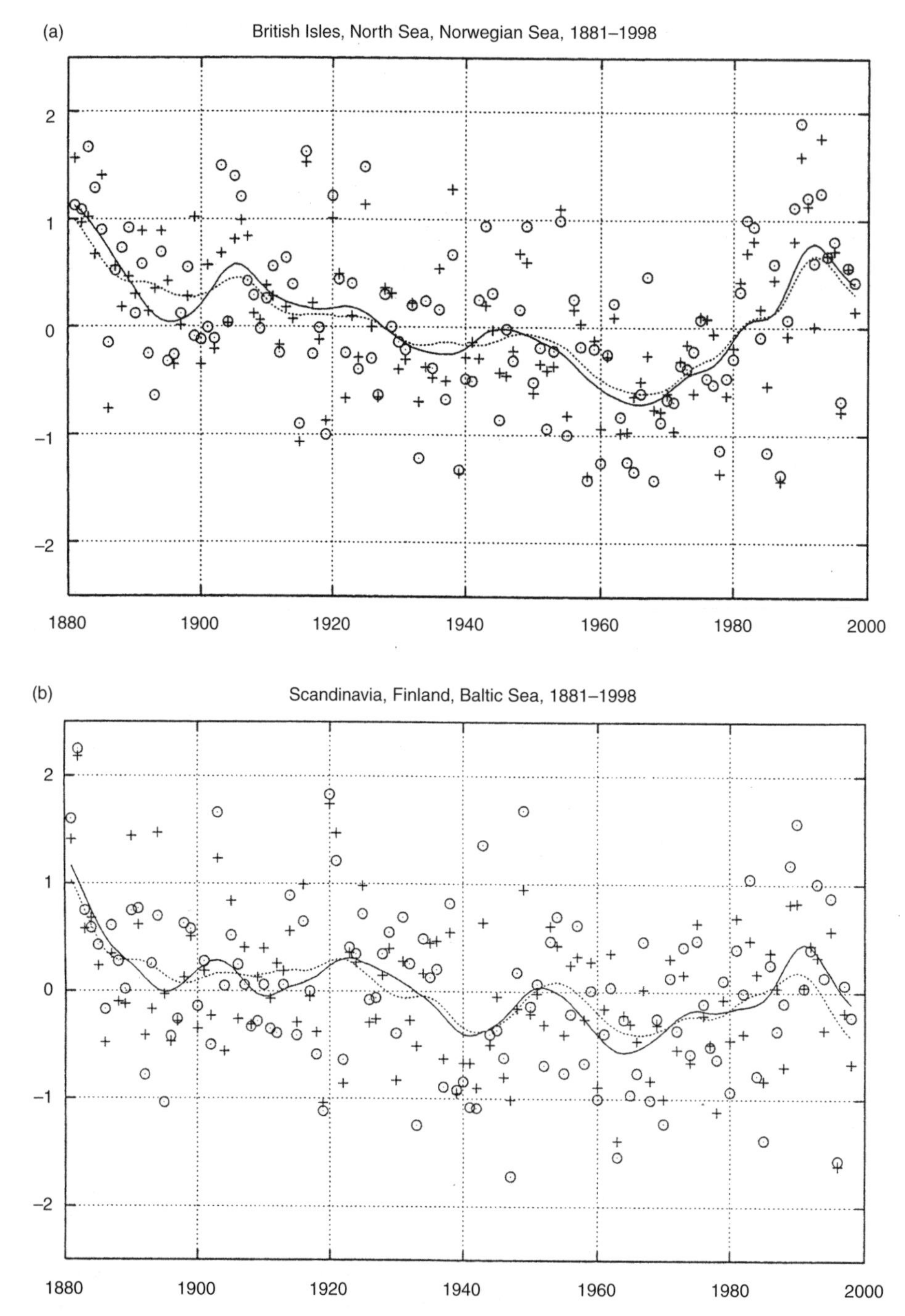

FIGURE 2.7 Standardised 95th (+, smoothed as solid line) and 99th (o, smoothed as dotted line) percentile values of yearly geostrophic surface wind speed distributions for the North Atlantic and north-west Europe from 1881 to 1998, calculated using surface pressure observations (see main text). Data are averaged for two areas: (a) British Isles, North Sea, Norwegian Sea; (b) Scandinavia, Finland, Baltic Sea. (Reproduced from Alexandersson *et al.*, 2000 with permission from Inter-Research Science Publisher)

This is in agreement with the generally rising trend in the NAO index over this period and the general NAO correlation noted by themselves (see section 2.3.1) and by Alexandersson *et al.* (1998) (see above). The rising NAO trend was in fact interrupted in 1996, and the lower values of the winter NAO index from 1996 to 1998 do seem to be reflected in reduced storminess in the WASA region (see Fig. 2.7 and Alexandersson *et al.*, 2000).

Although the NAO is part of the climate's natural variability, these results do not automatically rule out an anthropogenic influence in the past few decades. In either case, the recent high level of storminess in this region arguably involves different causal factors compared with the previous highest level of storminess in the record, which occurred in the 1880s, when the winter NAO index was not consistently high (Figs 2.1b and 2.7). That period is sometimes considered the tail end of the Little Ice Age (Lamb, 1977, p. 463), and Alexandersson *et al.* (1998) speculated that the many very cold years in northern Europe at that time may have been associated with a few particularly violent storms, so increasing the high percentile values. Lamb (1988, pp. 93–4) also found a high frequency of extreme wind events in the 1880s using pressure map data for part of the region, and noted the apparent association between extreme storms there and periods of cold climate and fewer westerlies.

Interesting as the most recent results are, these data are dominated by relatively frequent events with not especially hazardous wind speeds (see Section 2.2.3). Therefore they beg the question as to how representative they are with regard to very rare, very extreme, very hazardous wind events. First, the link between these proxy data and wind speed may not necessarily hold for the most extreme events (WASA Group, 1998). Second, these very rare events could in principle form a separate population with regard to causal factors, and so could be following trends that are not reflected in the less extreme event data. In the past few decades there have certainly been some extraordinarily severe European storms, two of which were noted earlier. Lamb (1991, pp. 33–4) speculated that their severity may be largely related to the general increase in SST leading to increased latent heat release within the storms (see Section 2.3.1).

There have also been a number of recent studies of extratropical cyclone trends in North America. Agee (1991) examined data from three previous investigations and found trends in the annual frequency of cyclones (of all intensities) over North America and surrounding oceans, namely an increase between 1905 and 1940 and a decrease between 1950 and 1980, and then a possible rise to 1985. For the same region Changnon *et al.* (1995) also found, for all seasons, a decrease in the frequency of all cyclones between 1950 and 1980, followed by an increase (of about 50 per cent) into the mid-1990s (Fig. 2.8). Again, the question arises as to how such results relate to *extreme* cyclones. For the Great Lakes region, Angel and Isard (1998) in fact found similar trends between 1900 and 1990 in the frequency of strong cyclones (those with minimum pressure Ł 992 mb). Also, correlations of data since the mid-1960s with the PNA index (Box 2.1) showed that strong winter Great Lake cyclones were more prevalent when the flow over the United States was zonal (i.e. with a negative PNA index; Fig. 2.2b) rather than meridional (Angel, personal communication). For the east coast of North America, Davis and Dolan (1993) found that the frequency of nor'easters was high from the mid-1940s to the mid-1960s, but declined to the mid-1970s. Since then, overall numbers have varied but have not reached pre-1965 levels, whereas the frequency of the strongest storms has greatly increased, with seven out of the eight most extreme storms between 1942 and 1992 occurring since 1960. These authors also suggested a link to the general circulation, associating the recent high frequency of extreme storms with the prevalence of meridional rather than zonal flow in winter over the United States (i.e. positive PNA index; Fig. 2.2a) over those decades. They argued that such flow promotes cyclone intensification in the region as large-amplitude north–south waves in the jet stream bring polar air southwards, and large temperature gradients develop between the cold land and the warm ocean off the east coast.

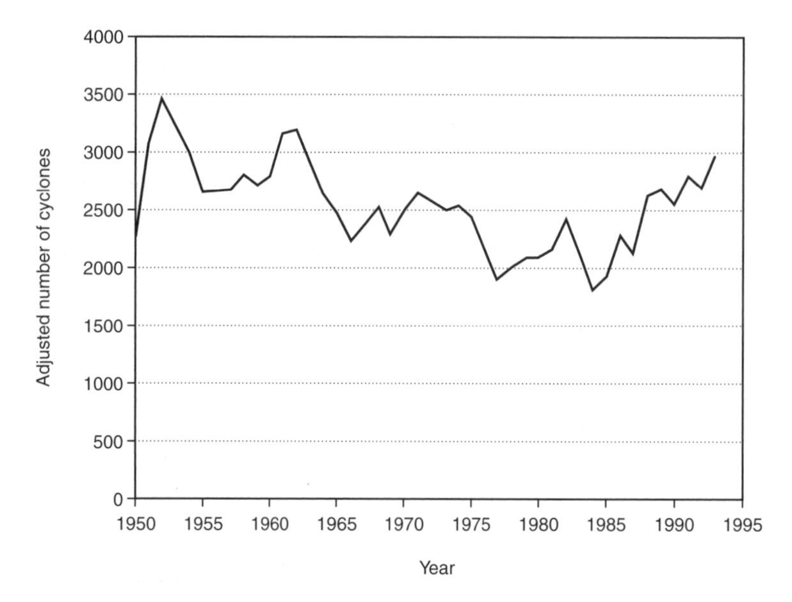

FIGURE 2.8 Annual cyclone counts for North America and the surrounding ocean between 1950 and 1993. The method used equal-area circles to sample a region from 25° to 70° N and 60° to 140° W. The totals were then adjusted for incomplete sampling. (Reproduced from Changnon *et al.*, 1995 with permission from the American Meteorological Society)

For Australia, Hopkins and Holland (1997) found that the frequency of east-coast cyclones nearly doubled between 1958 and 1992, but with considerable annual variability (ranging between zero and seven per year). There was a preference for the cyclones to form between extreme ENSO episodes.

As regards larger-scale studies using meteorological analyses, the data are again generally dominated by not particularly hazardous events, in terms of both intensity and location. Using central pressures, Lambert (1996) found decadal-scale fluctuations in the frequency of daily intense lows (<970 mb) for the Northern Hemisphere for the period 1900–90. In both North Atlantic and North Pacific sectors there was a rise in frequency in the past few decades. However, such data may be non-homogeneous, as noted earlier. Sickmöller *et al.* (2000) found very similar trends in the frequency of deep lows in similar longitude sectors in data from 1979 to 1997, but noted that these may be related to cyclone trajectories shifting northwards into a region of lower mean pressure. They found that the frequency of lows with high pressure gradients (i.e. high winds) actually *decreased* over the period 1979–97 in the North Atlantic sector, and showed no strong trend in the Pacific sector.

In conclusion, it is clear that the recent use of homogeneous data sets is now enabling genuine and consistent variations in the frequency of intense extratropical cyclones and high winds to be observed, particularly for the past few decades. There are clear decadal-scale variations, and apparent associations with oscillations in patterns of circulation as characterised by the NAO, PNA pattern and ENSO, in most cases involving changes in the numbers and preferred locations of cyclones. How such variations relate to the most extreme, hazardous events, however, is still uncertain, as is the extent of any anthropogenic influence on recent trends, particularly as the natural variability seems to be very large. To help clarify such issues, and understand the detailed mechanisms behind the observed variations, we must turn to modelling.

2.3.4 Predictions of changes in mid-latitude storminess

Hall *et al.* (1994) summarised the achievements and limitations of early dynamic models of the climate system in addressing the mid-latitude windstorm hazard. Typical models used in the 1980s had relatively poor spatial resolution, resulting in a poor simulation of the growth and decay of individual extratropical cyclones. Therefore they were unable to make accurate predictions of changes to the mid-latitude windstorm hazard with an enhanced greenhouse effect. However, they did all predict three global-scale changes that can affect storm frequency and intensity (see Section 2.3.1). The first is an increase in atmospheric water vapour, due to increased evaporation as the ocean surface warms. This tends to produce more intense storms through increased latent heat release within the cyclone. The second is an enhanced warming of the surface and lower troposphere at high latitudes in winter (due to a positive feedback, as a reduction in ice and snow cover leads to increased absorption of sunlight), though later coupled models generally show this effect restricted to the Northern Hemisphere (e.g. Carnell and Senior, 1998). This will reduce the surface meridional temperature gradient, leading to a weaker jet stream and less intense storms. The third is an enhanced warming of the upper troposphere at low latitudes, due to increase latent heating there, which provides a greater energy source for storms. Thus the expected changes are likely to have competing effects on the frequency and intensity of extratropical cyclones.

Another possible global-scale effect was pointed out by Trenberth (1999) (see also Section 2.2.2 and Chapter 3). He notes that there will be an increased poleward heat transport associated with the increased latent heat release in extratropical cyclones, and this may be balanced by there being reduced overall numbers of storms. The result of global warming may therefore be fewer but more intense extratropical cyclones.

In addition, local factors, such as the differential heating of land and ocean (affecting the temperature gradient), and the frequency and position of blocking features, are of crucial importance (Carnell and Senior, 1998). Any long-term trends in regional decadal-scale oscillations such as the PNA pattern or the NAO would also have an effect. There are also regional factors such as changes in the intensity and location of the shallow continental winter highs, or possible alterations to the North Atlantic Ocean circulation. In the far North Atlantic the surface water is dense because it has cooled and has increased in salinity following evaporation and sea ice formation. It therefore sinks and flows southwards at depth as it is replaced by warm northward-flowing surface water from the Gulf Stream. Most models predict a weakening of this thermohaline circulation (sometimes known as the Atlantic conveyor belt), mainly due to increased precipitation at high latitudes reducing the surface water salinity. This leads to reduced warming or even cooling at high latitudes.

More sophisticated models have since been used for predicting the effects of global warming on mid-latitude storms. Carnell and Senior (1998) noted that the most recent models are coupled ocean–atmosphere models with improved horizontal resolution. Such models not only can resolve individual extratropical cyclones and blocking anticyclones, but also can more accurately simulate aspects of natural variability such as ENSO. Nevertheless, they may still be inadequate to predict accurately the paths of individual cyclone tracks and the positions of blocking highs. Kattenberg *et al.* (1996) summarised some of the early results from the 1990s. The models simulated the storm track positions but tended to underestimate their strength. These results show little agreement in predicting changes in storminess due to enhanced CO_2, partly because of their different resolutions and methods of studying the storm tracks.

Since then, a number of different models have been used to investigate the effect of an enhanced greenhouse effect (in both equilibrium and transient experiments) on extreme wind events, either directly or via numbers of intense cyclones. In each case we note the predicted effects for CO_2 doubling, unless stated. In an equilibrium experiment, Lambert (1995) examined the frequency of winter lows (noted once per day, i.e. allowing many lows per cyclone system) as a function of central pressure. In the Southern Hemisphere he found a 5 per cent reduction in the total frequency of lows, with no significant increase in intense ones. Katzfey and McInnes (1996) investigated changes in Australian east-coast low systems in an equilibrium experiment. They found an overall reduction of almost 50 per cent in the frequency of such systems, but with a tendency towards lower minimum pressures. In the

Northern Hemisphere, Lambert (1995) found a 4 per cent decrease in the total frequency of lows, accompanied by a considerable increase in the numbers of more intense lows near the termination of the Pacific and Atlantic storm tracks. In a transient experiment, Carnell *et al.* (1996) too found fewer but deeper cyclones (each cyclone being noted only once) at the ends of these storm tracks. They also found an increase in wind speeds in the eastern North Atlantic area, with a 30–40 per cent increase in the numbers of gales over the British Isles. A similar wind speed trend was found by Lunkeit *et al.* (1996) in a transient experiment with a larger CO_2 increase. In an equilibrium experiment, Zwiers and Kharin (1998) examined global changes in wind speeds having a 20-year return period; in mid-latitudes the only (marginally significant) change is an increase in north-west Europe. They found that the number of strong-wind days per year also increases there, but decreases within the main storm tracks. Schubert *et al.* (1998), however, in a transient experiment, found little change in the frequency of lows or the intensity of cyclones in the North Atlantic, though there is a shift in frequency toward the north-east. Using ensembles of transient experiments, Carnell and Senior (1998) found an overall decrease of a few per cent in the number of Northern Hemisphere winter storms (each being counted once), but an increase in the number with minimum pressure <970 mb (by 11 per cent for runs including aerosol effects). They also found an increase (decrease) in the frequency of blocking highs in the eastern (central) Pacific, and noted a strong tendency for the positive PNA index pattern to become predominant in winter. Ulbrich and Christoph (1999), in a transient experiment, found a slight tendency for the positive NAO index pattern to become more prevalent, with a shift of its low-pressure centre to the north-east.

Thus, despite the problems outlined earlier, we find a consensus is beginning to emerge from recent models that with enhanced greenhouse warming there is likely to be an increase in the frequency of intense extratropical cyclones, particularly at the end of the storm tracks where the storms reach maturity. However, this will probably be accompanied by a reduction in the overall frequency of lows. Such changes accord with Trenberth's suggestion based on theoretical arguments (Trenberth, 1999), and can be explained in terms of the various competing mechanisms, as discussed earlier. In particular, most studies indicate consequent increases in the frequency of extreme wind events in the north-east Atlantic and north-west Europe. Positive phases of the PNA and NAO circulation patterns are also predicted to be more prevalent. Nevertheless, the predicted trends are relatively small compared with the observed decadal-scale natural variability. Because of this, and because of the difficulties in observing trends in the most extreme events, it is still unclear whether these predicted anthropogenic effects are actually beginning to occur in the observed record, despite a number of recent severe storms. Nevertheless, as models improve, particularly in their resolution, their predictions are likely to become more accurate and reliable, even without strong observational support.

2.4 Tropical cyclones, tornadoes, derechos and dust storms as the climate changes

Tropical cyclones, tornadoes and derechos produce damaging winds on a smaller scale than extratropical cyclones, which makes them more difficult to measure or model accurately. Nevertheless, they can be highly destructive, and so it is important to try to monitor and predict any changes in their frequency or intensity with global warming. In this chapter we will restrict ourselves to the wind hazard; the flood hazard from tropical cyclones is dealt with in Chapter 3. We also consider here the hazard from wind-generated dust storms. Only outlines of the phenomena are given here; most textbooks on meteorology give more detailed descriptions.

2.4.1 Tropical cyclones and hurricanes: an overview

Tropical cyclone (TC) is the generic name for all storms with a cyclonic circulation originating over warm tropical waters. Though there are various local names for TCs, for simplicity we will use here only the names and categories for TCs in the North Atlantic basin. TCs are categorised by maximum sustained (1 min average) surface (10 m) wind speed, using the gale-force and hurricane-force limits on the Beaufort scale (see Box 2.3): *tropical depressions* (<18 m/s), *tropical storms* (18–32 m/s), and *hurricanes* (ø 32 m/s). Hurricanes are in turn categorised by wind speed from 1 to 5 on the Saffir–Simpson disaster-potential scale (Table 2.1), with category 3, 4 and 5 hurricanes, having wind speeds of at least 50 m/s, being known as *major hurricanes*. Each year, worldwide, there are some 80–90 tropical storms and about 40–50 of these become hurricanes (Henderson-Sellers *et al.*, 1998). Hurricanes, and especially major hurricanes, can cause thousands of deaths and enormous amounts of damage. Over the ocean, where they develop, the associated extreme winds and waves cause a direct threat to shipping. If they make landfall, however, lives and property are threatened not only by winds and waves, but by flooding due to storm surge and heavy rainfall (see Chapter 3). In addition, the rainfall can cause landslides in some areas (see Chapter 4). Therefore it is of great importance to know whether the hurricane hazard is likely to increase under global warming.

Like intense extratropical cyclones, hurricanes have a very low central sea-level pressure and converging, cyclonically circulating surface winds, and their intensification also involves rising air near the centre being removed by divergence in the upper atmosphere more rapidly than it converges at the surface. However, their structure and mechanism are very different (Fig. 2.9). First, over the tropical ocean there is a single warm, humid air mass and no fronts. Thus when the converging air rises, the resulting cloud and rain are not frontal and asymmetrical in character, but involve deep convection in an unstable atmosphere, arranged symmetrically about the storm centre. Furthermore, with no upper-level jet stream, and little meridional temperature gradient, the energy source for hurricanes is almost entirely the warm tropical ocean. This prevents the surface air from cooling by expansion as it converges towards the central low pressure, and also provides moisture (i.e. latent heat) by evaporation. The rising and condensing of this warm, moist air into cumulonimbus clouds then increases the air temperature near the centre of the storm. The resulting 'warm core' at the centre creates (by vertical expansion of the air column) high pressure and hence divergence in the upper atmosphere. In some hurricanes, however, temperature gradients do play a role. These *baroclinically enhanced* hurricanes are either initiated or enhanced by divergence associated with nearby upper-tropospheric low-pressure features (Elsner and Kara, 1999). (A baroclinic atmosphere is one where density and pressure contours are not aligned, usually because of horizontal temperature gradients.)

Compared with intense extratropical cyclones, hurricanes generally have lower central sea-level pressures (often <950 mb, occasionally <900 mb), and they are usually smaller (typically several hundred kilometres in diameter), so the surface pressure gradient

TABLE 2.1 The Saffir–Simpson scale for hurricanes

Category	Damage	Central pressure (mb)	*V* (m/s)	Surge (m)
1	Minimal	≥ 980	33–42	1.0–1.5
2	Moderate	965–979	43–49	1.5–2.5
3	Extensive	945–964	50–58	2.5–4.0
4	Extreme	920–944	59–69	4.0–5.5
5	Catastrophic	< 920	> 69	> 5.5

V is the maximum (1-minute average) surface (10 m) wind speed. The storm surge limits are converted from feet and given to the nearest 0.5 m.

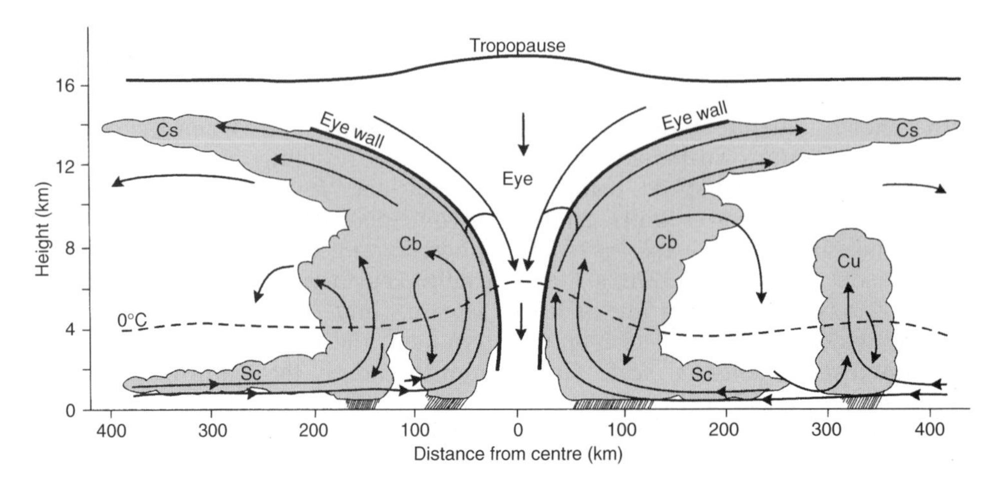

FIGURE 2.9 Schematic representation of the vertical structure of a mature hurricane, with arrows representing the air flows involved. (Reproduced from *Atmospheric Processes and Systems* by Russell D. Thompson, 1998 with permission from Routledge)

towards the centre is much greater, accounting for the stronger winds. Near the centre the resulting large centrifugal force (which dominates the weak low-latitude Coriolis force) balances the pressure gradient force, so that the circulating winds become tangential and cyclostrophic. The maximum surface wind speed, convection, atmospheric heating and rainfall all occur in the eyewall, a ring of towering cumulonimbus clouds with a typical radius of a few tens of kilometres. This encloses the eye of the hurricane, where winds are light and air is subsiding rather than rising. This air is thus compressed and warmed (helping the divergence aloft), with little condensation into cloud. The rest of the hurricane's convective cloud is organised into a series of spiral bands, giving the hurricane its characteristic appearance on satellite images. (The upper-level cirrus from the cumulonimbus flows out anticyclonically, so has the same spiral curvature as the cyclonic inflowing surface air.) Hurricanes travel with the regional winds (typically at around 10 m/s), so the wind speed will be increased by this amount on the left (right) side of the approaching hurricane in the Northern (Southern) Hemisphere, and reduced on the other side.

The hazard from hurricanes depends on their frequency, maximum intensity, size, duration and geographical distribution (particularly if and where landfall occurs), so we need to know whether any of these factors are changing, or are likely to change, as a result of global warming (Emanuel, 1997). Most studies have concentrated on frequency and intensity. Variations in geographical distribution and duration are to some extent incorporated within such studies, but variations in size distribution have been little studied. Hurricanes display a wide range of sizes, and damage is strongly dependent on size. (The radius within which gale-force winds have been recorded ranges from 50 km for Cyclone Tracey in 1974 to 1100 km for Typhoon Tip in 1979; Pielke and Pielke, 1997.) However, there is little currently known about the factors that determine hurricane size (Emanuel, 1997; Elsner and Kara, 1999). The factors governing the frequency of TCs (i.e. type A changes to the frequency distribution; see Section 2.2.3) are different from those governing their maximum intensity (i.e. type B or type C changes) (Emanuel, 1995). We will therefore review each of these aspects in turn.

The frequency of TCs in a given region is dependent on the frequency of suitable tropical disturbances such as meso-scale convective systems (MCSs) associated with easterly waves or the intertropical convergence zone (ITCZ), and on the presence of conducive environmental conditions. Few tropical disturbances actually become TCs, and though the suitability criteria are uncertain, modelling studies suggest that a major requirement is for the disturbance to have a sufficiently strong initial circulation, an independent 'kick-start' to initiate TC formation (Emanuel, 1991). As regards environmental conditions, empirically based studies led to the identification of six criteria necessary for TC genesis (Gray, 1979; Lighthill *et al.*, 1994; Henderson-Sellers *et al.*, 1998).

1. The latitude must be greater than about 5°. Equatorwards of this there is insufficient Coriolis force to provide the necessary initial cyclonic motion around the low pressure to prevent it filling.
2. There must be a substantial pre-existing amount of low-altitude cyclonic vorticity in the atmosphere to aid the cyclonic spin-up.
3. Values of vertical wind shear must be sufficiently low to prevent moisture and heat being blown away from the centre of the developing storm, and its symmetrical structure being disrupted.
4. The environmental lapse rate must be steep enough to allow free convection and the creation of cumulonimbus clouds.
5. Relative humidity in the middle troposphere must be high enough to avoid drying out of the air in the convective clouds by mixing with the surrounding atmosphere.
6. The sea surface temperature (SST) must be ø 26–27°C to a depth of at least 60 m. This warm surface water is the thermal energy source for the cyclone.

Gray (1979) quantified these criteria as predictor parameters of TC frequency. The product of the first three parameters is the 'dynamic potential' for TC formation (Gray, 1979), and the product of the last three is the 'thermal potential', or 'potential for cumulonimbus convection' (Gray, 1979). These last three criteria, in particular, are not independent. For example, the 26–27°C SST criterion is needed to allow a steep enough lapse rate for deep convection (Emanuel, 1986), and the 60-m depth criterion prevents excessive SST cooling as the hurricane winds mix cooler water from below (Emanuel, 1999). These thermal potential criteria together are fulfilled in the region of ascending air in the Hadley cells (Emanuel, 1988). They effectively set an upper latitude limit of 15–20° beyond which cooler waters prevent cumulonimbus convection, and Hadley cell subsidence of dry air reduces mid-tropospheric humidity and generates a shallow 'trade inversion' that limits free convection to the well-mixed lower troposphere (Lighthill *et al.*, 1994). In combination, the six criteria are generally satisfied in the regions shown in Fig. 2.10, in the summer and early autumn. Indices based on the product of all six parameters relate well to TC frequency maps in all basins (Gray, 1979; Watterson *et al.*, 1995), and also correlate with interannual variations in some basins (Watterson *et al.*, 1995).

As noted earlier, genesis of TCs can also be helped by the presence of a baroclinically related upper atmospheric feature causing divergence aloft. Interannual and intraseasonal TC frequency has also been linked to other local factors such as SLP (Gray *et al.*, 1993; Knaff, 1997), SST (Saunders and Harris, 1997; Shapiro and Goldenberg, 1998) and the Madden–Julian Oscillation (MJO), a modulation of tropical convection and winds (Maloney and Hartmann, 2000a, b; Sobel and Maloney, 2000). These are generally considered to affect hurricane genesis by influencing the above factors.

After formation, TCs may grow into hurricanes, typically over a period of several days. If the divergence at high levels removes the rising air faster than it converges at the surface, the sea-level pressure reduces and the storm intensifies. TC intensification involves a positive feedback whereby lower surface pressure and the consequent increased surface wind speed cause more heat transfer from the ocean, which leads via convection to more warming and divergence aloft. As they intensify, hurricanes usually move westwards in the easterly winds and then *recurve* polewards around the subtropical highs (Fig. 2.10). On reaching cooler water, hurricanes can rapidly decline as they lose their source of energy. They also decline soon after making landfall, owing to the reduction in surface

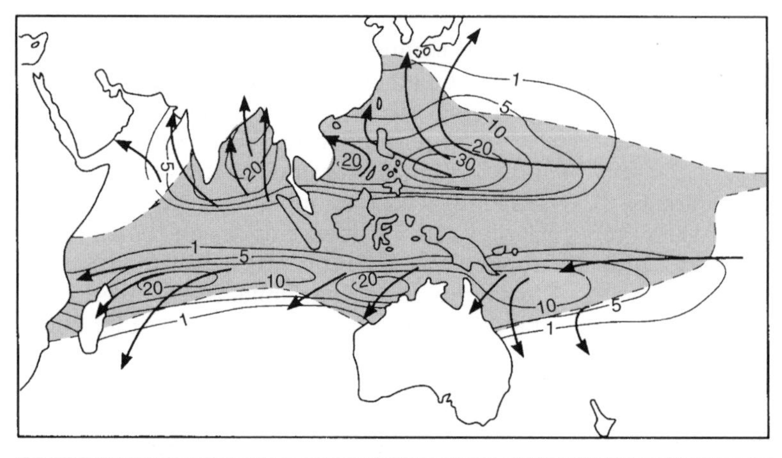

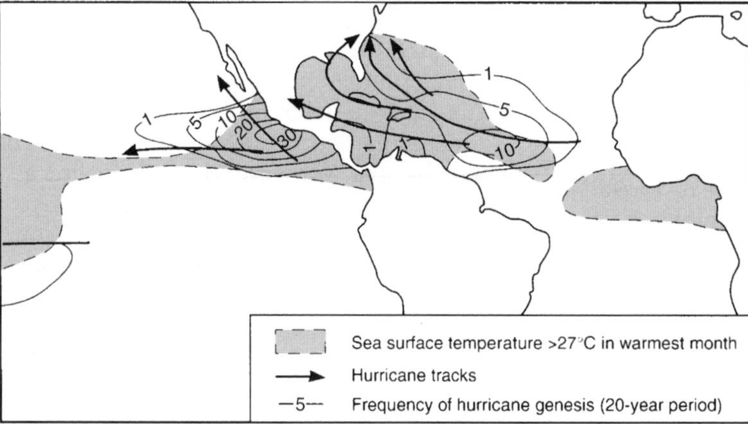

FIGURE 2.10 Contours of hurricane genesis frequency (in a 20-year period), showing the main formation basins. The main hurricane tracks, and areas that are warmer than 27°C in their warmest month, are also shown. (Reproduced from *Atmospheric Processes and Systems* by Russell D. Thompson, 1998 with permission from Routledge)

moisture as well as the increase in friction causing the central low to fill.

Hurricanes can intensify only up to a certain limit. Thermodynamic models can be used to predict the maximum potential intensity (MPI, calculated as the minimum possible pressure) for a given set of environmental conditions (Emanuel, 1986, 1988, 1991; Holland, 1997). The models show considerable ability to predict observed MPI, and both models show a strong dependence of MPI on SST (Fig. 2.11) which agrees reasonably well with observations (Tonkin *et al.*, 2000). However, even if hurricanes do not travel over cooler water or land, they generally do not achieve their MPI. Modelling of the intensification of individual hurricanes has shown that storm-induced ocean surface cooling by mixing cooler water from below is likely to be the main limiting factor (Emanuel, 1999; Bender and Ginis, 2000). Empirically, Emanuel (2000) found that the frequency distribution of hurricane intensities *relative to their MPI* is flat – that is, 'a given storm is equally likely to attain any intensity between hurricane force and its potential intensity'. (These data exclude hurricanes that are limited by declining MPI as they travel over land or cooler water.) Because of the variation in MPI values, however, the frequency distribution of *actual* intensities is peaked, with a long tail to high intensities (Elsner and Kara, 1999, p. 21).

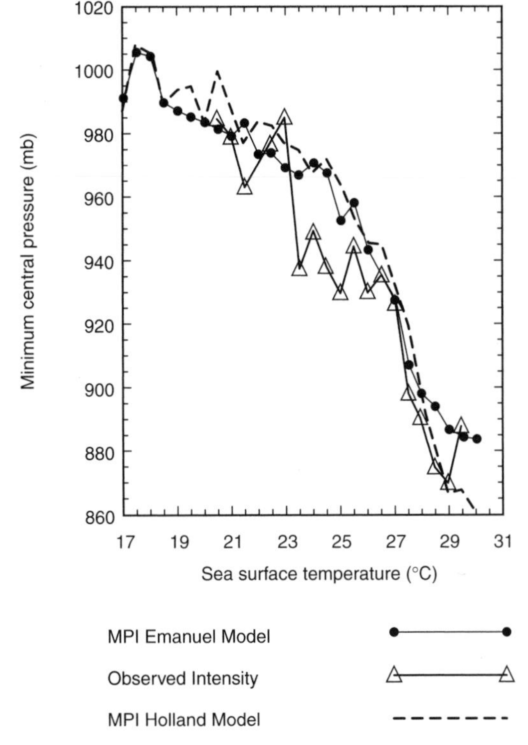

FIGURE 2.11 Hurricane maximum potential intensity (MPI, shown as minimum central pressure), as a function of sea surface temperature (SST), for the North Atlantic, north-west Pacific and Australian/south-west Pacific basins combined. Observed maximum storm intensities are compared with MPIs calculated using the thermodynamic models of Emanuel (1986, 1991) and Holland (1997). Below 26°C, the observed values are an overestimate of the MPI because the hurricanes originally formed in warmer waters. (Reproduced from Tonkin *et al.*, 2000 with permission from the American Meteorological Society)

2.4.2 Observations and predictions of changes in the hurricane hazard

Nicholls (1995), Henderson-Sellers *et al.* (1998) and Trenberth and Owen (1999) noted the problems of observing hurricane trends. Direct observations of hurricanes are difficult, and accurate records using reconnaissance aircraft or satellites are limited to between 25 and 55 years, depending on the ocean basin (Henderson-Sellers *et al.*, 1998). Artificial trends have been identified in both frequency and intensity due to changes in observation or analysis techniques. Nevertheless, both decadal-scale variations and trends have been observed in various basins, though there is no common pattern of variation, and no strong trend in hurricane frequency or intensity worldwide (Henderson-Sellers *et al.*, 1998; Saunders, 1999). Annual indices of hurricane activity used include frequency, mean and maximum intensity and hurricane-days (Landsea *et al.*, 1999).

In the North Atlantic basin, with reliable data since the 1940s, there has been a significant downward trend in major hurricane frequency, and in mean and peak intensity, but no significant trend in the overall frequency of hurricanes and tropical storms (Fig. 2.12) (Landsea *et al.*, 1996, 1999). However, there was a general shift in storm location from south to north of 23.5° N in the 1970s to early 1990s compared with the 1940s to 1960s, which corresponds to a shift from 'tropical-only to 'baroclinically enhanced' cyclones (Kimberlain and Elsner, 1998; Elsner and Kara, 1999; Landsea *et al.*, 1999). A similar inverse relationship between hurricane frequency at low and high latitudes is found with landfalling Atlantic hurricanes over the past two centuries (Elsner *et al.*, 2000). In the north-west Pacific basin, there was a decrease of about 30 per cent in tropical storm and hurricane frequency between 1960 and 1980, followed by a comparable rise since 1980 (Chan and Shi, 1996). In the north-east Pacific basin, over the past three decades there have been strong upward trends in TC and hurricane frequency in the extreme west of the basin (140° W–180° W; Chu and Clark, 1999) and in hurricane and major hurricane frequency in a broader western region (116° W–180° W; Collins and Mason, 2000). However, there is no significant trend for this basin east of 116° W (Collins and Mason, 2000). In the Australian region (105º E–160° E) since 1970, there has been a decrease of about 70 per cent in the frequency of moderate TCs (with minimum pressures between 970 and 990 mb), but a small increase in the frequency of intense (<970 mb) hurricanes (Nicholls *et al.*, 1998). Over the same period, the north Indian basin shows a notable

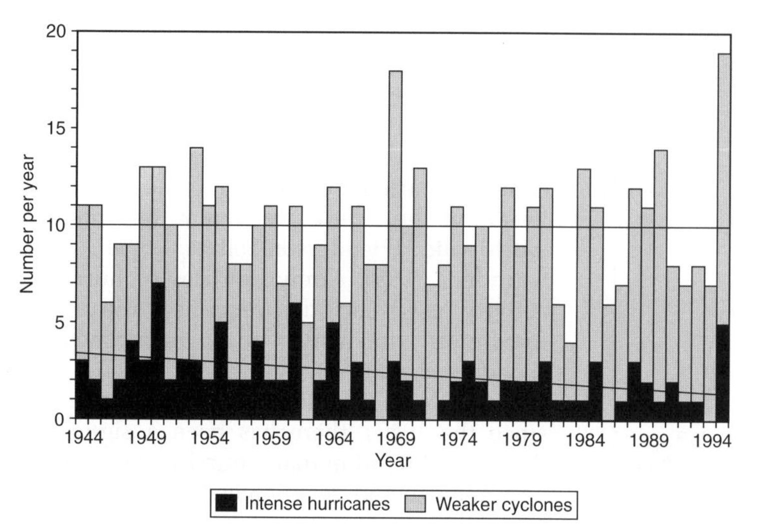

FIGURE 2.12 Annual frequency of tropical cyclones from 1944 to 1995 for the Atlantic basin. Dark bars are for major hurricanes only; light bars include all hurricanes, tropical storms and subtropical storms. Best-fit straight-line trends are also shown. (From Landsea *et al.*, 1996 with permission from the American Geophysical Union)

downward trend in frequency, but the south-west Indian basin and the south-west Pacific region (east of 165° E) show no long-term trends (Henderson-Sellers *et al.*, 1998).

Statistical studies of interannual variations suggest that these recent observed trends can at least partly be ascribed to decadal-scale natural variability, as large-scale circulation patterns modify the factors relating to TC genesis and intensification. Henderson-Sellers *et al.* (1998) reviewed the influence of the Quasi-biennial Oscillation (QBO; a variation in the direction of stratospheric winds) and ENSO. When QBO winds are westerly (easterly), there are more (fewer) hurricane-days in the west Atlantic, and an opposite, weaker relationship exists in the north-west Pacific. El Niño (La Niña) events are associated with fewer (more) hurricanes in the Australian/south-west Pacific region west of 170° E, the north-west Pacific west of 160° E, and the North Atlantic, and more (fewer) hurricanes in the south-west Pacific east of 170° E and the north-west Pacific east of 160° E. More recent results show that the latter relationship also holds for hurricanes in the north-east Pacific basin around Hawaii (Chu and Wang, 1997) and west of 116° W (Collins and Mason, 2000), but that there is no significant relationship with ENSO east of 116° W (Collins and Mason, 2000). Relationships with the NAO have also been discovered. The decadal-scale north–south shift in Atlantic hurricanes is related to July NAO values, probably via the influence of the subtropical high on hurricane tracks (Elsner *et al.*, 2000), and the annual NAO index is linked to correlated variations in global hurricane frequencies between different basins (Elsner and Kocher, 2000). Links with the Atlantic thermohaline circulation have also been suggested (Gray *et al.*, 1997, Elsner and Kocher, 2000).

As regards prediction of trends under global warming, frequency and intensity are generally considered separately. A number of semi-empirical studies have attempted to estimate frequency changes using GCM predictions of how the spatial and temporal distribution of the six TC genesis criteria may change (Ryan *et al.*, 1992; Royer *et al.*, 1998; Druyan *et al.*, 1999). However, the results are inconclusive because some of the parameter values are tuned to the present climate (Ryan *et al.*, 1992; Broccoli *et al.*, 1995; Royer *et al.*, 1998; Druyan *et al.*, 1999). For example, the area enclosed by the 26°C isotherm is predicted to increase under global warming, but the trade wind inversion high-latitude limit seems unlikely to change (Lighthill *et al.*, 1994), and thermodynamic modelling indicates that upper-atmospheric warming will lead to an increased SST threshold for TC formation (Henderson-Sellers *et al.*, 1998) (Fig. 2.13). Hence the geographical

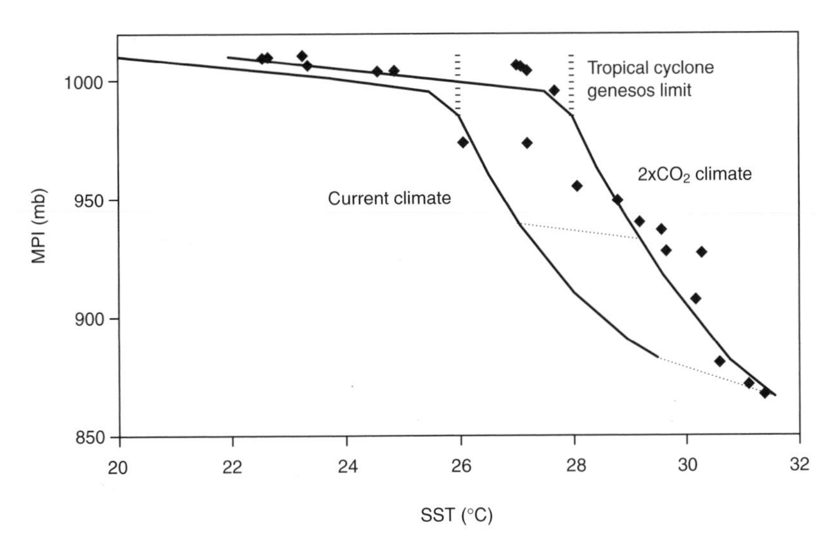

FIGURE 2.13 Modelled effect of global warming on the genesis and maximum potential intensity (MPI) of tropical cyclones. The thin line is based on an empirical curve of MPI as a function of sea surface temperature (SST) for the current climate. The data points (fitted by the thick curve) represent MPI values calculated using the thermodynamic model of Holland (1997) for atmospheric temperature and SST conditions after greenhouse warming. The thick dashed lines show an increase in the genesis threshold from 26°C to 28°C. The thin dashed lines indicate the sense of the MPI changes, with a slight increase in MPI predicted. (Reproduced from Henderson-Sellers *et al.*, 1998 with permission from the American Meteorological Society)

regions for cyclogenesis are expected to be unchanged (Henderson-Sellers *et al.*, 1998). In addition, such studies do not include possible changes in the frequency of suitable tropical disturbances, or in TC movement and decay.

The other main method used to predict TC frequency changes is modelling of the TCs themselves within GCMs. To determine TC frequencies accurately, such models need to be able to simulate the special atmospheric conditions and the frequency of occurrence of suitable initiating tropical disturbances, as well as the early stages of TC growth. Their potential problems are therefore a lack of spatial resolution and inadequate prediction of large-scale atmospheric circulation features such as ENSO (Kattenberg *et al.*, 1996; Henderson-Sellers *et al.*, 1998). However, since the mid-1990s, high-resolution (~100-km) GCMs and nested models (down to tens of kilometres resolution) have been able to reproduce current patterns of TC formation reasonably well, and the TC-like vortices (TCLVs) that are obtained have some of the features of TCs. Using high-resolution global models, Bengtsson *et al.* (1996) found a reduced TCLV frequency under global warming, and Krishnamurti *et al.* (1998) found a slight increase (no change) in TCLV frequency for simulated La Niña (El Niño) conditions under doubled CO_2. Using a nested model, Walsh and Katzfey (2000) found no significant change in TCLV frequency for the Australian region for CO_2 doubling, though the TCLVs tended to travel further polewards. Though these early results must be treated with caution (Henderson-Sellers *et al.*, 1998), it is expected that with increased resolution (down to a few kilometres) and excellent model physics, this approach should eventually provide good predictions of future TC statistics (Lighthill *et al.*, 1994; Walsh and Watterson, 1997). These would include not just frequency but distributions in location, duration, size and intensity.

As regards hurricane intensity, Henderson-Sellers *et al.* (1998) reviewed empirical and modelling studies that show hurricanes' MPI to be strongly dependent on SST (for example, see Fig. 2.11). Thermodynamic models predict that MPI values will increase by 10–20 per cent under GCM doubled-CO_2 scenarios (Emanuel, 1987; Henderson-Sellers *et al.*, 1998), although factors omitted from the models may act to reduce these increases (Henderson-Sellers *et al.*, 1998; Shen *et al.*, 2000). Provided that the distribution of hurricanes' intensities relative to their MPI values remains flat (see earlier), their actual intensities should increase by a similar percentage (Emanuel, 1999). These results have been compared with those of nested GCM experiments. For doubled CO_2, Knutson *et al.* (1998) found an increase in wind speed of 5–12 per cent for the north-west Pacific, in reasonable agreement with theoretical estimates of MPI changes. For the Australian region, Walsh and Ryan (2000) found some intensity increase, but less than MPI theory predicts. Thus current predictions of hurricane intensity changes are small compared with observed natural variability (Henderson-Sellers *et al.*, 1998). Nevertheless, and despite the uncertainty in the current predictions, it seems possible that a warmer world will experience hurricanes of a somewhat greater average intensity.

2.4.3 Tornadoes and derechos

Like tropical cyclones, tornadoes involve wind circulating cyclostrophically around a central low pressure, mainly cyclonically (though tornadoes are occasionally anticyclonic). Although they can have higher wind speeds than tropical cyclones, and generally move faster, their path of destruction is much more limited, as they are very much smaller in size and have a much shorter life. Nevertheless, they are a major windstorm hazard in that, unlike tropical cyclones, they are mainly land based, they can occur relatively frequently in some regions, and they are very difficult to forecast, either in their formation or their precise path. For example, major tornado disasters, each involving several hundred deaths, occurred on 3–4 April 1974 in the United States, on 9 June 1984 in the Soviet Union, and on 13 May 1996 in Bangladesh.

Tornadoes are mainly spawned from severe thunderstorms, though they can also form in association with ordinary thunderstorms or tropical cyclones. Their occurrence is highest where there are large numbers of supercell thunderstorms, with most forming in mid-latitudes and the majority (more than 800 per year on average) being found in the United States, mainly in the Great Plains. They occur in all seasons and at all times of day, but in the United States their incidence peaks strongly in spring and early summer, with most occurring in late afternoon when the near-surface atmosphere is most unstable.

For a supercell thunderstorm to generate a tornado, its updraught initially needs to be rotating as a *mesocyclone*, typically around 10 km in diameter. However, the cause of this rotation, and how it sometimes leads to tornado formation, are still matters of debate, though the main features of the supercell, namely strong updraughts and downdraughts, rapidly converging air at the base, and vertical wind shear, may all play a role. Typically a tornado consists of a funnel-shaped cloud extending from the base of the thunderstorm's cumulonimbus cloud, darkened by debris where it makes contact with the ground. Circulating wind speeds have a maximum at the edge of the tornado's 'core', and the most violent tornadoes are thought to have small but very damaging 'suction vortices' travelling around the main vortex. As with tropical cyclones, the velocity of the circulating winds adds to the travel velocity of the tornado, so the total wind speed (and damage) is less on one side of the vortex and greater on the other.

The median (mean) diameter and path length for tornadoes in the United States between 1950 and 1982, based on their damage paths, are 43 (117) m and 1.6 (7.1) km (Schaefer *et al.*, 1986), and they typically last for a few minutes, travelling at speeds of 10–20 m/s. However, the largest tornadoes are several kilometres in size, and some have speeds of more than 35 m/s. Moreover, some have been reported as travelling for several hundred kilometres and lasting for several hours. They often occur in families or outbreaks produced by a large cluster of severe thunderstorms known as a meso-scale convective complex (MCC). One of the most violent outbreaks was the 1974 event noted earlier, which involved 148 tornadoes.

The pressure drop at a tornado's centre has occasionally been measured, and it seems likely that it can be up to 100 mb, i.e. comparable to that seen in the most intense tropical cyclones, though over a much smaller distance (Bluestein, 1999a, pp. 13, 133–4). This sudden drop in atmospheric pressure as the vortex passes over a relatively well-sealed building may cause it to explode, but most damage is caused by the extremely high winds. The relatively few measurements of wind speed have all been indirect, by analysing movies of the tornado circulation or using Doppler radar. Estimates are therefore more usually based on the amount, type and appearance of the damage (Schaeffer *et al.*, 1986), using the Fujita scale (Table 2.2). Bluestein *et al.* (1993), using Doppler radar, have observed wind speeds as high as 120–125 m/s in an F5 tornado.

Even after several decades of storm-chasing and field programmes (Bluestein, 1999b) there are few such measurements, because tornadoes are relatively small, short-lived and rare in any individual location (for example, the annual probability of a tornado strike at any point has a maximum, in central Oklahoma, of 6.1×10^{-4} per year; Schaefer *et al.*, 1986). For the same reason, statistics on tornadoes' frequency and severity, and any trends with time, are likely to show inaccuracies. Frequency estimates are likely to be biased in favour of centres of population, and may show increasing trends owing to improvements in tornado-spotting. Estimates of severity based on damage are likely to be unreliable, as the effect of wind on structures is highly variable (even if they are present in a tornado's path) and path widths vary along the path. The parameter that can be estimated most accurately is path length (Schaefer *et al.*, 1986).

Nevertheless, there have been a number of statistical surveys of tornadoes in the United States (e.g. Schaefer *et al.*, 1980, 1986), and Niino *et al.* (1997) have done the same for Japan (though with only about 20 tornadoes per year, the statistical uncertainties are much greater). Generally speaking, the greater the wind speed (F-number) of a tornado, the greater the average path width, path length and damage area (Schaefer *et al.*, 1986). However, frequency distributions of all the variables are very skewed. In the United States between 1950 and 1978, for example, fewer than 2 per cent of tornadoes were in categories F4 and F5, though these accounted for 68 per cent of the deaths (Schaefer *et al.*, 1980).

TABLE 2.2 The Fujita scale for tornadoes

F-number	*V* (m/s)	Damage	Expected effects
F0	18–32	Light	Chimneys and signs damaged, tree branches broken
F1	33–49	Moderate	Lower limit is the start of hurricane wind speed. Roofs damaged, mobile homes overturned, moving cars pushed off the road
F2	50–69	Considerable	Lower limit is the start of major hurricane wind speed. Roofs torn off, mobile homes demolished, large trees uprooted
F3	70–92	Severe	Walls torn off buildings, cars lifted, most trees uprooted
F4	93–116	Devastating	Well-constructed houses destroyed, cars thrown
F5	117–142	Incredible	Buildings lifted off foundations and carried considerable distances to disintegrate, cars thrown more than 100 m, trees debarked

V is the wind speed.

As regards derechos, Bentley and Mote (1998) have investigated their spatial and temporal characteristics in the United States over the period 1986–95. They identify a derecho as a windstorm caused when downbursts (localised surface downdraughts) within the cells of a group of severe thunderstorms form a family of downburst clusters producing damaging surface gusts (>26 m/s) over a length of >400 km. Though less frequent and less damaging than tornadoes, derechos affect more people over a larger area. There were on average 11 per year during this period, mainly in the Central Plains, varying from 1 in the drought year of 1988 to 25 in 1995 when large amounts of warm, moist air were present.

The numbers of derechos are too small to show any meaningful trend, but the number of tornadoes reported in the United States has risen steadily from about 200 per year in 1950. However, since the frequency of F4/F5 storms has remained virtually constant, this trend is likely to be due to increased reporting of weaker storms (Schaefer *et al.*, 1980). In support of this, there has been no equivalent increase in the annual number of tornado days (Grazulis, 1993, referenced by Tobin and Montz, 1997). The same explanation has been suggested for the increasing trend in reported tornadoes in Japan prior to 1960, followed by a constant rate in more recent years (Niino *et al.*, 1997). Monfredo (1999) found a correlation between ENSO and the frequency of violent (F2 to F5) tornadoes in the central United States from 1955 to 1994, with more (fewer) tornadoes in La Niña (El Niño) years. He suggests that this might be due to airflow patterns producing a stronger mid-tropospheric inversion in La Niña years, allowing more supercell thunderstorms to form. Etkin (1995) showed that in the southern prairies of Canada, tornadoes from 1980 to 1992 occurred 11 days earlier (on average) than those from 1951 to 1979. His analysis suggests that this change may be related to a spring/summer warming that occurred in this region during the 1980s.

How are the frequencies of tornadoes and derechos likely to change with global warming? Using a physically based conceptual framework (Trenberth, 1999; see Chapter 3), Trenberth suggested that precipitating weather systems, including thunderstorms, are likely to become more severe. Other things being equal, this might lead us to anticipate more associated tornadoes (Saunders, 1999). However, other factors may not remain unchanged. For example, the overall frequency of storms may decrease (Trenberth, 1999). Predictions of the numbers and locations of supercell thunderstorms and MCCs, with their associated dynamic features (e.g. wind shear, inversions), ideally require numerical modelling. The difficulty in using GCM models to predict trends related to convective storms, however, is their very small scale, though high-spatial-resolution models are now beginning to be successful in forecasting convective events (Bernardet *et al.*, 2000). Griffiths *et al.* (1993) therefore suggested three alternative methods: using correlations with GCM-predicted general circulation indices such as ENSO; using GCM predictions of the necessary conditions for supercell storms, tornadoes, etc. to form (e.g. wind shear); and using GCM predictions of typical weather situations associated with the convective events. The first method could in principle be applied to the above correlations with ENSO for tornadoes in the central United States, but likely changes in ENSO under global warming are still not very certain. In a form of the third method, Etkin (1995) and White and Etkin (1997) have used the above correlation for Canadian tornadoes to suggest an increase in tornado numbers there with global warming.

2.4.4 Dust storms and environmental change

Dust storms are windstorms accompanied by suspended particles of dust (dry soil or sand), and are wide-ranging in their possible effects, as summarised by Goudie and Middleton (1992). Dust storms present various hazards to humans, both directly via air pollution and road accidents, and indirectly via economic effects such as transport disruption and soil erosion. They are also associated with environmental change in a number of ways:

1. Dust storms have many natural environmental consequences, for example relating to geomorphology (e.g. the formation of loess, stone pavements or desert varnish) and sedimentation (in rivers and oceans).
2. Dust storms are often a manifestation of environmental change, particularly *desertification* in semi-arid areas.
3. Dust storms may also be affected (in their frequency and intensity) by environmental change, for example climatic change, altered farming practices or lake bed exposure.

There are various kinds of dust event caused by wind (Goudie and Middleton, 1992). Technically, the term *dust storm* is restricted to situations where particles of dust are entrained into the air by turbulent wind systems, restricting visibility to <1 km. If the visibility is >1 km, the phrase *blowing dust* is used, and convective whirlwinds containing dust are known as *dust devils*. Suspended particles from a wind event that occurred some time earlier, or some distance away, are called *dust haze*.

Bryant (1991, pp. 44, 50–55) summarised the circumstances and formation mechanisms of dust devils and dust storms. Clearly, they can occur only where loose dry soil or sand is available – that is, in situations of low rainfall and low vegetation. They are most frequent in semi-arid regions where the annual rainfall is between 100 and 200 mm. Dust devils can occur when intense solar heating of the surface has warmed the lower levels of the atmosphere, causing it to be unstable to free convection. Localised surface heating then causes air to rise, and air converges at the surface to replace it. If this air rotates as it converges, it will increase its speed by conservation of angular momentum, causing a whirlwind that picks up any loose soil or sand from the surface. Most dust devils are only a few metres in diameter, but they can occasionally become large and destructive. Dust storms are associated with a wide range of wind systems. One common example is turbulent air spreading out from the downdraught of a large convective thunderstorm cell, giving rise to a dust storm known as a *haboob*. A haboob tends to have a sharp dome-shaped leading edge, as the dust is confined to the cool air behind the gust front pushing under the warmer in-flowing air. Another example is an intense cold front, the passage of which can generate an extensive dust storm in mid-latitude regions. As with haboobs, these storms often have a characteristic convex leading edge, as the dust is confined to the cold air region that is pushing under the warm air mass. Dust storms can also have other wind sources such as deep extratropical cyclones (Peterson and Gregory, 1993).

Increases in the frequency and intensity of dust storms at any location can occur either by changes to the nature of the land surface or by changes to the climate (McTainsh and Lynch, 1996), namely increased windiness, reduced precipitation or increased temperature (which, together with windiness, increases evaporation). Goudie and Middleton (1992) examined meteorological records to determine dust storm frequency variations over the past few decades for a large number of areas worldwide. They found upward trends in some areas (e.g. West Africa), downward trends in others (e.g. Mexico City), and cyclical variations in yet others. In many cases dust storm frequency has been governed by climatic variations, with runs of drought years being an important factor, as in the 1930s Dust Bowl in the United States. In some regions, however, human activities have also affected the frequency of dust storms by altering the land surface. Examples of adverse effects are inappropriate farming practices in the Great Plains of the United States, disruption of surfaces by human activity in Mongolia, and lake bed exposure due to excessive water abstraction in California and the Aral Sea. In Mexico, however, an exposed lake bed was stabilised by vegetation, leading to reduced numbers of dust storms.

What of future trends? An increase in the intensity of drought events is one of the likely consequences of anthropogenic global warming (see Chapter 3), so an overall increase in dust storm frequency seems possible. However, modelled changes in temperature and precipitation patterns, and in the frequency and location of the parent wind systems, would also need to be taken into account in assessing overall dust storm trends. As regards dust devils, these might be expected to increase in numbers and intensity if increased surface heating leads to increased atmospheric instability and more convective activity. Overall, however, future dust storm and dust devil trends at any location are likely to be affected not only by anthropogenic changes to the climate, but also, as in the past, by direct human modifications to the land surface. The overall result is likely to be decreases in some areas and increases in others.

3

Floods and other weather-related hazards in a changing climate

3.0 Chapter summary

This chapter follows on directly from Chapter 2, focusing on the other main weather-related hazards: floods, hailstorms, snow and ice storms, droughts, wildfires and temperature extremes. As in Chapter 2, we describe each hazard and then review any observations or predictions of change in its frequency and/or severity under global warming. The main emphasis is on floods, of both the river and coastal types. River flooding depends not only on climate but also on non-climatic factors involving the drainage basin, and we note how environmental change as a result of human activity has an important impact on flood severity and frequency. This also makes the effect of climate change alone difficult to assess, and most research has concentrated on changes in precipitation, with changes in rainfall extremes being widely observed and predicted. We review coastal flooding due to storm surges from tropical and extratropical cyclones, and note the consequences of sea-level change. Changes in temperature extremes such as heat-waves and cold waves are an inevitable consequence of a general rise in temperature, and we review the latest research in this area. We also briefly discuss the other weather-related hazards, though it is far less clear how these are likely to be affected by global warming.

3.1 River flooding and environmental change

The two main types of flood hazard are river flooding, when a river overflows its banks, and coastal flooding, when the sea inundates a stretch of coastline. As noted in Chapter 1, both types of flood can cause great damage and loss of life. There is therefore much interest in establishing whether flood events are likely to increase in frequency or magnitude as the world warms. We begin by discussing river floods.

The nature of river flooding is discussed, for example, by Smith and Ward (1998). At any point on a river, the relationship between its stage (the height of the water level) and its discharge (the volume of water passing per unit time) depends on a number of factors, including the river channel's cross-sectional area and gradient. Flooding occurs when there is an increase in discharge, and therefore in stage, sufficient to overflow the river channel. River floods are occasionally caused by exceptional events such as dam failure, but most are caused by rainfall that is unusually heavy (i.e. with a high rainfall rate) and/or prolonged, or by an unusually rapid and/or prolonged snow melt. At any point along the river, the flood hazard for any particular event is largely characterised by the flood hydrograph (the curve of discharge against time) – not only its peak value, but its detailed structure such as the lag between the rainfall and the peak, and the rates of rise and fall (Fig. 3.1). The shape of the flood hydrograph depends not only on the rainfall or snowmelt characteristics (e.g. duration, rate, and distribution within the drainage basin), but also on the size of the river and its location in the river network, and on various factors relating to the drainage basin itself. These include the size and shape of the basin, and surface processes such as evaporation and infiltration into the soil. In particular, if the infiltration capacity of the soil (the maximum rate at which the soil can absorb rainfall) is exceeded, the rainfall flows over the surface into streams and rivers as run-off. This can occur, for example, if the soil is already saturated from previous rainfall events. Hydrographs for hazardous floods vary greatly. At one extreme is a sudden-onset upstream flash flood produced by a local thunderstorm, like the Big Thompson, Colorado, flood of 1976 in which 135 people lost their lives. At the opposite extreme is a slow-onset downstream flood moving as a wave down a large river over a period of weeks, as a result of prolonged rainfall over a wide area of the basin. An example is the 1993 Mississippi flood. Moreover, even flood events at the same place with similar peak discharges can have significantly different hydrograph shapes, involving different total volumes of water (the area under the curve), as with the 1993 and 1995 Rhine floods (Fig. 3.1).

Peaks in flood hydrographs can in principle be used to observe long-term trends in flood frequency at a particular place, for example using a peaks-over-threshold approach (see Section 2.2.2). Some studies in the United States, for example, were reviewed by Kunkel *et al.* (1999). However, those authors pointed out that such trends can be difficult to interpret in terms of climate change because of human activity within river basins. For example, many river channels have been reduced in cross-section by human intervention (e.g. by the use of raised banks or levees), often so as to allow floodplain development. The effect on flood trends is illustrated by analyses of the 1993 Mississippi floods as reviewed by Kunkel *et al.* (1999) and Smith and Ward (1998, p. 139). Though the 1993 peak stage at St Louis was greater than any since 1840, the 1993 peak discharge was in fact exceeded in 1844 and

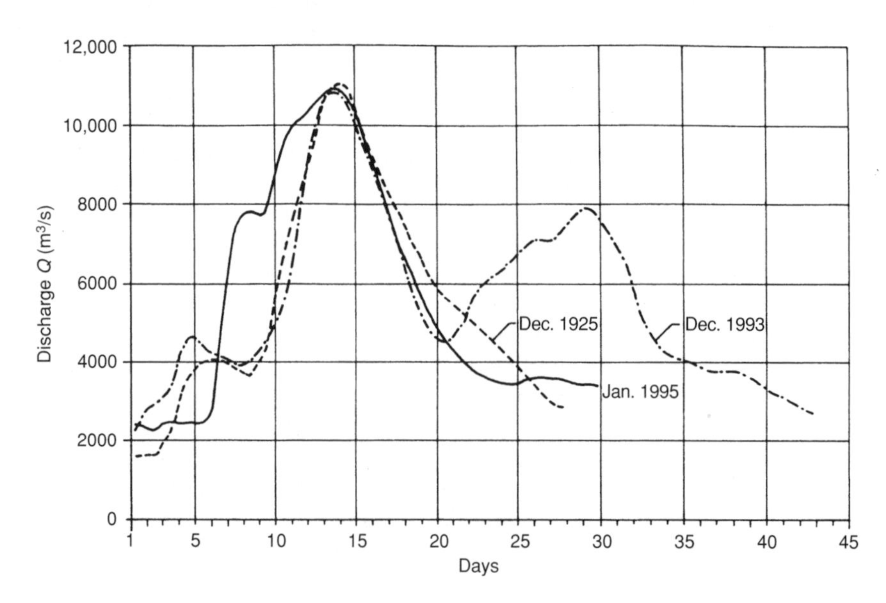

FIGURE 3.1 Comparison of discharge hydrographs of the flood events of 1925/26, 1993/94 and 1995 at the Cologne gauging station on the river Rhine. (From Engel, 1997 with permission from the International Association of Hydrological Sciences)

other years (Fig. 3.2). It is therefore important to use discharge values rather than stage values when assessing climate-related change (although discharge values may be much more difficult to measure during extreme flood episodes; Smith and Ward, 1998, p. 139). Even if discharge data are used, however, flood trends can be affected by human modification of basin characteristics such as infiltration, for example through urbanisation, deforestation or agricultural drainage. In some cases the effects on the flood hydrograph can be quite large, as noted in Chapter 1 (though in the case of deforestation the consequences are not always clear, with conflicting evidence in different regions; Smith and Ward, 1998, pp. 79–86). It is therefore essential to take into account such anthropogenic environmental changes, as well as climatic change, in any observations or predictions of flood trends.

To date, most observational studies on the effects of climate change have focused on rainfall extremes as these are the main cause of river flooding. The same is true of prediction studies, though for a number of individual river basins the impact of predicted climate changes on flood frequencies has also been assessed, by using GCM predictions as inputs to regional hydrological models (e.g. Kwadijk, 1993). Because of this focus on rainfall in the literature, our discussion of trends will also concentrate on precipitation extremes.

3.2 Trends in rainfall extremes that can cause river flooding

Precipitation needs to be considered in the context of the individual weather systems

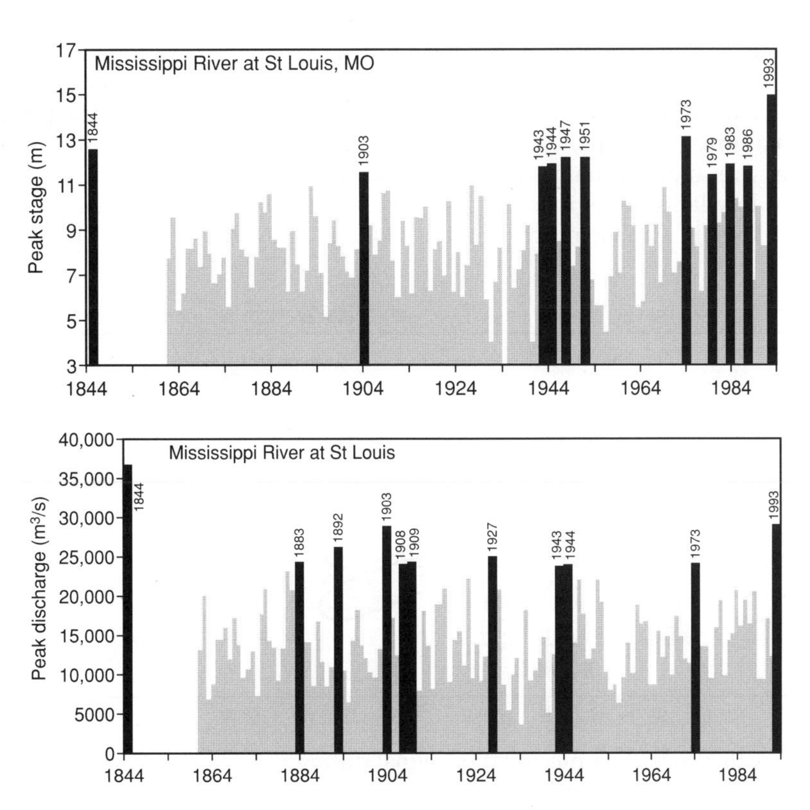

Figure 3.2 Annual values of (a) the peak stage and (b) the peak discharge of the Mississippi River at St Louis, for 1844–1993. The dark bars show the major flood events. (Reproduced from Kunkel *et al.*, 1999 with permission from the American Meteorological Society)

involved. These range in scale from simple convective cumulus shower clouds, through individual thunderstorms and groups of thunderstorms such as MCCs, to tropical and extratropical cyclones. Generally, the larger the scale of the system, the longer its lifetime. However, the duration of the rainfall in any location will also be affected by any movement of the system through the region. The global amount, distribution and seasonal variation of precipitation are determined by the occurrence and distribution of these weather systems within the general circulation of the atmosphere. In the tropics, much of the rainfall comes from convective systems associated with the ITCZ, and the migration of the ITCZ north and south with the seasons thus provides many low-latitude land areas with wet and dry periods during the year. In some regions, for example southern Asia, this process combines with other factors (such as the formation of a shallow thermal low over land in summer) to cause the *monsoon*, a dramatic reversal in wind direction from summer to winter associated with a large seasonal variation in rainfall. In the summer and autumn months, landfalling tropical cyclones are another major source of rainfall for some low-latitude regions. In mid-latitudes, particularly in winter, much rainfall is provided by the passage of extratropical cyclones and their associated fronts, which display a strong seasonal variation in intensity, frequency and location, as discussed in Chapter 2. In summer, the warming of the land means that many mid-latitude regions also experience substantial convective rainfall from shower clouds and thunderstorms, whether individually or in groups such as MCCs.

As noted in Box 2.1, oscillations of the climate system have a major effect on precipitation patterns, and therefore also on river flooding. ENSO episodes, of either phase, produce precipitation anomalies in a number of regions worldwide. Figure 3.3 shows areas of anomalous rainfall during El Niño events; the pattern is generally reversed for La Niña events. Rainfall teleconnections are also associated with the NAO (Hurrell, 1995). As noted in Chapter 2, a switch in its predominant phase in winter causes a seesaw in winter rainfall between northern and southern Europe as the main storm track alters its position, and summer NAO variations have also been linked to a shift in Atlantic hurricane tracks. The phase of the PNA pattern of winter circulation (Fig. 2.2) also helps determine rainfall patterns in North America, for example in the south-eastern United States, where the PNA phase is associated with

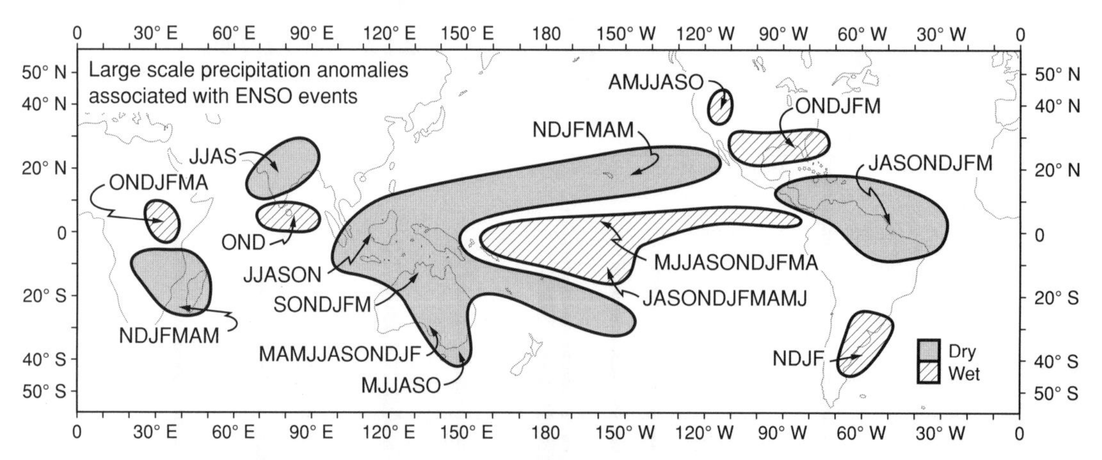

FIGURE 3.3 Areas with a consistent precipitation signal during El Niño events. For each region a listing is given of the months during which it is wetter or dryer than normal, in each case beginning in the year of the event. (Reproduced with permission from K.E. Trenberth, General Characteristics of El Niño–Southern Oscillation, in H. Glantz *et al.* (eds) *Teleconnections Linking Worldwide Climate Anomalies*, 1991, Cambridge University Press)

variations in the number, time sequence and size of rainstorms (Henderson and Robinson, 1994). It has also been linked to variations in extratropical cyclone frequencies (see Chapter 2).

What is the likely effect of climate change on precipitation, and particularly on the precipitation extremes that give rise to river flooding? To help interpret the observed and predicted changes in precipitation extremes, we start by outlining a physically based 'conceptual framework for changes of extremes of the hydrological cycle with climate change', as presented by Trenberth (1999) and shown schematically in Fig. 3.4. First, though, we note that for the global climate to be in equilibrium, the world's surface heat budget must balance, with annual average heat fluxes into and out of the surface being equal. A similar annual balance is required for the surface moisture budget, with average evapotranspiration (evaporation plus transpiration from plants) balancing average precipitation. As noted by Trenberth (1999), any increase in greenhouse gases will cause an increase in the global surface temperature by increasing the radiative heating of the surface by the atmosphere. This heating also causes extra evaporation of surface moisture, which helps to keep the surface heat budget in balance by providing an increased transfer of latent heat from the surface to the atmosphere. The need to balance the surface moisture budget as well means that globally, there must also be an increase in annual precipitation to match the increase in evaporation.

However, the extra evaporation, together with the fact that the warmer atmosphere can contain more water vapour before it condenses out as clouds, also means that there will now be more water vapour in the atmosphere. The result of this is that there will be a greater flux of moisture converging into each rain-producing weather system. Moreover, Trenberth (1998) has shown that all precipitating systems feed largely on moisture already in the atmosphere flowing into the systems, rather than on moisture recycled from their own rainfall. Thus the rainfall *rate* within each system will be increased, leading to a greater risk of flooding.

Thus on theoretical grounds it seems virtually certain that global warming 'will lead to a more vigorous hydrological cycle' (IPCC, 1996, p. 7), with an increase in annual global precipitation and an increase in rain rate for individual precipitation events. Trenberth (1999) also used the conceptual framework to predict other changes relating to precipitation, though the details are less certain. He noted how modelling shows that, other things being equal, the likely percentage increases in rainfall rate for individual events would provide a greater increase in total precipitation, globally, than is feasible for a balanced surface moisture budget. This implies that a compensating effect has to occur to limit the overall increase in global precipitation, such as a reduced overall frequency of rain events (though other compensating effects are also possible). In the case of extratropical cyclones, Trenberth (1999) argued that a reduced frequency of storms is possible as follows. The increased latent heat release as the water vapour condenses into cloud will tend to intensify the cyclones (though there are competing processes tending to reduce their intensity, as noted in Chapter 2). Arguments relating to the meridional heat transport by such storms then imply that if, on average, they *are* more intense, they could become less frequent. To clarify such issues, and to assess the detailed effect of climate change on the duration, size and location of precipitation events, as well as their rainfall rate and frequency, we need to turn to observation and numerical modelling.

Karl and Easterling (1999), Harvey (2000) and Easterling *et al.* (2000a, b) reviewed observations that appear to confirm that an enhancement of the hydrological cycle is indeed taking place as the world warms (see also Chapter 1). Observed trends include an increase in evaporation from the tropics, increased atmospheric water vapour, and an increase in convective clouds and continental cloud cover. Trends in annual precipitation are difficult to determine on a truly global scale but have been observed over land, with increases in most regions (especially in middle and high latitudes), though with some decreases in the subtropics. As regards the intensity of rain events, in many countries of the world there have been increases in the proportion of rainfall falling as 1-day and multiday heavy rainfall events, though there have also

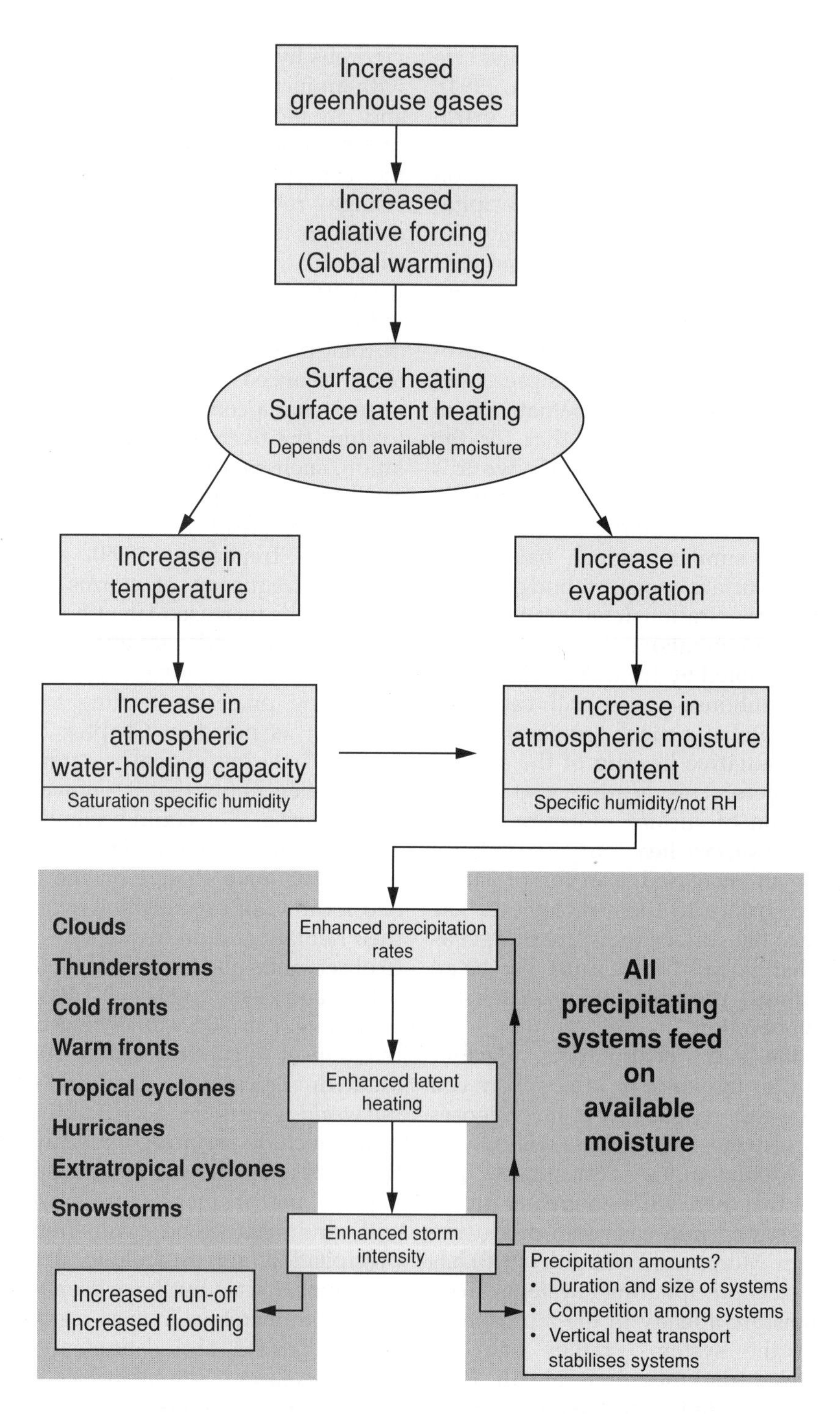

FIGURE 3.4 Schematic outline of a conceptual framework, described by Trenberth (1999), for how changes of extremes of the hydrological cycle are likely to occur with climate change. (From Figure 2 of Trenberth, 1999, p. 335, with kind permission from Kluwer Academic Publishers)

been decreases in some areas. In most regions, these trends in the frequency of intense rainfall events have the same sign as the trends in monthly or seasonal rainfall (Fig. 3.5).

Climate modelling studies predict a further intensification of the hydrological cycle as greenhouse gas concentrations increase, as reviewed by Harvey (2000), Meehl *et al.* (2000b) and Easterling *et al.* (2000b). Models are generally still of insufficient resolution to represent precipitation processes accurately (see Chapter 2). However, recent studies have confirmed the 1995 IPCC findings that under greenhouse warming the atmospheric moisture content increases, the global mean precipitation rate increases, the intensity of precipitation events increases in many regions, and there is an increase in the interannual variability of the south Asian monsoon. A new finding since the 1995 IPCC report is that extreme values of daily precipitation increase considerably more than mean values in most places. This accords with the results of modelling the measured frequency distribution of daily summer precipitation over land (Groisman *et al.*, 1999). Another new finding with some, though not all, models is a more El Niño-like state for the mean Pacific climate, with an associated change in the precipitation patterns. We also noted in Chapter 2 that some model results predict that the positive PNA and positive NAO circulation patterns may become more prevalent in winter, which would have regional effects on winter precipitation levels.

Thus it seems very likely, on both theoretical and modelling grounds, that in a warmer world there will be increases in annual global precipitation levels, and that these increases will be associated with increases in the intensity (i.e. rainfall rate) of individual precipitation events. It also appears that such changes are already being observed as the world warms. Both these trends will tend to lead to an increase in river flooding, though detailed predictions of the

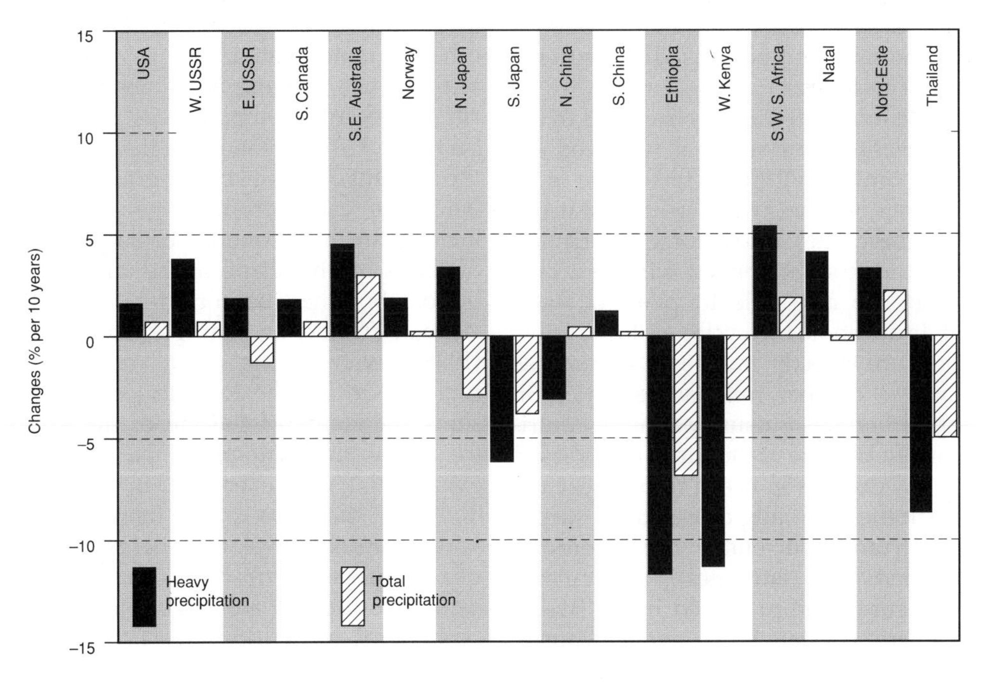

FIGURE 3.5 Linear trends in total seasonal precipitation (usually the season with maximum precipitation) and frequency of 1-day heavy precipitation events, for various countries. (Reproduced with permission from Easterling, D.R., Meehl, G.A., Parmesan, C., Changnon, S.A., Karl, T.R. and Mearns, L.O. 2000. Climate extremes, observations, modeling and impacts. *Science* 289, 2068–2074. Copyright (2000) American Association for the Advancement of Science)

effects of climate change in any region still await higher-resolution models that can accurately represent the precipitation processes. If the output of such models is used to drive hydrological models, the risk of increased flooding can then be assessed for any particular drainage basin, taking into account the effects of any other anthropogenic and natural environmental changes within the basins themselves.

3.3 Coastal floods and climate change

If a river already has a high level of discharge, a high tide can cause it to overflow as it reaches the coast in an *estuarine flood*. However, the most hazardous coastal floods (apart from those caused by tsunami) involve a *storm surge* (a regional increase in the height of the ocean generated by a cyclonic wind system). Such floods are responsible for much of the damage and loss of life associated with tropical cyclones, and can also be caused by extreme extratropical cyclones.

There are a number of factors involved in the creation of a large storm surge, as reviewed, for example, by Smith and Ward (1998). The main cause is strong onshore winds piling up water against the coast. The Coriolis force will tend to deflect the water, so prolonged winds parallel to the coast are also able to pile up water against it. In addition, the cyclone's low atmospheric pressure directly induces a rise in sea level (about 1 cm per millibar reduction in pressure), though this is usually a minor contribution. Finally, the size of the surge is increased if there is a reduction in ocean depth near the coast. Hurricanes typically cause storm surges of between 1 and 6 m (Table 2.1) where the onshore winds are strongest. Particularly devastating surges are produced in the Bay of Bengal, where coastal flood events caused 300,000 deaths in 1970 and 140,000 in 1991. Storm surges from extratropical cyclones tend to be worst in enclosed or semi-enclosed seas such as the North Sea, where in 1953 a surge of between 1 and 4 m, depending on location (Smith and Ward, 1998), caused severe coastal flooding in the Netherlands and England, with the loss of more than 2000 lives. In such cases the storm surge is strongly dependent on the location and track of the cyclone. In 1953 the surge was produced by strong north-westerly winds as the storm travelled southwards down the North Sea.

Regarding trends in coastal flooding due to climatic variation and change, the frequency and magnitude of storm surges are directly related to the frequency and intensity (in terms of both wind speed and pressure drop) of the associated cyclones. Therefore the observed and predicted trends in tropical cyclones and extratropical cyclones, as discussed in Chapter 2, can clearly be applied to this hazard as well. However, there is also another trend that is currently occurring and is virtually certain to continue, which is bound to make coastal flooding worse: sea-level rise (see Chapter 6).

To assess the effect of sea-level rise on the frequency of coastal floods, we need to consider the frequency distribution in flood level or magnitude (see Section 2.2). If we ignore any other climate-related changes in the distribution (such as those noted above), a sea-level rise represents a simple type B increase (see Section 2.2.3) in the mean of the distribution (as the amount of sea-level rise is just added to the height of each storm surge). The resulting increase in the frequency of hazardous floods depends on the shape of the tail of the distribution, but may be considerable (Smith and Ward, 1998). As an example, Smith and Ward (1998) reviewed the results of studies for a typical port on the English east coast. These show the shape of the tail of the exceedance frequency distribution to be approximately exponential, with every 0.15 m of sea-level rise doubling the exceedance frequency for any particular flood level. (In other words, a 0.15-m rise doubles the probability of a sea wall or dike being overtopped, a 0.3-m rise quadruples the probability, and a 0.45-m rise gives an eightfold probability increase.) The increase in overtopping probability for a given sea-level rise is even greater for the south of England. There will also be tidal changes as a result of sea-level rise, affecting estuarine floods, but these are difficult to predict without modelling (Smith and Ward, 1998). There are anyway likely to be other climate-related changes to estuary flooding, particularly due to precipitation changes, as discussed in the previous section.

3.4 Changes in other hydrological and temperature extremes

The other main hazards relating to precipitation extremes are hailstorms, snow and ice storms, and drought, with its associated wildfire hazard. Hail develops within thunderstorms, where particles of ice grow by accretion of supercooled liquid drops within the cumulonimbus cloud. Hailstones can grow to large sizes only if they remain in the cloud for some minutes, in the strong updraughts associated with severe thunderstorms. In the United States the main prevalence of hailstorms is in the western Great Plains. Though they generally cause relatively few fatalities, hailstorms are extremely damaging to property (e.g. cars) and to crops. Changnon and Changnon (2000) have carried out an analysis of hailstorm frequency in the United States over the past century. They find large multidecadal fluctuations with considerable regional variations. There are no consistent trends, with most stations reaching their lowest level over the last twenty-year period (1976–95), but some showing a peak at that time. The effects of climate change on hailstorms are not yet predictable, because individual thunderstorms cannot yet be resolved by GCMs.

Extratropical cyclones not only can cause a wind hazard (Chapter 2) and a direct flood hazard (this chapter), but in certain circumstances they can cause other precipitation-related hazards, namely severe snowstorms (including blizzards) and ice storms. The former can cause major disruption of transport, as well as potentially causing river flooding when the snow melts. Ice storms are associated with freezing rain, when rain (or cloud) droplets form thick accretions of ice on surfaces that are below freezing point. Ice storms can cause great damage to power lines and forests, and are a particularly important hazard in the Great Lakes region of North America. The general predictions of models regarding extratropical cyclones (see Chapter 2) should apply to these hazards, though more detailed regional modelling would be required to assess the effects of climate change, including regional warming, for any particular area.

The final precipitation-related hazard to consider is drought. Smith (1996) noted that a drought may be defined as 'any unusual dry period which results in a shortage of water'. Drought can often be a final trigger for a famine, and has therefore produced some of the world's worst disasters (Smith, 1996). When combined with poor land use practices, drought in dry areas can also lead to desertification (Smith, 1996). However, the drought hazard is outside the scope of this book in that it is a creeping (rather than rapid-onset) hazard, as the expected rainfall fails to occur day after day. Moreover, it is not a simple climate-related hazard, because, though deficiency of rainfall is always the initiator, the drought hazard depends on whether *useful* water is still available (e.g. in reservoirs and rivers) rather than whether there is simply an unusual lack of rainfall (Smith, 1996). Nevertheless, it is of interest briefly to touch on the immediate cause of drought, which is a prolonged negative extreme of rainfall. The reviews on positive extremes of rainfall, quoted earlier, also include some discussion of drought. Drought shows a large variability, but the areas of the world affected by both drought and excessive wetness are observed to have increased. Modelling studies suggest that the shift in precipitation towards heavier events may also be accompanied by an increase in the number of dry days in some areas. In particular, models predict an increased chance of drought in mid-continental areas in summer, owing to both decreased precipitation and increased temperature and therefore evaporation. The predicted change in the interannual variability of the South Asian monsoon would also increase the chance of drought there, as well as flooding.

Although drought is a creeping hazard, it can lead to the rapid-onset wildfire hazard, as summarised, for example, by Smith (1996). Wildfires tend to occur in regions with a

Mediterranean or continental climate where there is sufficient rain to produce vegetation for fuel, but there is also a dry period in the summer, making the vegetation easy to ignite. Lightning strikes from summer thunderstorms then provide the natural ignition source, though many wildfires are now started by human action, including arson. The wind also plays a major role both in helping to dry out the vegetation before it ignites, and in spreading the fire once it has started. Hence certain weather patterns are often associated with wildfires, such as warm, dry, unstable air with strong surface winds ahead of a cold front. Topography also plays a part, with fire driven upslope spreading particularly rapidly. Major wildfires are extremely common in Australia, where the vegetation and climate are especially conducive, but many other regions, such as southern Europe and the western United States, are also very prone to wildfires. The main hazard is to forestry and to buildings in the fire zone, as well as to the lives of people caught in the fire's path, including firefighters. The effect of climate change on wildfire frequency is complex, operating via weather variables (particularly temperature, relative humidity, wind speed and precipitation, which together determine 'fire weather') as well as the frequency of cloud-to-ground lightning strikes, and the length of the fire season (Flannigan *et al.*, 1998). For example, Flannigan *et al.* (1998) presented evidence that in much of the world's northern boreal forest, despite an increase in temperature there has been a decrease in fire frequency since the end of the Little Ice Age, because of increased precipitation. They predicted a further decrease in fire frequency there under global warming, by using GCM modelling of conducive fire weather, though without considering lightning strike changes. Their study also showed large regional variations in the effects of climate change, with significant increases in the frequency of wildfires predicted for central continental areas, in accord with the drought predictions above. Beer and Williams (1995) have also modelled fire weather using GCMs and predict an increase in wildfires over much of Australia, with slight decreases in relative humidity being the main determining factor. Modelling of fire frequency in the south-western United States by Price and Rind (1994), again using a GCM but including changes in the lightning regime, also predicted an increase in the frequency of fires there under global warming.

The final hazard we will consider briefly is that of temperature extremes. An average global surface temperature rise is the best-known, and arguably the most certain, observation and prediction relating to climate change. However, there has been less research on changes in extremes of temperature. This is despite the fact that unusually hot days in summer or cold days in winter, especially if forming a multiday period (a heatwave or a cold wave), are an important hazard with significant socio-economic impacts (for example, see Kunkel *et al.*, 1999). Temperature extremes can affect human health and cause many deaths through heat stress or hypothermia, though these are often linked to humidity and wind chill respectively (for example, see Tobin and Montz, 1997, pp. 53–55). Temperature extremes can cause property damage too (e.g. through damage to frozen pipes). Frost is also a well-known hazard for agriculture, and crops can be damaged by prolonged high temperatures as well (Mearns *et al.*, 1984).

Observations of changes in temperature extremes have been reviewed by Karl and Easterling (1999), Harvey (2000) and Easterling *et al.* (2000a, b). Over the period 1950–93 there were larger increases in mean surface air temperature over land than the global average, and the increase in (mainly night-time) mean minimum temperatures was twice as great as the increase in (mainly daytime) mean maximum temperatures, so that the average diurnal range has been reduced. These asymmetrical changes in the daily mean values are also reflected in changes in extreme temperatures (Table 3.1). One consistent and widespread peaks-over-threshold result is an increase in the number of frost-free days per year, with a longer frost-free season, which has been reported for several regions of the world. Furthermore, for a number of regions that have been examined, the lowest daily minimum temperatures (within certain periods, e.g.

TABLE 3.1 Summary of analyses of temperature extremes around the world (Source: Easterling *et al.*, 2000a. Reproduced with permission from the American Meteorological Society)

Country	Frost days	Warm minimum temperatures	Warm maximum temperatures	Cold waves	Heatwaves
Australia	Fewer		Up		
China	Fewer	Up	Down	Fewer	
Central Europe	Fewer				
Northern Europe	Fewer				
New Zealand	Fewer		Up		
United States	Fewer	Up	No trend	No trend	No trend

month, season or year) have a strong increasing trend, whereas the highest daily maximum temperatures show little or no trend.

Harvey (2000), Meehl *et al.* (2000b) and Easterling *et al.* (2000b) reviewed predicted changes in temperature extremes due to greenhouse gas increases. In many regions, models do indeed predict greater increases in mean (nighttime) minimum temperatures than in mean (daytime) maximum temperatures, with a reduced diurnal range. One reason for this appears to be that cloud reduces the amplitude of the diurnal cycle in regions where cloudiness increases. As expected from statistical considerations (see Section 2.2.3), these type B changes in the mean values are also associated with changes in the frequency of extreme values. This leads to an increase in the frequency of very warm days in summer (as the mean maximum temperature increases), and an even greater decrease in the frequency of very cold days in winter (as the mean minimum temperature increases still more). However, some models show type C changes in the spread of the daily temperature as well as the mean. For example, in northern temperate mid-continental regions, some models predict increases in variance in summer (further increasing the likelihood of very warm days) and decreases in variance in winter (further decreasing the likelihood of very cold days). There are also direct predictions of extreme temperature events. For example, the 20-year return maximum daily temperature values (in summer) show the largest increases where soil moisture is most reduced, and 20-year return minimum daily temperature values (in winter) show the largest increases at high latitudes where snow and ice have retreated. All these predictions also imply that the frequency of cold waves will generally reduce, and the frequency of heatwaves will increase.

A final point of interest regarding weather-related hazards, of particular relevance to precipitation and temperature extremes, is the recent development of the Climate Extremes Index (CEI; Karl *et al.*, 1995a) (Box 3.1). This currently combines indicators of precipitation and temperature extremes for the United States. The century-long record of the CEI for the United States shows large decadal-scale variations, and suggests that the US climate has become more extreme in recent decades. This is probably related to variations in the PNA pattern and ENSO (Karl *et al.*, 1995a). The authors suggested that new indicators relating to other extreme weather (e.g. heatwaves, cold waves, freezes, strong winds, tropical cyclones, hail, tornadoes, etc.) could be added to the CEI as the data improve, particularly with regard to their homogeneity. Easterling and Kates (1995) concluded that the CEI has several of the necessary attributes to become 'usable knowledge', widely understood by users and the general public in the same way as, for example, wind chill factor or the Palmer Drought Severity Index (PDSI). Particularly if extended to other regions and other parameters, the CEI thus promises to be useful for summarising in a quantitative way the results of investigations into the effects of climate change on weather-related hazards, at a time when their impact is increasingly being felt and noted worldwide.

Box 3.1 The Climate Extremes Index (CEI)

The Climate Extremes Index (CEI) was introduced by Karl *et al.* (1995a) to quantify observed changes in the extremes of climate in the United States. It is based on an aggregate of existing climate extreme indicators, though currently only those concerning temperature and precipitation extremes. The following definition of the CEI was given by Karl *et al.* (1995a), who also discussed the reasons for the choices of indicator. In each case, occurrences 'much above (below) normal' are defined as those falling above (below) the upper (lower) tenth percentile of the local, century-long period of the record, and the percentages refer to the conterminous US area.

Definition of the United States Climate Extremes Index (CEI)

The US CEI is the annual arithmetic average of the following five indicators:

1. The sum of (a) the percentage of the USA with maximum temperatures much below normal, and (b) the percentage of the USA with maximum temperatures much above normal.
2. The sum of (a) the percentage of the USA with minimum temperatures much below normal, and (b) the percentage of the USA with minimum temperatures much above normal.
3. The sum of (a) the percentage of the USA in severe drought (equivalent to the lowest tenth percentile) based on the Palmer Drought Severity Index (PDSI), and (b) the percentage of the USA with severe moisture surplus (equivalent to the highest tenth percentile) based on the PDSI.
4. Twice the value of the percentage of the USA with a much greater than normal proportion of precipitation derived from extreme (more than 2 in. or 50.8 mm) 1-day precipitation events.
5. The sum of (a) the percentage of the USA with a much greater than normal number of days with precipitation, and (b) the percentage of the USA with a much greater than normal number of days without precipitation.

4

Landslides and environmental change

4.0 Chapter summary

Landslides are the most widespread and undervalued natural hazard on Earth. They are undervalued because the most common landslides are too small to pose a threat to human life and so rarely attract popular attention. Unlike many other geophysical events, however, even small landslides can incur considerable economic loss. Thus a small slope collapse of a few thousand tonnes blocking a motorway can cost several million pounds in repair work, as well as indirect losses that might result, for example, from reduced tourism. At the other extreme, giant catastrophic landslides (of, say, 10^9–10^{10} tonnes), although rare, can release within seconds the energy of a major volcanic eruption or of an earthquake with Richter magnitude between 6 and 8. Such events can wipe out entire communities without warning.

The main factors controlling how landslides develop are the weight of the unstable mass (which drives a landslide downhill), the landslide's rheological properties (which resist movement and depend on what type of material is unstable) and the local environment (such as ground slope and topography). A dominant factor triggering instability is the effective weakening of a slope by the accumulation of water below the surface. A pervasive theme of this chapter, therefore, is how water circulation and its variation with environmental change may affect the frequency of small and giant landslides alike.

4.1 Impact of landslides

Annual losses due to landslides across the world amount to tens of billions of US dollars (Fig. 4.1), a figure that includes the costs of destroyed property, of repair work and of maintaining defence measures, such as barriers and ground-anchoring systems (Schuster, 1996). In the United States alone, land movements cause more deaths (25–50) and losses ($2 billion) each year on average than all other natural hazards put together. In developing nations, the cost to human life and misery is much greater. This is largely because developing communities depend heavily on local resources and have only limited access to emergency support from outside. As a result, the loss of a road or critical communal facility, such as a school or a hospital, not only severely disrupts the community, but may also induce years of hardship since restricted outside help increases the time needed to recover.

Mass movements also have insidious consequences on the environment (Sidle *et al.*, 1985). In zones of landsliding themselves, movement enhances surface erosion and can strip fertile soils from agricultural land. This in turn increases the amount of sediment being dumped into rivers, lakes and the sea and, by causing conditions in the water to deteriorate, can significantly increase aquatic mortality rates. Fish, for example, are vulnerable because an increase in suspended sediment can (a) damage their gills, (b) reduce the transmission

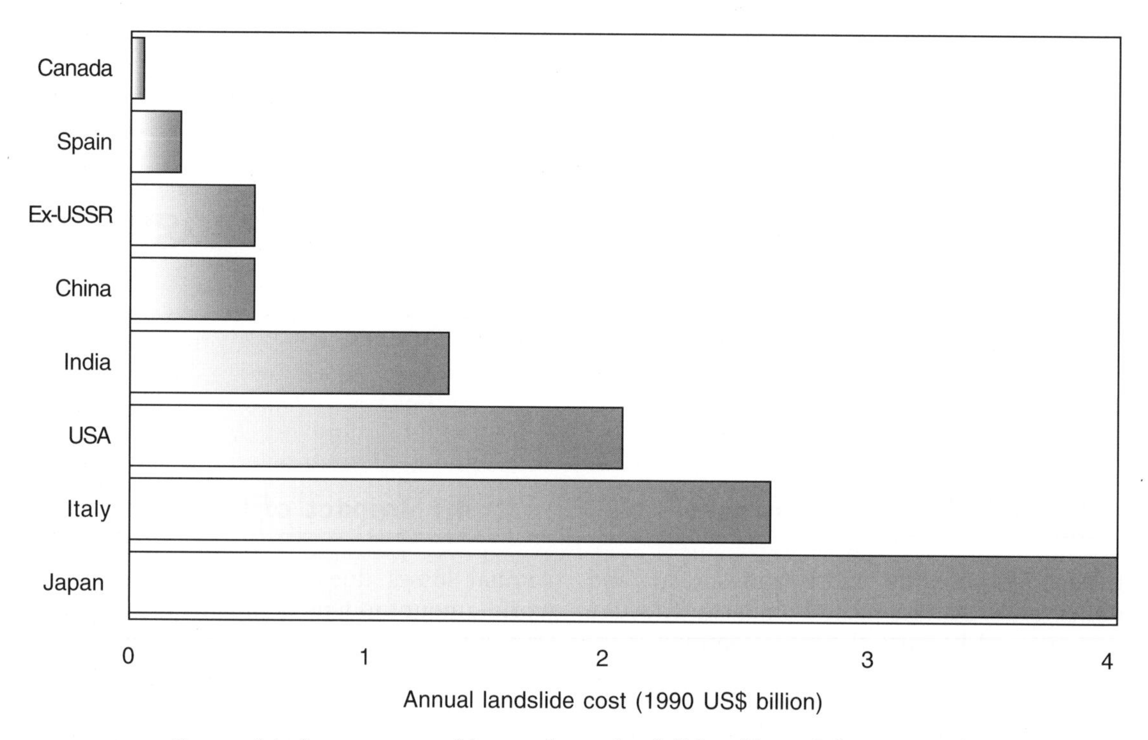

FIGURE 4.1 Average annual losses due to landslides. (From Schuster, 1996)

of light and, hence, rates of photosynthesis and oxygen production in water, and (c) mingle with eggs in spawning grounds and so reduce the amount of oxygenated water reaching the eggs, as well as blocking the escape of newborn fry (Herbert and Merkens, 1961; Phillips, 1971; Meehan, 1974). At the same time, increased sediment loads in rivers can damage purification filters for drinking water, as well as pumps for irrigation systems.

At a larger scale, slope collapses can alter the paths of rivers or dam them completely. For example, an earthquake in February 1911 triggered the collapse in Tajikistan of 2.9 km^3 of mountainside, which destroyed the village of Sarez and blocked the river Murgab, creating Lake Sarez, a reservoir of some 19 km^3 of water. Since then, the river has continued to erode the natural dam. If a major breach should open catastrophically, about 4 million people will be at risk from flooding in Tajikistan, Uzbekistan, Turkmenistan and Afghanistan (Vorobiev, 1998).

Giant, catastrophic collapses can also trigger other forms of natural hazard, the impact of which is greater than that from the landslide alone. Especially pernicious are flank failures at active volcanoes and giant landslides into the sea. On 18 May 1980, an earthquake of magnitude 5.1 triggered the collapse of about 3 km^3 from the northern flank of Mount St Helens in the Washington State section of North America's Cascade Mountains. The landslide exposed the gas-rich magma that had been accumulating inside the volcano for at least two months. Exploding laterally, the magma disintegrated into a cloud of gas and volcanic debris that, at temperatures of several hundred degrees Celsius, raced downslope at 300–500 km/h. This cloud, together with the blast from the air pushed ahead of it, laid waste some 600 km^2 of land within two minutes (Lipman and Mullineaux, 1981).

Even larger collapses (~10^2 km^3) have been recorded geologically from the flanks of ocean islands, including the volcanic archipelagos of the Canary Islands and Cape Verde Islands in the Atlantic, the Hawaiian Islands in the Pacific, and Réunion in the Indian Ocean (Keating and McGuire, 2000; Labazuy, 1996; Moore and Moore, 1984). These collapses have triggered landslides extending as much as hundreds of kilometres along the sea floor, and may also have initiated giant tsunami energetic enough to destroy coastlines on opposite sides of an ocean basin.

In addition to their impact on human activity, therefore, mass movements can severely disrupt the environment at scales from the microscopic (small fry) to intercontinental (megatsunami). To mitigate the future hazard from landslides, it is crucial to identify and to quantify consistent patterns in their behaviour. Only then will it be possible to evaluate how landslide behaviour might alter in response to environmental change.

4.2 Types of landslide

Landslides are unstable mixtures of soil and rock. They occur when the pull of gravity overcomes natural slope resistance. This situation typically arises when slope resistance is reduced below a critical value by some combination of chemical weathering, increased fluid pressure in rock and soil (e.g. after persistent rainfall), and the undercutting of the base of a slope, either by natural erosion or by human activity. Bulk soil resistance may also vary significantly according to the amount and type of overlying vegetation, since root systems can act as efficient anchors, while the vegetation itself draws water out of the ground.

Weakening processes are acting on slopes all the time. Such weakening is gradual and tends to trigger landslides individually, as isolated collapses. However, violent external events may cause rapid weakening and initiate numerous landslides simultaneously. Heavy rainstorms are especially dangerous and, as described in Section 4.5, can trigger destructive landslides across entire nations.

Instability also occurs when an additional force assists gravity. The assistance may be gradual, as when persistent tectonic movements slowly steepen a slope. More dramatic, however, is when seismic accelerations combine with gravity during an earthquake (in addition to the fact that violent ground-shaking may cause liquefaction). The stronger an earthquake, the greater is the furthest distance at which it can trigger a landslide. Earthquakes of magnitude 5 can destabilise slopes up to a kilometre away, whereas those of magnitude 6 can reach to 50 km and those of magnitude 8 to about 400 km (Keefer, 1984). The nature and size of each collapse depends on whether the landslide consists of soil or rock, as well as on how close the slope is to the earthquake's source. At one extreme, entire mountainsides can be dislodged (e.g. the Sarez landslide, Section 4.1). At the other, whole districts can be dissected by tens of thousands of soil-rich landslides, each with a volume the size of a house. On 13 January 2001, a magnitude-7.6 earthquake struck beneath the Pacific Ocean close to the coast of El Salvador. The shaking triggered landslides throughout the country. One of the collapses occurred from the Las Colinas hillside in Santa Tecla, just south of San Salvador. Although a modest collapse (first estimates gave volumes of less than 10^5 m^3), the landslide crashed through a residential district of Santa Tecla, burying more than 400 homes and claiming about 1000 lives.

Landslide volumes and rates of movement vary from cubic metres to tens of cubic kilometres and from centimetres a year to hundreds of kilometres per hour (Fig. 4.2). The wide ranges in material composition and emplacement behaviour have spawned numerous descriptive schemes for classifying landslides by type of material (rock, debris or earth) and style of emplacement (sliding, flowing, toppling or falling). Even the simplest scheme, due to Varnes (1978), proposes fifteen categories of mass movement involving only sliding or flowing (Table 4.1), and it has become common to assume that each category is distinguished also by a particular dynamic condition. As we see in Section 4.7, however, only small differences may in fact exist between many of the landslide categories. In addition to Varnes (1978), landslide classification is amply

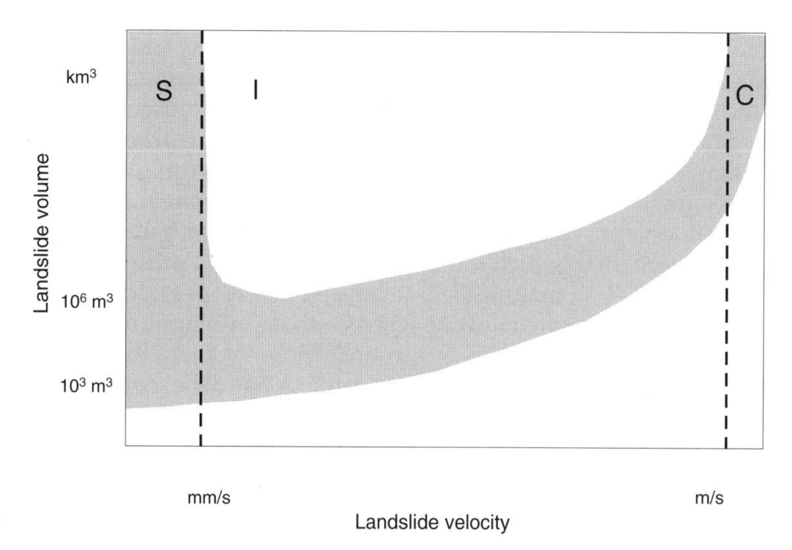

FIGURE 4.2 Schematic division of landslides according to volume and velocity (note the non-linear scales on each axis). The shaded area shows the preferred volume–velocity conditions for sluggish (S), intermediate (I) and catastrophic (C) movements. *See also* Table 4.2.

TABLE 4.1 Landslide classification scheme (Simplified after Varnes, 1978)

Type of movement	Bedrock	Debris (coarse soil)	Earth (fine soil)
Falls	Rock fall	Debris fall	Earth fall
Topples	Rock topple	Debris topple	Earth topple
Slides, rotational	Rock slump	Debris slump	Earth slump
Slides, translational	Rock block slide*	Debris block slide*	Earth block slide*
	Rock slide	Debris slide	Earth slide
Lateral spreads	Rock spread	Debris spread	Earth spread
Flows	Rock flow	Debris flow	Earth flow
Complex	Combination of two or more types of movement		

* Block slide refers to slides relatively undeformed away from the slide plane.

discussed by Hutchinson (1988), as well as the general landslide texts by Selby (1993), Turner and Schuster (1996) and Dikau *et al.* (1997).

From a hazards point of view, masses that slide or flow are by far the most dangerous type of ground movement. Masses that topple and fall are normally a modest threat in comparison. This chapter will thus concentrate on slides and flows. With this proviso, it is convenient first to allocate slides and flows to categories according to their typical rates of movement. Velocity is a measure of a landslide's destructive potential (recall that the square of velocity is proportional to kinetic energy per unit mass) and has been used to define seven classes of landslide intensity, covering a range of velocities from less than metres per century to more than metres per second (Table 4.2, after Cruden and Varnes (1996)). Derived from empirical observation (Varnes, 1958, 1978), each class embraces a hundredfold variation in velocity and is associated with a qualitatively distinct damage potential. Damage potential, in turn, can be used to gather the classes into three loose groups (Fig. 4.2; Table 4.2): sluggish movements that show obvious deformation over intervals of days to years and which commonly pose only a minor threat to structures (classes 1–4); catastrophic collapses that trigger movements of ~10^2 km/h and for which destruction is total (class 7); and intermediate movements

TABLE 4.2 Landslide intensity and damage (After Cruden and Varnes, 1996)

Landslide velocity class	Description	Velocity limits (m/s)	Destructive impact
Catastrophic			
7	Extremely rapid	> 5 (~5–50 m/s)	Violent catastrophe. Escape unlikely, many deaths. Buildings destroyed.
Intermediate			
6	Very rapid	5×10^{-2}–5 (~m/min to m/s)	Some lives lost. Buildings destroyed.
5	Rapid	5.10^{-4}–5.10^{-2} (~m/h to m/min)	Escape possible. Buildings damaged.
Sluggish			
4	Moderate	5×10^{-6}–5.10^{-4} (~m/week to m/h)	Insensitive structures (roads, strong buildings) can be maintained temporarily.
3	Slow	5×10^{-8}–5×10^{-6} (~m/y to m/week)	Some damaged structures can be repaired during movement.
2	Very slow	5×10^{-10}–5×10^{-8} (~m/century to m/y)	Minor damage to permanent structures.
1	Extremely slow	$< 5 \times 10^{-10}$ (< m/century)	Movement perceptible only by instruments. Construction possible with precautions.

that involve velocities from metres per hour to ~10 km/h (~m/s), sufficient to endanger life and cause significant damage to buildings (classes 5 and 6).

Although the classes are grouped by damage potential, it happens that the sluggish, intermediate and catastrophic categories are also crudely associated with different preferred styles of emplacement. Thus, while sliding is common to sluggish behaviour, flow is more important for faster movement, the distinction between intermediate and catastrophic events depending on a change in some combination of landslide resistance and the force per unit mass (i.e. acceleration) driving collapse. This should not be surprising, since extreme differences in rates of movement are often a clue to significant changes either in the controlling style of deformation, or in the imbalance between the forces driving and resisting motion. (As an analogy, consider two journeys between London and Brighton, one taking about an hour, the other a day. Possible explanations for the time difference are (a) different modes of transport, such as a train and a bicycle, or (b) similar modes of transport but with different capacities, such as by car in a modern Ferrari or in a Model-T Ford.)

Another feature of the threefold classification is that the different groups have preferred maximum volumes. While sluggish and catastrophic behaviour can occur among very large masses, intermediate movement appears to be restricted to intermediate maximum sizes (Fig. 4.2). In addition, while sluggish movements can also occur at very small volumes, the minimum volume for deformation appears generally to increase as velocity becomes greater (Fig. 4.2). Such volumetric influences are not intuitive. Velocity depends on the forces imposed *per unit mass* (or, roughly, per unit volume, because most landslides have similar bulk densities). A volume control on velocity must therefore reflect the influence of one or more hidden factors, among which we will find the most important to be landslide geometry, material resistance, and the structure of the upper kilometre of the Earth's crust.

4.3 Sluggish deformation

Significant movement (decimetres to metres) over timescales of days to years almost invariably involves deformation of soil or clay-rich rock. Slower movements (metres per 10–100 years or more) often involve also the deformation of apparently hard crustal rock, whether sedimentary, igneous or metamorphic. The second group of movements occurs mostly on mountains (Jahn, 1964) and volcanoes (Borgia, 1994), but though important geomorphologically, they do not constitute a major hazard. Accordingly, we will focus attention on the 'faster sluggish' movements associated with soils and clays, a collection broadly encompassing the earthflows, mudflows and mudslides of common classification schemes (Table 4.1).

Sluggish earth deformation is the most common form of mass movement in nature (Keefer and Johnson 1983). Movements may involve volumes from a few cubic metres to hundreds of millions of cubic metres. They occur on slopes from 2° to 45°, but appear to be most frequent (or most obvious) on slopes around 20°–25°. Morphologically they can be subclassified into two groups according to the shape of their basal failure surface.

Rotational slides are those for which movement occurs over curved surfaces, concave upwards (Figs 4.3 and 4.4), that may resemble the bowl of a spoon or part of a cylinder lying across the slope. The sliding masses tend to have widths similar to or greater than (by as much as 10–20 times) their downslope lengths, and these, in turn, are about 3–10 times greater than their average thicknesses. Maximum thicknesses are of the order of 10 m, and maximum volumes about 10^5–10^6 m^3.

Translational slides, in contrast, are those for which movement away from the headscarp occurs over planar or weakly undulating surfaces (Keefer and Johnson, 1983). Seen from above, the moving masses commonly have a tongue or teardrop shape (Figs 4.3 and 4.5), with lengths normally greater than their widths (but also smaller, especially if truncated downslope by a cliff face) and at least ten times larger than their thicknesses. As for rotational slides, their maximum thicknesses are normally measured in tens of metres, but by increasing their surface area maximum volumes may reach 10^7–10^8 m^3.

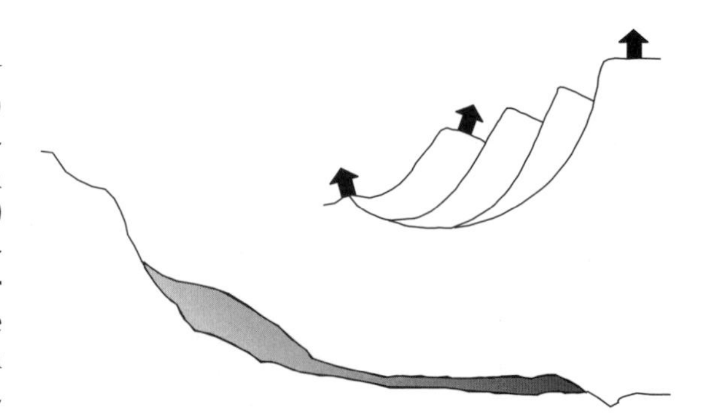

FIGURE 4.3 Common modes of slope failure. Rotational slides (*top*) collapse along shear surfaces that are concave upwards (approximately cylindrical or spoon-shaped). The unstable mass may itself break into smaller rotational units, separated by curved shear planes. Translational slides (*bottom*) move away from the curved headscarp along planar or undulating surfaces.

FIGURE 4.4 Rotational slide in the Barranco de Tirajana of Gran Canaria in the Canary Islands. The slide (*centre*) has slipped towards the bottom left along a concave surface. The steep headwall scar can be seen on the right. The steep surface on the left is another shear surface, revealed by a second collapse along the lower portion of the initial landslide. (Photo: C.R.J. Kilburn)

The head of an instability is marked by a curvilinear scarp, often surrounded by parallel cracks (*crown cracks;* Figs 4.5 and 4.6). The scarp bounds a natural amphitheatre (Fig. 4.5), in which the surface becomes lower than the surrounding hillside as material escapes downslope. The main body of the slide moves along lateral and basal planes of shearing, often with rough and grooved slickenside surfaces, and its toe may show a marked bulge, a result either of upward rotation or of material piling up behind the stalled or slowly moving front. In either case, the toes normally advance only a small distance from the source of the landslide.

In addition to forming shear planes, movement invariably cracks the landslide surface, so that the unstable mass can readily be infiltrated by precipitation. Easy access of rainwater and snowmelt is important because sluggish flows frequently become mobile during and after periods of heavy precipitation. Indeed, during a single, widespread storm, several small landslides may occur at the same time on neighbouring slopes. Initial movement may be abrupt, especially when a scarp is first formed, but then settles to the slow velocities that characterise sluggish landslides (Table 4.2).

FIGURE 4.6 Crown cracking around the scarp of the Black Ven landslide in Dorset, England. The crack is about 5–10 cm wide. (Photo: W. Murphy)

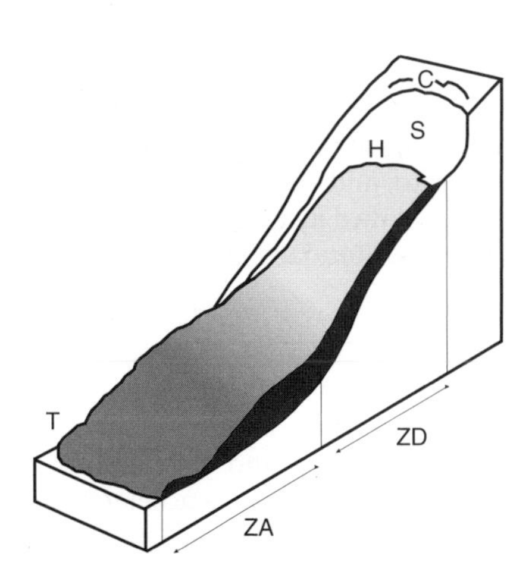

FIGURE 4.5 Simplified landslide structure, showing crown cracks (C) around the scarp (S) of the collapse zone. Movement produces a net loss of material from the region near the head (H) of the landslide, or zone of depletion (ZD). This material accumulates near the toe (T) of the landslide, in the zone of accumulation (ZA). (After Keefer and Johnson, 1983)

Sudden motion is commonly restricted to the scarp and head of the slide, and becomes slower downslope as the upper material pushes upon and activates the main bulk of the mass.

Once the whole mass is in motion, it tends to move as a block at nearly constant velocity, deformation being concentrated along the basal and lateral zones of shearing (Fig. 4.7). The Lugnez slope in Switzerland, for example, is an area of some 25 km^2 that has been advancing for more than a century down a 15° slope at several decimetres per year (Huder, 1976). The slide is carrying six villages along with it, but because differences in velocity across the surface are tiny, virtually no damage has been caused to any builidngs, even those more than 300 years old. However, buildings have been less fortunate on other inhabited sluggish slides, including those moving Ancona in Italy (Coltorti *et al.*, 1985), Ventnor on the Isle of Wight in England

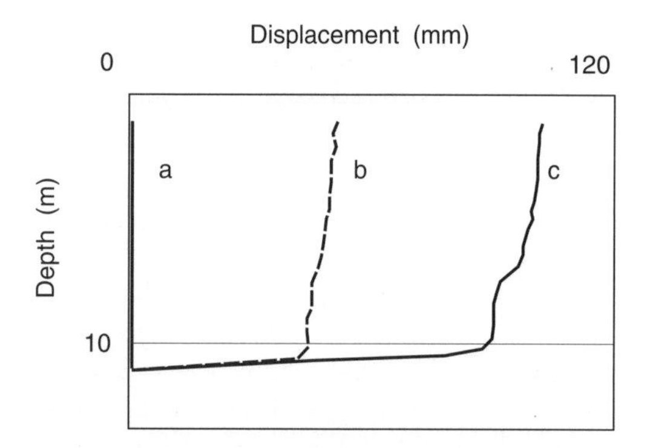

FIGURE 4.7 Block-like creep of the Rio Roncatto landslide, northern Italy, in July 1991. Following initial measurements on 3 July (a), movement of the landslide was concentrated within a narrow horizon, about 1 m thick and 10 m below the surface. Measurements on 12 July (b) and 17 July (c) show that creep was continuing at an almost constant rate of 7 mm/day. (Data from Panizza *et al.* presented in Brunsden and Ibsen, 1996)

(Brunsden, 1995), Rosiana on Gran Canaria in the Canary Islands (Lomoschitz and Corominas, 1996; Solana and Kilburn, 2001) and Villerville in France (Flageollet, 1989).

4.3.1 Analysis

Shape and thickness, block movement at nearly constant velocities, and a sensitivity to water content are essential clues to understanding the behaviour of sluggish landslides. Whether rotational or translational slides, sluggish masses have widths much greater than their thicknesses; as a result, lateral spreading is usually much less important than downslope movement to deforming a mass. Each slide category also has its preferred geometry – spoon-shaped and cylindrical for rotational slides and box-shaped, elongated downslope (unless truncated by topography) for translational slides. These shapes reflect preferred stress directions along which failure planes are formed: spoon shapes are favoured in homogeneous and isotropic materials (whose properties are similar throughout), while planes are favoured along weak horizons within a layered structure. The fact that all types of slide have maximum thicknesses of about 10 m further suggests a limit imposed by the Earth's crust. Only the upper few metres of the crust normally contain the soils and weak rocks of which sluggish masses are composed; deeper levels are dominated by stronger rock, whose failure, as we see below, also results in faster rates of movement.

Sluggish slides travel almost as blocks, focusing movement along discontinuities around their bases and edges, and tend to settle down to a constant velocity. The attraction to constant velocity is surprising. It shows that some of a landslide's resistance must increase with velocity, so that movement begins by accelerating to, and then staying at, the velocity at which resistance just balances the downslope pull of gravity. Standard sliding models cannot

Box 4.1 Modelling landslide resistance

Observation suggests that landslides tend when possible to advance at a nearly constant velocity. A simple force balance gives

$$F_g = \tau A + \rho V a \qquad (1)$$

where F_g is the downslope pull of gravity, τA is the landslide's resisting force (τ is basal shear resistance and A is the basal area of the landslide), and ρVa is the inertial force related to acceleration and deceleration (ρ is density, V is volume and a is acceleration).

For constant velocity, inertial forces can be neglected ($\rho Va = 0$). The gravitational force F_g is $\rho g V \sin \beta$, where g is gravitational acceleration and β is the mean angle of slope down which the landslide is travelling. Since a velocity-dependent term must operate to maintain a constant velocity, the only candidate left in eqn (1) is the basal stress τ.

Assuming that deformation is concentrated within a thin basal layer, three simple models are immediately available to describe resisting stress. The first follows traditional landslide studies and treats a landslide as a rigid block (Hoek and Bray, 1981), for which friction is determined by irregularities between the landslide and the ground. In this case, basal friction can be described by the Mohr–Coulomb relation for sliding:

$$\tau = \mu(1 - U/\sigma)\sigma \quad (2)$$

where μ is the coefficient of friction, σ is the normal stress and U is a fluid pressure (see also Box 4.2).

Alternative bulk models consider that material in the basal layer deforms as a fluid. In this case, landslide resistance is controlled by the fluid rheology. The simplest class of fluid has a Newtonian rheology, for which

$$\tau = \eta \, du/dy \quad (3)$$

where η is the viscosity of the material and the deformation rate du/dy is the change in forward velocity (du) over a thickness dy. To a first approximation, du/dy can be replaced by u/δ, where u is the mean landslide velocity and δ is the thickness of the basal layer. Thus, for a material with constant velocity deforming in a basal layer of constant mean thickness,

$$\tau = Ku \quad (4)$$

where $K = \eta/\delta$.

Equation (3) implies that a landslide will halt ($du/dy = 0$) when the resisting stress and, hence, also the driving stress are zero. Unless the slide is contained by topography, the net stress on a viscous boundary layer approaches zero only when the layer becomes infinitesimally thin. Hence, either eqn (3) holds until the basal layer is exhausted (i.e. it has been left as a deposit trailing behind the main landslide), or the basal layer itself possesses a strength that must be exceeded for flow to continue. In the second case, the simplest bulk model is that for a Bingham fluid, which yields

$$\tau = \tau_y + \eta \, du/dy \quad (5)$$

where τ_y is the yield strength below which no motion can occur. An important feature of eqn (5) is that at sufficiently large deformation rates (du/dy), the effect of the yield strength becomes negligible and eqn (5) approximates to eqn (3), but when du/dy becomes small, landslide resistance depends on the yield strength τ_y alone. In other words, provided that $du/dy >> \tau_y$ for most of a landslide's advance, the resisting stress can again be approximated to $\tau = Ku$, as for the Newtonian case.

For the force balance in eqn (1) to yield a constant velocity, τ must depend on velocity. Thus, for $\tau = Ku$, eqn (1) leads to

$$u = F_g/(KA) \quad (6)$$

More complex rheological models for material in the basal layer may also provide a basal stress that varies with velocity. However, without evidence to the contrary, experience favours the simplest behaviour consistent with observation, encouraging the view that most landslide materials can be treated as effectively viscous fluids.

For simplicity, the present analysis has assumed that the entire basal layer deforms as a fluid. This assumption is reasonable for intermediate and catastrophic movements, but is less obviously valid in the case of sluggish advance. Keefer and Johnson (1983) suggest that the movement of sluggish landslides may indeed involve irregular surfaces moving against each other, but that the rate of advance is actually controlled by a film of viscous mud squeezing between asperities. Another possibility is that surfaces scraping against each other persistently break existing surface asperities and create new ones. In this case, the rate of frictional energy loss may depend on the rate of breaking asperities which, in turn, depends on the landslide's rate of advance.

account for such behaviour (Box 4.1, p. 72). They assume that sliding resistance is controlled by a surface roughness that remains the same at all velocities. Alternative explanations are discussed in Box 4.1 and Section 4.7. For now, it is important to note that even sliding masses have a resistance that depends on velocity.

Slope movement is also favoured by heavy precipitation and the accumulation of water in the ground. Since masses behave as blocks, the influence of water must be to change conditions along the planes of sliding. First, the mass becomes heavier and so more likely to move. Second, the pressure of accumulated water can help push apart adjacent sliding planes, previously interlocked by irregularities along their surfaces, and so reduce frictional resistance (Section 4.6). Whichever effect is dominant, the fact that slope stability is sensitive to water content means that the frequency of sluggish landsliding in a district may alter as climatic variations induce changes in seasonal precipitation.

4.4 Catastrophic deformation

Catastrophic landslides occur when huge volumes (from ~10^6–10^7 m^3 to ~10–100 km^3) of rock collapse and shatter with accelerations of the order of 1 m/s^2. They can travel kilometres within minutes and erase everything in their paths. The result of sudden slope failure, they occur at least twice a decade in regions of high relief, from young mountain chains to volcanic edifices. They may thus consist of all major rock types, from carbonates, through metamorphic rocks, to volcanic material. For instance, the largest sub-aerial catastrophic landslides on record have occurred in the metamorphic rocks of the Alps (Flims landslide in Switzerland, 12 km^3), the limestone of Kabir Kuh in Iran (Saidmarreh landslide, 20 km^3) and young volcanoes in Mexico (Colima, 12.5 km^3, and Popocatépetl, 28 km^3) and the north-western United States (Shasta, 26 km^3).

Apart from a few examples triggered by large earthquakes (Richter magnitude >6.5; Hadley, 1978; McSaveney, 1978; Plafker and Ericksen, 1978; Blodgett *et al.*, 1998), catastrophic slope failure is normally preceded by months to decades of accelerating creep (Voight, 1978). Moments before 10–20 million m^3 of rock crashed through the Swiss village of Goldau in 1806 and claimed about 475 lives, the last words of one victim were (Hsü, 1969) 'For thirty years we have waited for the mountain to come, well, now it can wait until I finish stuffing my pipe.'

Giant catastrophic collapse normally occurs on slopes greater than 20°. Failure scars may be rotational or translational and have geometries similar to scars associated with sluggish landslides. An essential difference is that for catastrophic collapse, the zones of failure have *minimum* thicknesses of tens of metres. A plethora of names have been attached to the resulting landslides, including rock slide-avalanches, rock or debris avalanches, and landslides of long run-out. Here we will follow Hsü (1975) and describe giant, rapid landslides as *sturzstroms*.

Sturzstroms form lobate, finger-like sheets that follow topographic depressions and have lengths about 3–10 times greater than their widths and 10–100 times greater than their thicknesses (Hsü, 1975, 1978). Unlike sluggish masses that tend to stay within the zone of failure, sturzstroms spread to areal extents much greater than those of their collapse scars. Their deposits consist of collections of fragments, from fine grains to blocks the size of a house, the fine material tending to be concentrated either within the lower layers of a deposit or along narrow horizons at different levels throughout its thickness (Fig. 4.8).

Another notable feature of sturzstrom deposits is that they preserve their large-scale pre-failure stratigraphy, such that a pre-failure sequence of, say, carbonates over gneiss would yield a sturzstrom deposit with disrupted carbonate levels sitting on broken gneiss. Indeed, the exposed upper portions of some giant sturzstrom deposits, such as at Köfels and Flims in the Alps, preserve fine structures at scales of 0.1–1.0 m (Erismann, 1979). These latter deposits are remarkable when seen in outcrop, in that they appear from a distance to consist of massive rock faces (~10–100 m

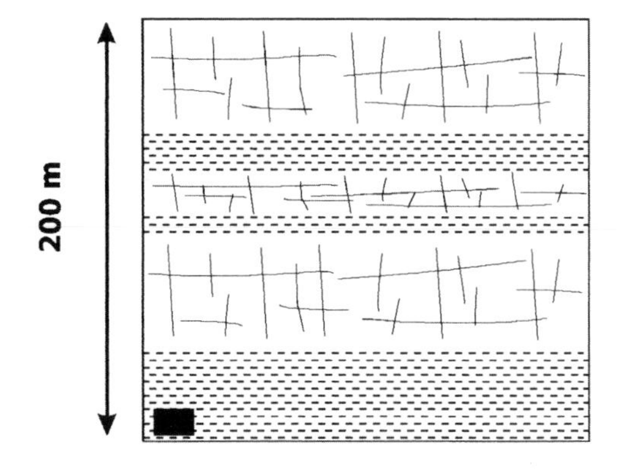

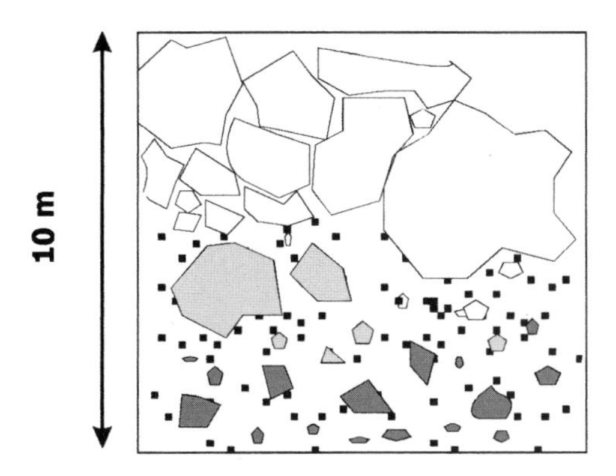

FIGURE 4.8 Sturzstrom deposits consist of broken rock. Near the source (*top*), heavily fractured hard rock (*white*) overlies a thinner base of gravel and small debris, possibly mixed with mud or clay (*stipple*), along which deformation is initially concentrated; secondary clay-rich layers within the body of the landslide may later develop as preferred horizons for deformation. Away from the source (*bottom*), the whole sturzstrom has become much thinner, and a larger basal fraction consists of a small debris supported by a fine-grained matrix (*stipple with grey and black debris*). This layer is overlain by giant angular blocks (*white*). Notice the poor mixing between different stratigraphic horizons (*white, grey and black levels*). The black box in the top diagram shows the relative thickness of the lower section. Although idealised, these sections are based on observations at Vajont, Italy (Hendron and Patton, 1985; *top*) and in the Karakoram Himalayas (Hewitt, 1988; *bottom*).

across), when they in fact consist of rock masses which have been shattered at the centimetre scale, but which have experienced no *relative* deformation during emplacement. They are the equivalent of a three-dimensional jigsaw puzzle that has travelled several kilometres without any of the pieces becoming dislodged.

Few direct observations exist of sturzstrom behaviour. Eyewitnesses were either victims or survivors running for their lives at the time. Remarkably, fragmentary accounts do remain of a handful of events, mostly from those in the Alps (notably Goldau (1806), Elm (1881) and Vajont (1963)) and concerning collapses with volumes of 20–200 million m^3 – the smaller end of the sturzstrom spectrum. These accounts emphasise the suddenness of slope failure, which takes place in less than a minute, and the remarkable speed of a sturzstrom, advancing at average rates of 15–30 m/s (~60–100 km/h). The suddenness of collapse indicates a low initial frictional resistance, during which most of the landslide's energy is converted from potential to kinetic and used to accelerate the unstable mass. Simple calculations show that within 30 s or less, a sturzstrom could be travelling at up to 30 m/s, comparable to its mean rate of advance during emplacement. It is conceivable, therefore, that the few minutes of a sturzstrom's advance occurs at almost the same velocity (Hsü, 1975, 1978). Indeed, accounts of the Elm sturzstrom describe the front of the landslide halting abruptly. This suggests that most deceleration occurred during the final seconds of emplacement, consistent with previous advance at velocities that changed only slowly with time.

The staggering speed of a sturzstrom has long been considered anomalous. The high velocities have been taken to imply unusually low rock friction, and several exotic mechanisms have been proposed, from trapped cushions of air (Kent, 1966; Shreve, 1968) and vaporisation of water in the landslide (Goguel, 1978) to rock melting (Erismann, 1979). However, these mechanisms appeal to special physical conditions that subsequent studies have shown cannot apply in all cases (Howard, 1973; Melosh, 1987; Campbell, 1989). Other explanations appeal to particular varieties of granular flow, but all these remain theoretical and have yet to be confirmed in the field (Melosh, 1979, 1987; Dent, 1986; Campbell, 1989; Campbell *et al.*, 1995; Kilburn and Sørensen, 1998). An alternative view, discussed in Section 4.6, is that high speeds are not necessarily the result of low friction, but instead reflect an impetus larger than expected during collapse.

As is the case for sluggish landslides (Section 4.3), the approach to catastrophic collapse can be hastened by water accumulating within an unstable mass. Unlike the shallow failure triggering sluggish landslides, sturzstrom collapse involves failure at depths of at least several tens to hundreds of metres. Since it has to percolate through an increasing volume of overlying rock, groundwater has greater difficulty in accumulating at deeper levels before escaping. As a result, the conditions for water to accelerate sturzstrom collapse are achieved less frequently than is the case for shallower landslides. Key requirements appear to be an unusually large supply of water and a mechanism for blocking the escape of water from deep levels. Favourable conditions thus include the rapid melting of snow caps or glaciers, as well as extreme and persistent rainfall, but most favourable of all is the filling of a water reservoir in a naturally rainy area.

The classic example of catastrophic failure into a reservoir was the tragic collapse in 1963 of Italy's Mount Toc (Müller, 1964; Hendron and Patton, 1985; Kilburn and Petley, 2001). In the foothills of the north-eastern Alps and just 80 km north of Venice, Mount Toc was an unexceptional mountain peak before 1960. Anonymity was lost when the deep Vajont river valley along Toc's northern flank was chosen as the site for a reservoir 5 km long. One hundred and fifty million cubic metres of water were to be contained by a pioneering dam which, 265.5 m high, was the tallest double-arched dam in the world. Trial flooding of the reservoir began in February 1960. Soon afterwards, a massive crack, nearly 2 km long, opened above the reservoir on Toc's upper flanks. During the next three years, some 200 million m^3 of rock were to shudder downslope by almost 4 m until the evening of 9 October 1963 when, at 10.39 pm, the mountainside plunged into the reservoir (Figs 4.9 and 4.12). Enormous waves lapped the sides of the reservoir before overtopping the dam and sending a wall of water crashing into Longarone, 2 km downstream. Minutes later, Longarone ceased to exist and the life had been crushed from more than 2000 victims.

The final failure of Mount Toc occurred after more than two months of slowly accelerating slope movement. During these months the reservoir had been filled almost to capacity, the depth of water reaching 220 m above the valley floor. Rainfall had also been heavy during September and the start of October. Blocked by the reservoir water, percolating rainwater was unable to escape through the lower flanks of the mountain, as it had been able to do before the valley was flooded. Increasing quantities of water became trapped within the mountain, and the natural conclusion is that, as in the case of shallow landslides, the accumulated water reduced slope resistance by opening gaps already present in the rock (Hendron and Patton, 1985). An alternative view is that the water catalysed *new* rock-cracking (Kilburn and Petley, 2001). These contrasting roles of water are discussed in Section 4.6. The key issue for

FIGURE 4.9 The scar in Mount Toc produced by the catastrophic landslide into the Vajont reservoir in 1963. Trails from small collapses can be seen down the main scar. Most of the deposit lies out of view below the photograph, filling in the Vajont river valley upstream from the dam, beyond the view to the right. (Photo: C.R.J. Kilburn)

now is that water can stimulate even deep-seated slope failure.

4.4.1 Analysis

Compared with sluggish landslides, giant catastrophic collapse involves similar failure geometries but minimum depths of tens of metres, similar to the maximum values for the slower movements. Because failure is deep-seated, it invariably involves strong, coherent rock that shatters during run-out. An immediate inference is that rock-cracking is important to the formation of sturzstroms.

Sudden collapse is commonly preceded by long intervals of accelerating creep. The rate of creep can be increased by the accumulation of water in rock for a suitably long period of time. Although the destabilising influence is commonly attributed to increased water pore pressure (Section 4.6), it happens that water is also extremely efficient in corroding rock under stress (Atkinson, 1984). This second mechanism and its potential for forecasting catastrophic failure are discussed in Section 4.6.

Catastrophic failure provides enough excess momentum to drive material several kilometres from the collapse scar. The resulting deposit is an elongated sheet much wider than it is thick, suggesting that more momentum may have been lost in moving downslope than in spreading. Although composed of broken rock, sturzstrom deposits tend to preserve their pre-failure stratigraphy, a feature that can occur only if deformation is concentrated along narrow horizons roughly parallel with the ground. The simplest interpretation is that sturzstroms can be treated as collections of jostling blocks carried along by strongly deforming basal boundary layers; otherwise a greater mixing between blocks would be expected (Erismann, 1979; Kilburn and Sørensen, 1998).

The few observations available indicate that both slope collapse and the arrest of sturzstroms can occur rapidly, in tens of seconds. Since emplacement normally takes much longer (a few minutes) and involves mean velocities similar to those calculated for collapse alone, it is plausible that most of a sturzstrom's travel occurs at nearly constant velocity. If this idea is correct, then sturzstrom behaviour is strongly influenced by a velocity-dependent resistance and so cannot be controlled by simple sliding (Box 4.1).

4.5 Intermediate deformation

Intermediate deformation broadly gathers together fast slides and the flow categories in Varnes's classification (Table 4.1). Movement typically occurs in poorly consolidated material, ranging from masses of mud, silt and clay to collections of debris (up to 80–90 vol.%) within a fine-grained matrix. The resulting landslides are produced by rotational or translational failure on slopes normally greater than 20°–25°. Failure is commonly shallow (depths of metres) and, whereas sliding may dominate initial failure, flow processes usually control motion away from the scar. In either case, deformation is focused along the base and lateral margins of the landslide, within horizons that may deform as a collection of sliding planes (as in a deck of cards) or as a true fluid.

Most intermediate landslides have volumes in the range 10^3–10^6 m^3 and, for all sizes, they show two broad styles of movement. The first style (sliding according to Varnes) resembles sluggish behaviour, except that it occurs at mean velocities of metres per hour to metres per minute. The mass slips as a block and advances only a short distance beyond its scar. The second style (flow according to Varnes) tends to occur at mean velocities of decimetres to metres per second and produces deposits that extend from their scars as elongate tongues, reaching lengths from 10^2 m to 10–30 km as volume increases (Fig. 4.15; Iverson, 1997; Iverson *et al.*, 1998). During advance, faster flows may scour the ground and gather up new material. Their extreme lengths reflect their tendency to follow narrow valleys that restrict their widths (Fig. 4.10).

Flows may be triggered directly or they may evolve from rapid slides. For both styles of movement, increased water pore pressure is a

FIGURE 4.10 Between 1998 and 2000 (date of photograph), numerous volcanic mudflows filled in large sections of the Belham river valley on the flanks of the active Soufriere Hills volcano on Montserrat, West Indies. A house, originally raised on the banks of the valley, has been surrounded and buried to depths of about 1 m. The mudflows are triggered by rainfall destabilising thick layers of volcanic ash that have accumulated on the mountain's upper flanks during eruptions. As of 2001, the eruption was still in progress, and further mudflows were expected. (Photo: C.R.J. Kilburn)

key factor initiating failure. The most spectacular examples occur during regional rainstorms. Such storms can rapidly drench hundreds of square kilometres. Frequently, much of the affected terrain has shared a common geological and geomorphological history and so enjoys a similar surface condition (e.g., amount and type of soil cover and typical angles of slope). Thus, when one slope becomes unstable, it is likely that many of its neighbours are in the same condition.

In January 1982 the San Francisco area was deluged by a Pacific storm that released 440–600 mm of rain within 32 hours. As it passed, it left behind the scars of 18,000 flows of mud and debris that cost 33 lives and about US$300 million in damage to property (Ellen *et al.*, 1988; Bell, 1999). In November 1994, western Piemonte in Italy was struck by an intense two-day storm that dumped as much as 250 mm of rain. In addition to severe flooding, the rainfall triggered tens of thousands of landslides (mostly intermediate flows with volumes of about 10^3 m^3), some slopes being dissected by 170–180 slides per square kilometre (Casale and Margottini, 1996). Four years later in southern Italy, a sudden downpour on 5–6 March 1998 dispersed 100 mm in 48 hours over western Campania. Twelve kilometres inland from Vesuvius, the rain mobilised the remains of poorly consolidated deposits of broken pumice, ejected by one of the volcano's prehistoric eruptions. Sixteen mudflows were triggered along a crest 5 km long, most of which ploughed through the villages of Sarno, Quindici, Bracigliano and Siano, 3 km downslope. At least 159 people were killed and nearly 200 buildings destroyed or rendered useless (Campagnoni *et al.*, 1998).

The most dramatic regional events are inevitably associated with hurricanes, cyclones and typhoons. During the last week of October 1998, Hurricane Mitch, moving west from the Atlantic, stalled for six days over Central America. As much as 2 m of rainfall fell over mountainous areas, triggering floods and landslides across Honduras and parts of Guatemala, Nicaragua, Belize and El Salvador (Fig. 4.11). At Las Casitas volcano in north-western Nicaragua, partial collapse of the summit crater's walls on 30 October fed a debris flow 16 km long and up to 8 km wide. The landslide destroyed at least four villages and claimed over 2000 lives. Across the affected

FIGURE 4.11 A building in the Honduran capital, Tegucigalpa, badly damaged by mudflows triggered by intense rainfall during the passage of Hurricane Mitch in 1998. (Photo: M.C. Solana)

countries together, the hurricane triggered about a million landslides, killed over 10,000 people, left another 1.5 million homeless and caused more than US$5 billion worth of damage. It had been the most destructive Atlantic hurricane for 200 years (Long, 1999; NOAA, 1999).

Investigations of regional landsliding (focusing on flows) have sought to identify critical combinations of mean rainfall intensity (I) and duration (D) above which slopes become unstable. Following the pioneering studies of Caine (1980), the critical condition is expressed in the form

$$I \geq CD^{-B}$$

where the constants B and C depend on regional soil and slope conditions. Thus, for D in hours and I in mm/h, Caine's original analysis yielded $B = 0.39$ and $C = 14.8\,\text{mm/h}^{0.61}$; other examples are given in Table 4.3.

The transition from sliding to flow occurs as the fine-grained component (mud, silt and clay) of a landslide liquefies, at least at the base of the moving mass. Advance may occur as a series of pulses or surges superimposed on a steadier mean velocity. Surging is especially common among flows, and peak speeds of as much as 25–30 m/s have been observed (Nash *et al.*, 1985). Some of the largest examples occur on volcanoes, where they are termed 'lahars', and grade into sturzstroms with volumes of 25–30 km^3 (Stoopes and Sheridan, 1992; Siebert, 1996; Iverson *et al.*, 1998).

4.5.1 Analysis

In the transition between sluggish and catastrophic movements, intermediate landslides share the basic characteristics of the other two categories. They frequently start with rotational or translational slope failure; they concentrate deformation along their basal and lateral margins; they tend to a steady rate of advance, even if this is punctuated by episodic surging; and they are readily triggered by water accumulation in an unstable slope.

Intermediate landslides are also associated with shallow failure, typically metres deep. As a result, they are normally composed of loose, fine-grained material that may contain boulders and smaller debris. Faster examples may erode the underlying surface and so grow in volume during emplacement. Since their volumes and essential components are similar to those of sluggish landslides, the greater velocities of intermediate landslides must be the result of different external conditions and material state. Indeed, intermediate slides are commonly found on steeper slopes and are associated with liquefied matrix material.

At the other extreme, intermediate landslides laden with debris grade into sturzstroms. The key distinguishing feature here is that sturzstroms begin as deeper-seated failures (tens of metres or more) and so involve the collapse of consolidated rock. However, the gradational nature of the transition strongly suggests a dynamic continuity between the intermediate landslides and sturzstroms.

TABLE 4.3 Threshold rainfall conditions for the onset of regional landsliding (flows)

B (dimensionless)	C (mm/h$^{(1-B)}$)	Limits of duration	Location and source
0.39	14.8	1 min–10 days	Caine (1980)
0.21	14.7	10 min–1 day	Hope, Coast Mtns, British Columbia, Canada; Church and Miles (1987)
0.47	35.4	3–23 hours	San Francisco, USA, wet districts; Cannon and Ellen (1985)
0.76	32.4	10–20 hours	San Francisco, USA, dry districts; Cannon and Ellen (1985)

4.6 Water and slope failure

We have seen that a slope becomes less stable when it contains more water. Four key mechanisms by which water can induce instability are by reducing the effective normal stress, liquefying mixtures of soil and fine-grained rock, chemically weathering a rock, and enhancing rock cracking.

4.6.1 Reducing the effective normal stress

The overlying surfaces of a failure plane are rough and uneven and can resist movement by interlocking their small-scale surface irregularities (like the teeth of interlocked cogwheels). Water within a rock exerts a hydrostatic pressure on the walls of enclosing pores and cracks. This pressure is the product $\rho_w g h_w$ of water density, gravity and the thickness of the water column above the level of interest. Since it is hydrostatic, the water pressure acts against the normal pressure σ on a given level due to the weight of the overlying mass. As a result, opposing faces of a plane can be prised open, much as the pressure in a tyre can raise the wheel of a car off the ground. Fewer irregularities remain in contact, and so sliding resistance is reduced. (As a kitchen experiment (Hubbert and Rubey, 1959; Hoek and Bray, 1981), place an opened beer can filled with water (it would be a shame to waste the beer) on an inclined surface, and tilt the surface until the can just begins to slide. Now puncture a hole in the base of the can, so that water can enter the space between the can and the incline. The hydrostatic pressure due to the water in the can raises the can a small distance above the incline. The rough surfaces of the can and incline come into contact over a smaller total area than before. The effective sliding resistance of the can is decreased and it can move over a much smaller gradient compared to its unpunctured state.)

Increased pore pressure is the mechanism most commonly invoked to explain the destabilising effect of water, be it along the base of sluggish and intermediate movements (Hutchinson and Bhandari, 1971) or of deep-seated movements, both slow (Hubbert and Rubey, 1959) and catastrophic (Voight and Faust, 1982; Hendron and Patton, 1985). In terms of the Mohr–Coulomb failure law (Box 4.2, eqn (1)), sliding resistance is $\tau_{c0} + \mu_0 \sigma (1 - U_0/\sigma)$, where τ_{c0} is material cohesion (due to chemical forces holding rock together), μ_0 is the coefficient of sliding friction (due to surface irregularities), U_0 is the pore pressure of water, and σ is the normal pressure due to the weight of overlying rock. Increasing water pressure (U_0) reduces the effective normal pressure to $(1 - U_0/\sigma)\ \sigma$ and, by reducing the number of surface contacts per unit area, it may in addition reduce μ_0; indeed, water pressure may even overcome some of the cohesive forces binding rock grains together, thereby reducing also material cohesion (Hsü, 1969).

4.6.2 Liquefying soil and fine-grained rock

A sufficient supply of water can transform a mass of interlocked grains into a fluid, either by carrying the grains along in suspension or by dissolving them chemically. Increased pore pressures are required to separate adjacent grains. However, liquefaction independently reduces resistance by reducing the friction coefficient, μ. This process is normally associated with intermediate landslides rich in mud, silt and clay (Selby, 1993).

4.6.3 Chemical weathering

Water can enter the molecular framework of minerals, distorting their lattices and reducing their structural integrity (Selby, 1993). Macroscopically this can weaken a slope by reducing material cohesion.

4.6.4 Enhanced rock cracking

Water can break the bonds between atoms, especially when they are already stretched under an applied load. Such breaking is further favoured where stresses are locally concentrated at the tips of flaws and cracks (Lawn, 1993). The net result is a reduction in material cohesion. In addition, the shape of the new

Box 4.2 Slope collapse and landslide runout

A slope remains stable while the gravitational stress τ_g pulling a mass downslope is smaller than or equal to the slope's peak sliding resistance τ_P (Hoek and Bray, 1981):

$$\tau_g \leq \tau_P = \tau_{c0} + \mu_0(1 - U_0/\sigma)\,\sigma \qquad (1)$$

where, at critical stability along the future plane of failure, τ_{c0} is material cohesion, μ_0 is the coefficient of sliding friction, and the effective normal stress, $(1 - U_0/\sigma)\,\sigma$, is the difference between actual normal pressure σ and fluid pore pressure U_0.

After failure, slope resistance drops from its peak to residual value τ_R (Fig. 4.13) and the unstable mass accelerates, for which a force balance yields

$$\tau_g - \tau_R = \tau_g - [\tau_c + \mu(1 - U/\sigma)\,\sigma + Ku] = \rho ha \approx \rho hu^2/x \qquad (2)$$

where the residual stress consists of a sliding stress $\tau_c + \mu(1 - U/\sigma)\,\sigma$ and a velocity-dependent stress Ku (Box 4.1); K is a dynamic resistance (material property), u is the mean velocity downslope, achieved when the mass has moved a distance x at a mean acceleration a, and ρ and h are the mean density and thickness of the landslide. The suffix 0 has been dropped from μ and U to illustrate that both these quantities might differ from their values before failure.

Subtracting eqn (2) from the limiting case of eqn (1) gives

$$\Delta\tau_s + Ku \approx \rho hu^2/x \qquad (3)$$

where $\Delta\tau_s$ is the difference in sliding resistance between the critically stable and moving states, given by $\Delta\tau_s = \Delta\tau_c + \sigma\,\Delta[\mu(1 - U/\sigma)]$, where Δ denotes 'a change in' the associated quantity. For simplicity, the normal pressure σ is assumed to be constant.

At the start of collapse, velocities are small and the resistance Ku is negligible compared with the change in sliding resistance (Ku ! $\Delta\tau_s$). For this simplified analysis, we assume (a) that the landslide has a block geometry with the ground and failure surfaces parallel, and (b) that changes in cohesive stress are much smaller than changes in the frictional sliding stress ($\Delta\tau_c$! $\sigma\,\Delta[\mu(1 - U/\sigma)]$). More complex analyses relaxing these assumptions yield substantially similar results.

From assumption (a), the normal stress $\sigma = \rho gh \cos\alpha$, where α is the mean slope angle of the failure surface. Using this expression and assumption (b), eqn (3) simplifies to

$$u = \{gx\,\Delta[\mu(1 - U/\sigma)]\cos\alpha\}^{1/2} \qquad (4)$$

If x^* is the distance travelled by the mass before it leaves the collapse scar, then the landslide velocity u^* after collapse is

$$u^* = \{gx^*\,\Delta[\mu(1 - U/\sigma)]\cos\alpha\}^{1/2} \qquad (5)$$

Equation (5) assumes that dynamic frictional stresses (Ku) become important only during the last stages of collapse when the landslide is about to leave its collapse scar. This is a good first approximation in the case of sturzstroms (Kilburn and Sørensen, 1998) and is assumed, by extrapolation, to be reasonable also for intermediate landslides. However, the assumption may break down in the case of sluggish movements.

As the landslide moves away from the collapse scar, it spreads over the surface while trying to maintain its initial velocity u^*. Changes in velocity are small (a approaching zero) so that eqn (3) can be approximated to $\Delta\tau_s \approx Ku^*$. If we note that (a) $\Delta\tau_s = \tau_{s0} - \tau_s = \tau_g - \tau_s$, (b) $\tau_s \leq Ku$, and (c) $\tau_g \approx \rho gh_1 \sin\beta$, where h_1 is the mean landslide thickness *beyond* the collapse scar, where the ground has a mean angle of slope β, the simplified force balance yields

$$h_1 \approx Ku^*/(\rho g \sin\beta) \qquad (6)$$

The important feature here is that the landslide is assumed to adjust its thickness to maintain its velocity at u^*. The factors controlling velocity can thus be linked directly to conditions during collapse (eqn

(5)). Clearly, the assumption is not valid if spreading is restricted by topography.

A landslide moves under gravity. Its energy for motion is the potential energy released as it collapses and advances downslope. This energy is converted to kinetic energy for accelerating the mass and finally lost in overcoming friction. Equating the total potential energy released with the total frictional energy consumed, a global energy balance gives

$$\rho g V H \sim \tau_f A L \qquad (7)$$

where V and A are the volume and mean basal area of the landslide, τ_f is mean frictional stress across the base of the mass, and H and L are the vertical and horizontal distances between the top of the collapse scar and the toe of the landslide deposit (Fig. 4.14). Strictly, H and L should refer to displacements of the landslide's centre of mass; however, the form in eqn (7) is good to a first approximation.

Equation (7) can be rearranged to give the travel distance $L \sim \rho g V H / \tau_f A$. We have seen that a landslide's travel distance increases with volume (Fig. 4.15), so the form of the energy equation looks promising as an explanation of the L–V trend. Since ρ and g are independent of volume, the next step is to examine the volumetric dependences of the terms H, A and τ_f.

Observation shows that collapse scars have preferred geometries. If z_0 is the vertical distance between the top and bottom of the scar, and x^* is the length of the failure surface, then geometric similarity implies that $z_0 \propto x^* \propto V^{1/3}$. Similarly, we can set $A \propto V^{2/3}$ in the first instance. The mean resisting stress during runout from the scar is Ku^*. From eqn (5), $u^* \propto x^{*1/2} \propto V^{1/6}$. Using these volumetric dependencies, eqn (7) simplifies to

$$L = C^* \phi(\rho g^{1/2}/K)(H/z_0)V^{1/2} \qquad (8)$$

where $\phi = \{\Delta[\mu(1 - U/\sigma)] \cos \alpha\}^{-1/2}$ and C^* describes landslide geometry (here assumed to be constant). The ratio $H/z_0 \geq 1$ and its value depends on local topography below the collapse scar ($H/z_0 = 1$ when a landslide collapses directly onto a horizontal surface).

Equation (8) indicates that the distance a landslide travels increases with the square root of its volume. This indeed is the trend observed (Fig. 4.15), albeit with a scatter induced by variations in the other terms.

crack surfaces may decrease the friction coefficient, while water migrating through new cracks may increase local pore pressures.

Slow cracking requires unbroken rock and is favoured by large applied stress. It is thus most likely to operate among deep-seated landslides for which failure occurs beneath the upper crustal layers of soil and weathered rock. Conventionally, it is assumed that water destabilises even deep-seated landslides by increasing pore pressure. A drawback of this interpretation is that it requires the persistent saturation of large thicknesses of rock in order for pore pressure to become a significant fraction of the normal pressure. Slow cracking is a ready alternative mechanism and has the advantage of being able to describe observed accelerations in slope movement before catastrophic failure.

Microcracking is dominated initially by the formation of new cracks, but later by crack growth. Eventually, the growing cracks link together to produce a major plane of failure (McGuire and Kilburn, 1997; Kilburn and Voight, 1998). While rates of crack nucleation increase exponentially with time (Main and Meredith, 1991), rates of crack growth increase exponentially with crack length (Lawn, 1993). In addition to single cracks, the exponential relations also hold for a population of cracks with a fractal size–frequency distribution (using mean values of nucleation rate, growth rate and crack length; Main *et al.*, 1993). Since small cracks must interlink before producing a new plane of failure, it is the onset of deformation dominated by crack growth that provides the essential precursor to slope collapse.

In the case of unstable slopes, cracking occurs owing to the downslope stress acting on the deforming horizon. The opening and growth of a population of cracks will thus be recorded macroscopically by a combination of bulk dilation and of downslope displacement. Assuming that the bulk movements are proportional to the total rate of cracking, rates of downslope displacement (dx/dt) shortly before collapse will be given by

$$dx/dt = (dx/dt)_0 \, e^{\psi(x-x_0)} \qquad (4.1)$$

where the suffix 0 denotes conditions when crack growth first dominates crack nucleation, and ψ is an inverse length scale that depends on the applied stress, rock properties and the geometry of the crack array (Kilburn and Voight, 1998).

Manipulation shows that eqn (4.1) is equivalent to a linear decrease with time in the *inverse* rate of displacement:

$$(dx/dt)^{-1} = (dx/dt)_0^{-1} - \psi(t - t_0) \qquad (4.2)$$

Equating the time of failure with the condition that $(dx/dt)^{-1}$ becomes zero (i.e. rates of deformation become infinitely large), eqn (4.2) can be used to forecast the onset of catastrophic collapse by linearly extrapolating measured inverse rate trends to the condition $(dx/dt)^{-1} = 0$ (Voight, 1988; Kilburn and Voight, 1998).

Few precursory data are available for catastrophic slope collapses. A notable exception is the set of observations compiled before the collapse of Mount Toc into the Vajont reservoir on 9 October 1963 (Müller, 1964). Final failure occurred after more than two months of slowly accelerating slope movement. When the inverse deformation rate during this interval is plotted against time (Fig. 4.12), a clearly linear decrease is observed, as expected from the slow-cracking model (eqn (4.2)). Notice that the trend is evident by day 30 so that, had the model been available at the time, catastrophic failure might have been forecast almost a month before it occurred (Kilburn and Petley, 2001).

4.6.5 Cracking and the speed of collapse

In addition to bedrock, clay-rich layers can break, rather than flow, when under loads of at least 1–10 MPa, corresponding to depths of ~10^2 m (Petley, 1999; Kilburn and Petley, 2001). The approach to brittle failure occurs in three stages (Fig. 4.13; Jaeger, 1969; Hallbauer *et al.*, 1973): (a) an elastic deformation for applied shear stresses smaller than a peak value τ_p; (b) continuous creep at τ_p; and (c) beyond a critical strain, an abrupt decrease in shear resistance to a residual value τ_r. This behaviour is consistent with microscopic cracking (Jaeger, 1969; Hallbauer *et al.*, 1973; Main and Meredith, 1991; Lockner, 1995; Petley, 1999). Initial elastic deformation involves the stretching of molecular bonds and frictional sliding between the surfaces of any existing small cracks. New cracks appear as the peak stress is approached and continue to open and grow at τ_p until they link together to form a major plane of failure. The shear resistance abruptly decreases to a

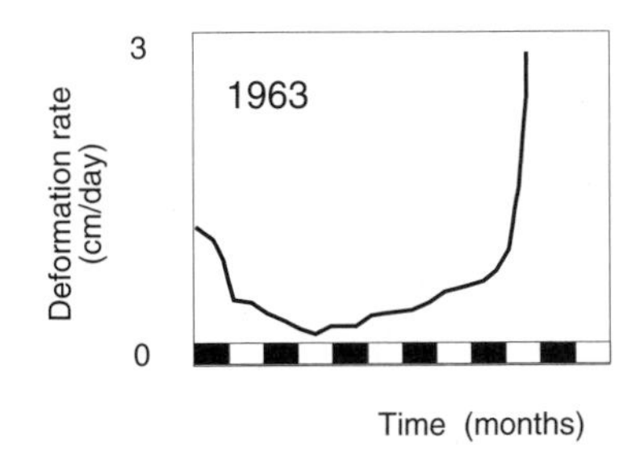

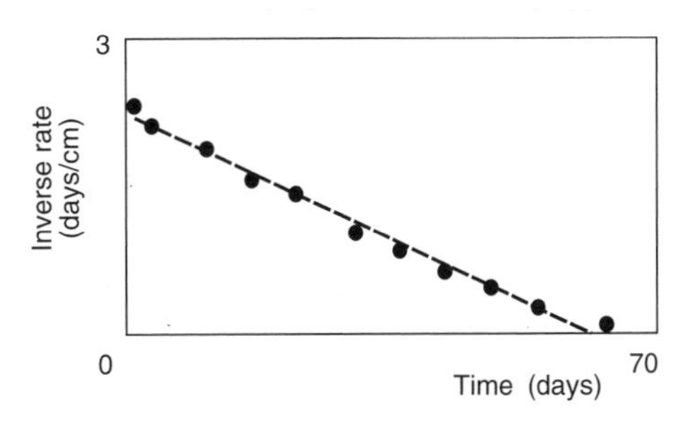

FIGURE 4.12 Rates of movement (*left*) of the Mount Toc landslide before it collapsed catastrophically into the Vajont reservoir on 9 October 1963. The inverse rate of deformation (*right*) decreased almost linearly with time for nearly 60 days before the catastrophe occurred. The linear inverse-rate trend is that expected from the slow-cracking equation (eqn 4.2).

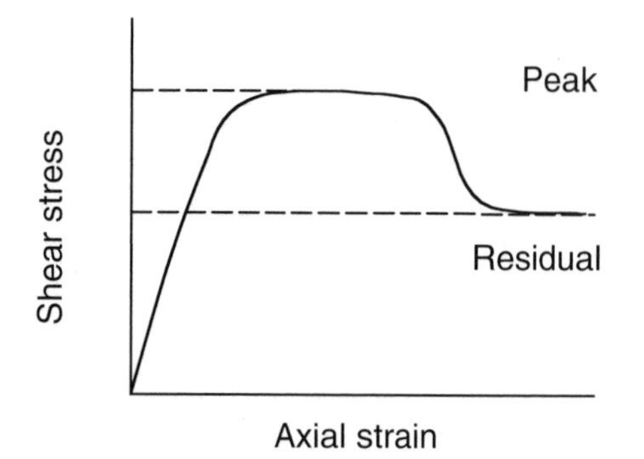

FIGURE 4.13 Idealised deformation of rock under loads of 1–10 MPa (depths of 50–500 m). The rock initially deforms elastically for applied stresses smaller than its peak resistance (τ_P). It continues to deform at τ_P as microcracks nucleate and grow. When the cracks link to form a major failure plane, the shear resistance of the rock rapidly decreases to a residual value (τ_R), the stress at which final collapse takes place.

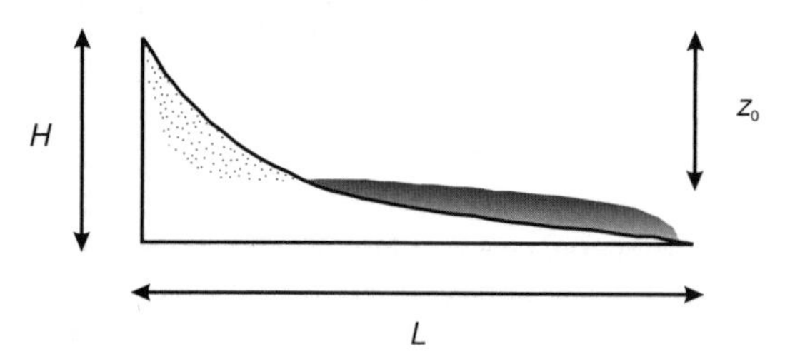

FIGURE 4.14 Definition sketch of key landslide dimensions. The landslide deposit (*grey*) is derived from material that has collapsed from the source volume (*stipple*). *H* and *L* represent the vertical and horizontal distances between the top of the collapse scar and the base of the landslide toe. Collapses, especially those of very large volume, frequently occur near the base of a slope, such that the vertical height of the collapse scar (z_0) is only slightly smaller than H.

residual value τ_R determined by sliding friction along the new failure plane.

Since the gravitational pull on a slope remains unchanged, the sudden decrease in resisting stress from peak to residual values leaves the unstable mass with an unbalanced force downslope. Collapse occurs with a mean acceleration *a*, given by (Box 4.2, eqn (2))

$$a = (\tau_P - \tau_R)/\rho h$$

where the peak resistance τ_P must equal the gravitational driving stress τ_g, otherwise failure could not begin, and ρ and h are the mean density and thickness of the landslide.

Taking clay as the most probable weak material in a crustal sequence, then $(\tau_P - \tau_R) \approx \gamma\tau_P$, where γ is about 0.1–0.2 (Kilburn and Petley, 2001). Setting $\tau_P = \tau_g = \rho g h \sin\alpha$ (where α is the mean angle of slope along the plane of failure), collapse accelerations are expected to be $\sim\gamma g \sin\alpha$, which, for slope angles of 25°–40°, yields mean accelerations of 0.4–1.2 m/s. In other words, a sturzstrom could accelerate to velocities of 15–30 m/s in less than a minute. Faster accelerations would follow the sudden failure of stronger rock, the peak strength of which is larger than that for clay (Jaeger, 1969; Hanks, 1977).

The cracking model can thus account for slope accelerations both before *and* during catastrophic collapse. Importantly, the model does not require any special mechanism (such as melting or the vaporisation of trapped water) for reducing landslide friction. It has thus the attractive quality that it can operate among normal rocks under ordinary conditions. However, cracking does not mean that pore water is unimportant. On the contrary, not only does the presence of water catalyse slow, corrosive cracking, but the associated pore pressures, even if small, may be just sufficient to reduce the effective peak resistance below the gravitational pull, and so initiate the onset of cracking.

4.7 Water and landslide run-out

As well as destabilising a slope, accumulated groundwater reduces the resistance of a moving mass and, hence, increases its potential velocity. Since faster initial velocities favour longer travel distances, groundwater accumulation is also important to landslide run-out. A first-order analysis is outlined in Box 4.2. It has been simplified by assuming that (a) both collapse scars and landslide deposits have restricted variations in geometry, (b) landslides accelerate to their peak velocity during

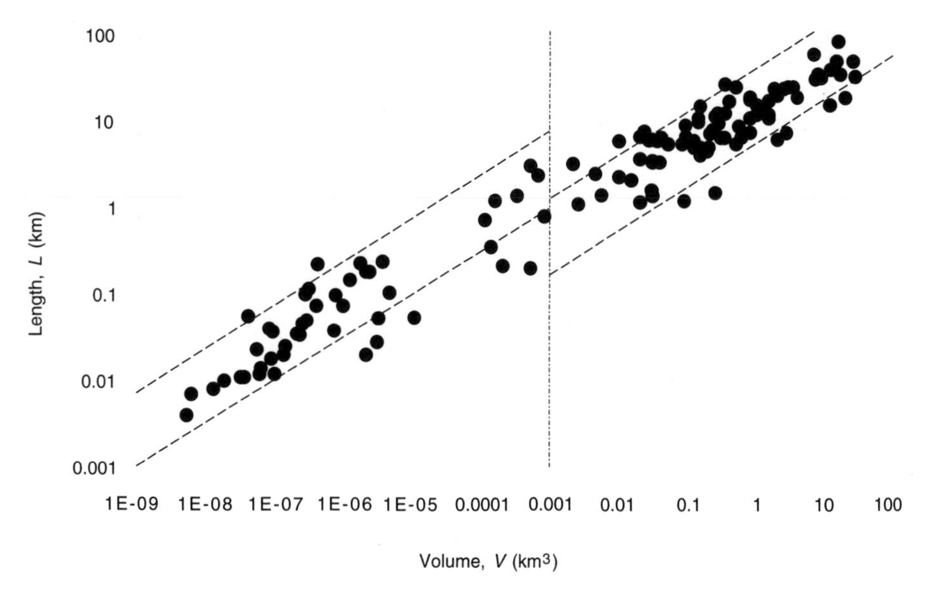

FIGURE 4.15 Volume is an important control on how far a landslide can travel. The general trend (*dashed lines*) is consistent with landslide length increasing with the square root of landslide volume (Box 4.2, eqn (8)). This trend appears to shift at volumes of about 10^6 m^3 (10^{-3} km^3, *dot-dashed line*), larger volumes corresponding to sturzstroms. (Data from Hayashi and Self, 1992, and Corominas, 1996)

collapse, and (c) resistance beyond the collapse scar is concentrated along the base of the landslide and increases with velocity, so that most of the run-out occurs at a nearly constant rate of advance.

The analysis (Box 4.2, eqn (8)) indicates that run-out length L depends, among other factors, on the square root of both volume V and water-induced changes in effective friction (in the term ϕ):

$$L = C^* \phi \, (\rho g^{1/2}/K) \, (H/z_0) \, V^{1/2}$$

where $\phi = \{\Delta[\mu(1 - U/\sigma)] \cos \alpha\}^{-1/2}$ and C^* describes landslide geometry (see Box 4.2 for other terms).

As shown in Fig. 4.15, the actual variation of runout length with volume closely follows two L–$V^{1/2}$ trends, a transition from one trend to the other occurring at volumes of 10^6–10^7 m^3. The transitional volumes correspond to the minimum values for sturzstroms to form, and reflect conditions when initial failure becomes sufficiently deep-seated to disrupt strong crustal rock, as opposed to the weaker rock and soil involved for smaller landslides.

The trend for smaller landslides is remarkable both for the huge range of volumes covered, from cubic metres to 10^6 m^3, and for the fact that the data involve movements belonging to several different conventional categories: translational slides, earthflows, mudflows, mudslides, debris flows and debris slides (Corominas, 1996). In addition to their restricted geometries, therefore, these different categories evidently share similar dynamic constraints. Two obvious constraints are the focusing of deformation along thin basal layers, and the nature of the deforming material. These landslides invariably contain a matrix of fine-grained material (such as soil, mud, silt and clay), and so the rheology of the matrix material (measured by K) is the most likely second feature to favour a similar dynamics among different types of landslide.

Similarly, the sturzstrom data (Hayashi and Self, 1992; Corominas, 1996) include landslides conventionally classified as rock avalanches, debris avalanches and debris flows. Again, their essential dynamics must be similar despite the variety of names applied and, by analogy with smaller-volume landslides, it is feasible that sturzstrom run-out is controlled by deformation within basal layers of fine-grained, comminuted rock (whose rheology determines K).

Irrespective of classification, therefore, landslides behave as slowly deforming masses that are carried on top of more rapidly deforming basal layers rich in fine-grained material. The analysis in Box 4.2 assumes that the basal material has a resistance that increases with velocity, a feature characteristic of simple viscous fluids. Such behaviour is indeed possible for muddy and granular suspensions within particular ranges of solid concentration and deformation rate (Bagnold, 1954; Keefer and Johnson, 1983; Pierson and Costa, 1987). The agreement between model and observed $L–V^{1/2}$ trends thus implies, but does not prove, that common matrix materials in landslides deform as viscous fluids. New studies are required to verify this interpretation.

The $L–V^{1/2}$ equation also implies that the jump in trends at $10^6–10^7 m^3$ is favoured by a combination among small-volume landslides of smaller effective viscous resistance (K), larger relative run-out over steep ground (H/z_0), and larger relative water pore pressures (U/σ). Giant, catastrophic landslides tend to occur towards the base of a mountain, whereas smaller collapses can occur from any level. Thus H/z_0 is normally about 1 for sturzstroms, but often greater than 1 for smaller landslides. In addition, it is traditionally assumed that sturzstroms have a lower effective viscous resistance (smaller K) than smaller landslides. This assumption has never been proven, but if it is correct, then changes in K and H/z_0 will tend to counterbalance each other in controlling the jump in $L–V^{1/2}$ trends. Possibly, therefore, the jump in trends is due to smaller masses acquiring larger relative pore pressures before they collapse. Such a condition is feasible since it is easier for circulating water to saturate rock to depths of 1–10 m than to 100 m or more. Hence, at least among landslides smaller than $10^6 m^3$, water may serve not only to induce slope failure but also to favour larger landslide velocities and travel distances.

4.8 Effect of environmental change on global rates of landsliding

A recurring theme of this chapter is the importance of water in destabilising slopes and in influencing how quickly and how far a landslide can travel. A key feature is the length of time for which rock and soil remain saturated with water. The amount and duration of water retention increase as the rate of water supply becomes ever greater than the rate of water escape from a slope. Of these two factors, it is the rate of water supply that appears generally to be more susceptible to environmental change. This is especially the case when addressing the potential consequences of climate change on landslide activity.

By altering atmospheric and surface temperatures, climate change will affect the locations and rates of annual precipitation, as well as the amount of water stored in liquid and frozen form. If recent climate changes continue unrestricted, model forecasts suggest that the Earth will be on average 3°C warmer in 2080 than it is today (UK Meteorological Office, 1999). The implications for water and its distribution are (UK Meteorological Office, 1999):

1. Thermal expansion of sea water and the melting of ice sheets and glaciers globally will raise mean sea level by about 0.4 m.
2. Annual precipitation will increase by about 100–300 mm at latitudes of 45° or greater, covering Antarctica and the tip of South America as well as Canada, the northern and central United States, northern Europe, the former Soviet Union and China. At lower latitudes (about 5°–45°), annual precipitation will decrease by about 100–700 mm, affecting Australia, southern Africa and western South America, the southern United States, Central America and the Caribbean, northern Africa and the Mediterranean, the Middle East and India. Close to the equator itself, reduced rainfall (by 100–600 mm a year) is expected in the Americas and South-East Asia, whereas an annual increase of 100–300 mm is expected across central Africa.

3. Vegetation cover will decrease in the tropics, especially the forests and grasslands of northern South America and central southern Africa. In contrast, vegetation growth will be favoured at higher latitudes, notably in the Northern Hemisphere.

The increase in mean sea level will favour a general increase in the frequency of coastal landsliding. Much of the movement is expected to have only a local effect. However, many of the world's active volcanoes have also a coastal (and island) setting, and any increase in the frequency of edifice failure may contribute to an increase in global rates of volcanic eruption. Inland, meanwhile, the melting of glaciers and permafrost in high mountain ranges will favour greater water accumulation in rocks immediately underneath and downstream. Recent studies in the Alps (Brown, 2001; PACE, 2001) have revealed increases in ground temperature of 0.5–1°C between 1985 and 2000. Such increases are large enough to favour melting of the uppermost layers of permafrost, typically tens of centimetres below the surface. The permafrost horizon marks the level below which rock has been filled with ice for millennia, in contrast to shallower levels in which trapped water freezes and thaws with seasonal variations in temperature. The melting of glaciers and permafrost will thus favour an increase in the annual number of landslides in mountainous areas. Moreover, since these areas of high relief are also the most vulnerable to giant, catastrophic collapse, global warming is expected in addition to increase the mean frequency of sturzstroms.

The forecast changes in precipitation suggest that significant mass movements may occur more often at high latitudes and in Central Africa, but might generally diminish towards the equator. However, increased vegetation cover may counteract the effect of rainfall (by absorbing rainfall and binding soil with roots), especially in the Northern Hemisphere.

4.9 The future

At the start of the twenty-first century, landslide studies are poised to enter a new and exciting phase for understanding landslide dynamics and for developing reliable and quantitative models for forecasting slope failure and landslide run-out. The results will invariably inspire also new methods for stabilising slopes and otherwise reducing landslide hazard. However, the fact that the Earth's crust is constantly deforming means that mass movements will always present a threat to human activity. The most efficient means of reducing future losses, therefore, will be to raise public awareness of landslide hazards and so encourage more prudent land-use policies and emergency strategies, such as evacuation plans, before the next disaster strikes.

5

Volcanoes and environmental change

5.0 Chapter summary

Here we take a look at the various ways in which volcanic activity and environmental change have been linked, both during recent times and in the geological record. Following a brief introduction to some of the relevant literature and consideration of the main issues, the characteristics of volcanic eruptions are addressed, as they relate to environmental impact. The volcanic contribution to the atmosphere is examined, together with the range of resulting physical and chemical processes that occur within the stratosphere. We look in some depth at the complex, and to some extent still controversial, impact of volcanic eruptions on climate. Historical evidence recording the aftermaths of major volcanic eruptions from the past millennium, together with detailed geophysical observations of the effects of recent large eruptions, supports the idea that such events have a clear, if short-term impact on weather and climate. Much of the detail remains to be resolved and clarified, however. Brief discussion of the potential cultural impacts of large eruptions is followed by consideration of the devastating effects of past volcanic super-eruptions on the global environment and their potential to harm modern society. We conclude the chapter by evaluating the complex apparent relationship between volcanism and glaciation–deglaciation cycles and addressing the possibility that the largest eruptions of all might have been responsible for mass extinctions within the geological record.

5.1 Volcanism and environmental change: a brief history of research

The 1783 eruption of Laki (alternatively Lakagigar or the Skaftár Fires) in Iceland and the 1815 explosion of the Indonesian volcano Tambora (Box 5.1) provide strong evidence that volcanic eruptions are capable of causing dramatic changes to the environment, and in particular to the weather and climate (although read Section 5.7 for a more cautionary perspective). For some time now, volcanic activity has been recognised as being linked with environmental change, most commonly as a trigger, but also through major changes in the environment triggering or enhancing volcanism itself. On the grandest scale, large basaltic flood eruptions have been implicated in mass extinctions, while the most voluminous explosive super-eruptions have been held responsible for initiating the episodes of severe global cooling known as *volcanic winters* (Rampino and Self, 1992, 1993a), and even leading to the decimation of the human race (Rampino and Self, 1993b). Volcanic activity has been put forward by some (e.g. Bray, 1976, 1977; Rampino and Self, 1992, 1993a) as a potential initiator, or at least accelerator, of Quaternary glaciation and by others (e.g. Matthews, 1969; Rampino *et al.*, 1979; Hall, 1982; Wallmann *et al.*, 1988; Nakada and Yokose, 1992; Sigvaldason *et al.*, 1992; McGuire *et al.*, 1997) as being triggered by dramatic changes in Quaternary climate. As evidenced by the Laki and Tambora events, and more recently by climatologically significant

Box 5.1 Laki and Tambora: a tale of two eruptions

On 8 June 1783, following a period of intense seismic activity, a 27-km-long fissure system opened up to the south-west of the Vatnajökull ice cap in Iceland, producing one of the largest basaltic lava flows in historic times (Thordarson and Self, 1993). Effusion rates of up to $9 \times 10^3\,m^3/s$ produced lava flows that travelled 35 km in 4 days, and by the eruption's end the following year, nearly 15 km^3 of lava covered an area of over 600 km^2. Lava fountains generated during the eruption reached heights of almost 1.5 km and convecting eruption columns from the vents reached heights of around 15 km. Over a period of 8 months the atmosphere was loaded with 122 million tonnes of SO_2 – sufficient to form ~250 Mt of H_2SO_4 aerosols, together with huge volumes of chlorine and fluorine gas (Thordarson *et al.*, 1996). The aerosol cloud spread over Europe and North America, causing one of the greatest atmospheric pollution events in recent history (Grattan and Charman, 1994; Grattan and Brayshay, 1995), and had a considerable environmental and climatic impact on the Northern Hemisphere. In Iceland, a low-altitude acid haze stunted grass growth, and over 50 per cent of the island's livestock was lost to fluorine poisoning in less than a year. Both fresh and salt water fish stocks also appear to have fallen dramatically as a consequence of the eruption. The resulting *haze famine* caused over 10,000 fatalities – over 20 per cent of the island's population (e.g. Hálfdánarson, 1984). In Europe, fine ash and a dry acid fog, containing noxious or toxic materials from the eruption, caused damage to plants and crops across the continent (e.g. Grattan and Brayshay, 1995), and dry fogs were also reported from North America (e.g. Franklin, 1784). The sulphate aerosol cloud is widely held to have been responsible for significantly depressed Northern Hemisphere temperatures during the period 1783–85, estimated to have been 1–2 degrees C below average (e.g. Angell and Koshover, 1988).

A little over thirty years after the Laki event, what is generally considered to have been the largest and most violent volcanic eruption in recorded history occurred on the Indonesia island of Sumbawa. The Tambora blast of 1815 ejected around 150 km^3 of tephra, with ash falling nearly 1500 km away and the sound of the climactic explosion carrying for 2600 km (Sigurdsson and Carey, 1989). The impact on the planet's climate appears to have been almost as spectacular as the eruption itself. Within weeks, gases and fine ash in the stratosphere were causing unusual atmospheric phenomena over Europe, including brilliant sunsets and twilights, blue moons and brown suns. As around 150 Mt of sulphate aerosols spread out across the planet, weather patterns were so disrupted that the following year – 1816 – became known as *the year without a summer* (Stothers, 1984). In that year, much of the north-eastern United States lay under a persistent 'dry fog', and temperatures throughout the spring and autumn were anomalously low, with both frost and snow reported during the summer of 1816. Similarly, the summer following the eruption was unusually cold and wet in Europe, causing low harvest yields, and even ruining Mary Shelley's continental vacation to such an extent that she stayed indoors and penned the gothic classic *Frankenstein*. The Northern Hemisphere temperature fall resulting from the aerosol cloud generated by the eruption is estimated by Stothers (1984) to have been of the order of 0.7 degrees Celsius, the largest drop recorded in historic times.

eruptions at El Chichón (Mexico) in 1982 and Pinatubo (the Philippines; Fig. 5.1) in 1991, volcanism is also implicated in short-term variations in weather and climate, and is linked to brief (2- to 3-year) cooling events following individual eruptions (e.g. Pollack *et al.*, 1976; Kelly and Sear, 1984; Robock and Mao, 1995). Longer-term interdecadal episodes of climate change such as the medieval cold spell known as the *Little Ice Age* have also been attributed to multiple eruptions occurring over a limited time span (e.g. Robock, 2000).

Somewhat surprisingly, the link between volcanoes and weather was recognised very early on in human history. Two thousand years ago Plutarch reported cooling, crop loss and famine resulting from the huge 44 BC eruption of Mount Etna. Much more recently, Benjamin Franklin attributed the cold summer of 1783 and the following winter to the Laki basalt eruption in Iceland (Franklin, 1784; Thordarson and Self, 1993; Dodgshon *et al.* 2000). Although Humphreys (1913, 1940) proposed that cooling following large volcanic eruptions was the result of the formation of volcanogenic sulphate aerosols in the stratosphere, the first serious analysis of the link was not undertaken until the early 1960s (Mitchell, 1961). It was probably not until 1970, when Hubert Lamb published his seminal paper 'Volcanic dust in the atmosphere, with a chronology and assessment of its meteorological significance', that the notion that volcanic eruptions could have a major influence on climate and weather became widely accepted. A plethora of papers followed that looked in some detail at the likely nature of links between volcanic eruptions, weather and climate, including Toon and Pollack (1980), Robock (1989) and Kondratyev and Galindo (1997). Research and publication in the field was further stimulated in the 1980s and 1990s following the El Chichón and Pinatubo explosive eruptions, both of which had a measurable effect on global temperatures. The eruptions coincided fortuitously with the realisation that the satellite-based *Total Ozone Mapping Spectrometer* (TOMS) was also capable of monitoring atmospheric sulphur dioxide, so permitting its use in estimating the mass of volcanogenic SO_2 injected into the stratosphere and the global tracking of SO_2 clouds from the eruptions (e.g. Bluth *et al.*, 1992; Robock and Matson, 1983). Following intense El Niño-Southern Oscillation (ENSO) events after both the El Chichón and Pinatubo eruptions, speculation arose concerning a possible causal link between the eruptions and the climatic signal. A number of studies (e.g. Robock *et al.*, 1995; Self *et al.*, 1997) have since, however, revealed that the sea surface warming associated with the ENSO had already started prior to both eruptions. Most recently, interest has focused on the potentially far-reaching impact on the global environment of infrequent but devastating *super-eruptions* (e.g. Chesner *et al.*, 1991) capable of ejecting sufficient volumes of sulphur gases into the stratosphere to trigger severe global temperature falls (e.g. Rampino and Self, 1992, 1993a; Bekki *et al.*, 1996; Zielinski *et al.*, 1996).

FIGURE 5.1 The 1991 eruption of Pinatubo in the Philippines ejected 20 million tons of SO_2 into the stratosphere, sufficient temporarily to reduce global surface temperatures by almost half a degree Celsius. (Image courtesy of USGS/Cascades Volcano Observatory)

5.2 Eruption characteristics and environmental impact

The potential for a volcanic eruption to have a significant impact on the environment is dependent upon a number of factors, the most important of which are size, explosivity, magma chemistry and geographical location. In very broad terms, volcanoes can be divided into *effusive* and *explosive* types, with eruptions of the former being dominated by lava flow formation and the latter by the more energetic expulsion of disrupted magma (*pyroclastic activity*) in the form of *tephra* (volcanic debris ejected into the atmosphere) and *pyroclastic flows*. Generally speaking, explosive volcanic eruptions are the most efficient at transmitting mass through the troposphere (the lowest 10 km or so of the atmosphere) and into the stratosphere, where it has the greatest potential to affect the climate on a regional or global basis. Large, predominantly effusive eruptions, such as that at Laki (Iceland) in 1783 (Thordarson and Self, 1993), may also, however, be capable of lofting sufficiently large volumes of gases into the stratosphere to cause severe short-term climatic effects on a continental scale (e.g. Grattan and Pyatt, 1999; Dodgshon *et al.*, 2000). As addressed later in this chapter and in Chapter 7, very large basaltic flood eruptions have also been implicated in major global climate change through the production over hundreds of thousands or millions of years of huge volumes of gases and their transmission into the atmospheric envelope.

5.2.1 Factors determining the style of eruption

In reality, effusive and explosive eruptive behaviour represent two extremes of a continuum, and many volcanoes are characterised by both lava flow production and explosive pyroclastic activity. The explosivity of a volcano often varies over time and is influenced by a number of factors including the gas content of the magma (mainly water vapour, but other common gas phases include sulphur dioxide, carbon dioxide, chlorine and fluorine), the rate of magma ascent in the crust, and magma chemistry. Broadly speaking, the higher the *silica* (SiO_2) content of a magma, the more explosive its eruptions are likely to be, primarily because the polymerising nature of the silicon and oxygen atoms binds the magma together and raises its viscosity. Higher-viscosity magmas, such as andesites, dacites and rhyolites, are more effective at trapping gas within the rising magma column. Consequently, high gas pressures tear the magma apart explosively when it encounters ambient atmospheric pressure on reaching the surface. Low-silica basaltic magmas, in contrast, tend to lose much of their gas during the ascent phase through the crust, and the arrival of magma at the surface is characterised by less energetic *lava fountaining* and the quiet effusion of lava flows.

The rate of magma ascent is also an important factor in determining the explosivity of an eruption. If magma ascent is slow, there is more time for dissolved gases to exsolve and form bubbles as it rises into lower-pressure regimes. As a result, when the magma reaches the surface, a relatively small gas overpressure exists and the consequent eruption is typically either effusive or mildly explosive. Where, however, magma ascent is rapid, there may be insufficient time for significant amounts of the contained gases to exsolve, particularly for high-viscosity magmas. When the magma encounters ambient pressure at the surface, therefore, the high overpressure causes the contained gases to tear the magma apart in a violent explosion. Similarly, at open-vent volcanoes such as Mount Etna in eastern Sicily, the pressure acting on a rising magma column falls steadily as it approaches the surface, allowing dissolved gases – provided magma viscosity permits – to exsolve and form bubbles that rise with the magma and are free to expand and coalesce. Where no open vent exists, or where the vent has become plugged by the products of a previous eruption, gases often remain dissolved in the magma as it enters the volcano from below, exerting considerable pressures on the magma itself. When the surface is breached, the large differential between ambient pressure and the internal pressure exerted by the dissolved gas tears the magma apart, generating an explosive eruption column. Sometimes, as demonstrated by the climactic eruption of Mount St Helens (Washington State, USA) in 1980, such an explosive decompression of magmatic gases can be triggered by collapse of a volcano's flank due to a giant landslide, thereby instantly unroofing and exposing the magma body to atmospheric pressures.

5.2.2 Measuring the size of a volcanic eruption

The explosivity of a volcanic eruption is measured on the *Volcanic Explosivity Index* (Newhall and Self, 1982; *see* Box 5.2) and is of critical importance in determining the degree to which an eruption can affect the environment on a large scale. While less powerful explosions still have the potential to devastate the immediate vicinity through the generation of heavy ashfall, often metres deep, and destructive pyroclastic and debris flows, the affected area rarely amounts to more than a few hundred square kilometres. Serious environmental impact is restricted to large explosive eruptions that pump sufficient sulphur gases into the stratosphere to cause surface cooling, and even damage to stratospheric ozone (e.g. Brasseur and Granier, 1992; Yang *et al.*, 1996).

5.2.3 Influence of magma chemistry, geographical location and eruption column height

Magma chemistry, in terms of the sulphur content, is also important in determining the potential for a particular eruption to have a significant impact on weather and climate. Consequently, sulphur-rich eruptions such as those of El Chichón (Mexico) in 1982 and

Box 5.2 The Volcanic Explosivity Index

The *Volcanic Explosivity Index* (VEI) (Newhall and Self, 1982; Table 5.1) uses a number of parameters, including height of the eruption column, volume of material ejected and eruption rate, to determine the scale of an eruption. The index starts at 0 and is open-ended, although nothing larger than a VEI 8 has yet been either observed or identified in the geological record. This may, perhaps, reflect the fact that crustal properties will not support a magma chamber large enough to supply greater volumes of magma to the surface in a single eruptive episode. Like the Richter Scale of earthquake magnitude, the index is logarithmic, so that each value on the scale represents an eruption ten times larger than the previous value. The lowest value on the VEI is reserved for non-explosive eruptions that involve the gentle effusion of low-viscosity basaltic magmas such as characterise eruptions of the Hawaiian volcanoes such as Kilauea and Mauna Loa. VEI 1 and 2 eruptions are described as small to moderate explosive eruptions, which eject less than 10 million cubic metres of debris. VEI values 3 to 7 designate progressively more violent explosive eruptions of andesitic and dacitic magma, capable of ejecting greater and greater volumes of debris and gas to higher levels in the atmosphere. Eruptions registering 8 on the scale are very rare and involve the explosive ejection of thousands of cubic kilometres of rhyolitic magma. Such *super-eruptions* are capable of triggering dramatic climate change through the emplacement of huge volumes of debris and gas into the stratosphere and their distribution across the planet.

While eruptions in the 0–4 range are commonplace, the larger events have progressively lower frequencies. VEI 5 eruptions, of which the 1980 Mount St Helens blast is an example, occur – on average – every decade or so, while VEI 6 events (e.g. the 1991 eruption of Pinatubo, the Philippines) have return periods of around a century. The only historic eruption to merit a 7 occurred at Tambora (Indonesia) in 1815, and the frequency of such large events is likely to be at most once in 500 years and probably considerably lower. Over the past 2 million years, VEI 8 *super-eruptions* have demonstrated an average return period of 50,000 years (Decker, 1990), the last occurring at Toba (Sumatra, Indonesia) around 73,500 years ago (Chesner *et al.*, 1991).

TABLE 5.1 The Volcanic Explosivity Index (VEI) (Sources: Newhall and Self, 1982; Simkin and Siebert, 1994)

VEI	Tephra volume (m^3)	Eruption column height (km)	Tropospheric injection	Stratospheric injection	General description
0	$<10^4$	<0.1	Negligible	None	Non-explosive
1	10^4–10^6	0.1–1	Minor	None	Small
2	10^6–10^7	1–5	Moderate	None	Moderate
3	10^7–10^8	3–15	Substantial	Possible	Moderate to large
4	10^8–10^9	10–25	Substantial	Definite	Large
5	10^9–10^{10}	>25	Substantial	Significant	Very large
6	10^{10}–10^{11}	>25	Substantial	Significant	Very large
7	10^{11}–10^{12}	>25	Substantial	Significant	Very large
8	$>10^{12}$	>25	Substantial	Significant	Very large

Pinatubo (the Philippines) in 1991 lofted sufficient SO_2 into the stratosphere – 7 and 20 million tons respectively – to cause significant falls in global temperature ranging from a few tenths to half a degree Celsius (e.g. Jones and Kelly, 1996). Geographical location also has a controlling influence on the environmental impact of a volcanic eruption, as it determines the likely distribution pattern of the aerosol cloud. While aerosol clouds generated by high-latitude eruptions are rarely transported beyond the mid-latitudes of the same hemisphere, those produced by eruptions in the tropics are far less geographically confined (Jakosky, 1986). Although the precise path followed by a volcanic aerosol cloud generated at low latitudes is strongly dependent on stratospheric wind patterns at the time of eruption, the general circulation pattern in the stratosphere will lift the aerosols in the tropics, transport them towards the poles and then return them to the troposphere at high latitudes. Aerosol clouds generated by tropical eruptions such as El Chichón and Pinatubo are therefore much more likely to have a global impact on temperature and weather patterns than similar-sized eruptions, such as Mount St Helens, occurring at higher latitudes.

Provided that an eruption column reaches an altitude sufficient to penetrate the stratosphere, there appears to be no relationship between column height and climatic impact, and eruptions that eject gases higher into the stratosphere are not, as a consequence, more effective at lowering global temperatures. In fact, Jakosky (1986) notes that the injection of sulphur gases into the stratosphere is more important in determining an eruption's effect on the climate than either the total explosivity of an eruption or the actual height reached within the stratosphere. Although less capable of subsequent dispersion across the globe, sulphur gases from high-latitude eruptions – for example in Alaska and Kamchatka (Russia) – have a greater chance of reaching the stratosphere than comparably sized eruptions in the tropics. This is because the stratosphere becomes progressively lower with increasing latitude, its base falling from ~15–17 km over the equator to less than 11–12 km at mid-latitudes and only 8–9 km in polar regions (Pyle *et al.*, 1996). In these circumstances, the height differential between the volcano summit and the tropopause (the upper boundary of the troposphere) becomes a critical factor in determining whether or not the gas clouds from a less powerful eruption are capable of penetrating the stratosphere. Pyle *et al.* (1996) estimate, in fact, that over 50 per cent of eruption columns that reach the stratosphere are due to moderately sized VEI 3 events at volcanoes located north of 40° N.

5.3 Contribution of volcanic eruptions to the atmosphere

Volcanic eruptions result in the injection of a range of particles and gases into the atmosphere (Fig. 5.2). Particulate matter may be derived either from the disruption of fresh magma (*juvenile ejecta*) as it reaches the surface or from older, solidified volcanic material ripped from the inside of the vent owing to the violence of eruption (*lithic ejecta*). While larger – centimetre- to metre-scale fragments – eventually accumulate at the surface within the vicinity of the volcano, the finer ash and dust fractions may be deposited thousands of kilometres from the eruption site. The very finest volcanic dust particles may be buoyed up in the atmosphere for longer, but residence times for such material are rarely more than a few weeks in the troposphere and a few months in the stratosphere. Because of these short residence times, particulate volcanic debris, from all but the largest explosive eruptions, rarely has a significant impact on the climate. The temporary loading of the atmosphere with large volumes of fine volcanic material can, however, reduce the amplitude of the diurnal cycle of surface air temperature in the region of the tropospheric cloud, leading to local warming. This effect was observed immediately following the climactic eruption of Mount St Helens in May 1980 (e.g. Mass and Robock, 1982), which locally loaded the troposphere with huge quantities of volcanic ash. At Yakima (Washington State), 135 km east of the volcano, the surface air temperature remained constant at 15°C for 15 hours, independent of the diurnal

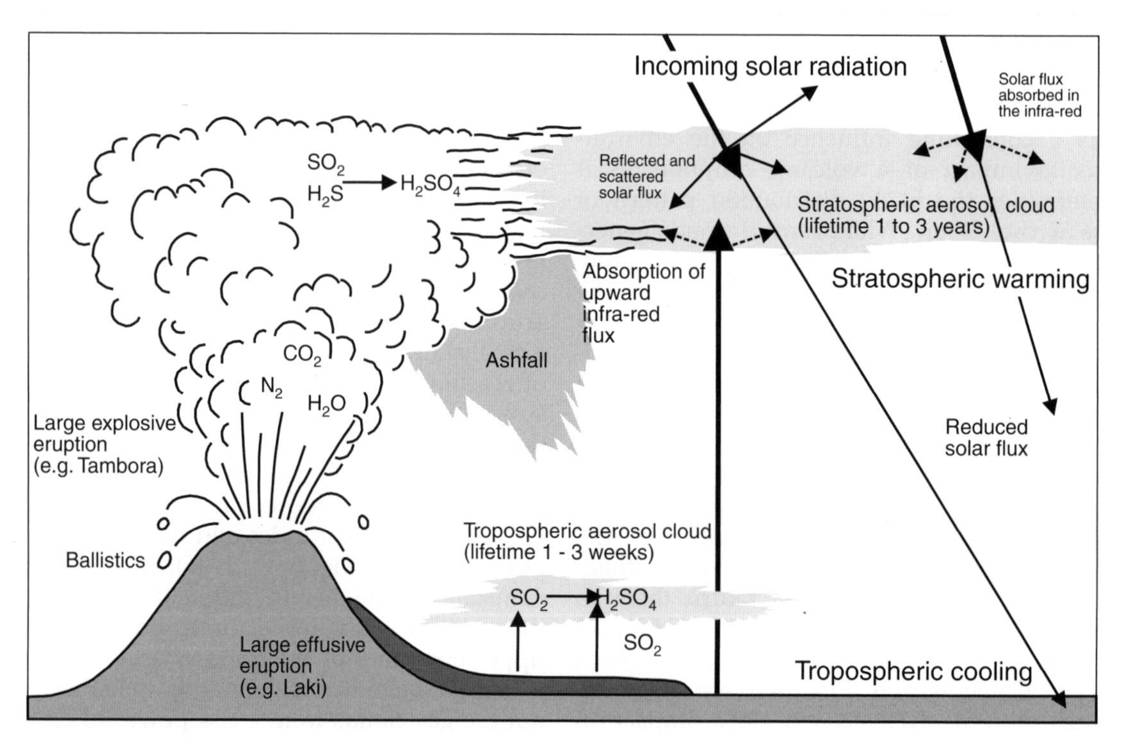

FIGURE 5.2 Volcanic eruptions eject both gas and particulate matter into the atmosphere. Most of the latter is in the form of volcanic ash that is soon removed by gravity. Gases, particularly SO_2, can, however, combine with atmospheric water to form aerosols that may remain in the stratosphere for up to 3 years. Here, their effects on the atmosphere and incoming solar radiation can result in stratospheric warming accompanied by significant cooling of the surface. See text for explanation. (Adapted and simplified from Robock, 2000)

cycle. The same effect was noted in the Indonesian capital Batavia (now Jakarta) after the cataclysmic 1883 eruption of Krakatoa.

In addition to particulate matter, volcanic eruptions also release large volumes of gases, of which H_2O, N_2O and CO_2 are the most abundant. Although both H_2O and CO_2 are greenhouse gases, the volumes released by erupting volcanoes are insufficient to augment anthropogenic planetary warming. From the perspective of volcanogenic climate change, it is the injection of volcanically derived sulphur into the stratosphere – primarily in the form of SO_2, although sometimes accompanied by H_2S – that is the main culprit (e.g. Pollack *et al.*, 1976; Rampino and Self, 1984). Within weeks, these sulphur species are oxidised through reaction with atmospheric H_2O and OH to form sulphate aerosols that effectively lead to a cooling of the lower atmosphere and the surface through reducing the level of solar radiation reaching it (Table 5.2). The conversion of SO_2 to H_2SO_4 is described by the generic reaction (Bekki, 1995)

$$SO_2 + OH + 3H_2O \Rightarrow H_2SO_4 + HO_2$$

The stratospheric aerosol cloud typically forms over a period of several weeks, the length of this period reflecting a so-called *e-folding* time of between 30 and 40 days (Bluth *et al.*, 1997). Comparable to the half-life of a radioactive isotope, *e-folding* refers to the time needed for an exponential decay process to remove 1/e of the initial amount. In other words, following an eruption, 4–6 weeks is sufficient for half of the volcanogenic stratospheric SO_2 to be converted to sulphate aerosol. Following formation of the aerosol cloud, its residence time in the stratosphere is of critical importance in determining the longevity of any climatic impact. Sulphate removal is dominated by gravitational settling but is also dependent on other factors, including the season and the size of the aerosol particles. For example, the net removal rate of aerosols generated by the El Chichón eruption was approximately 20 per cent slower during the winter (Jäger and Carnuth, 1987). The larger aerosol particles generated by the El Chichón eruption also had atmospheric residence times around 2 months shorter than smaller particles (Hofman and Rosen, 1987). Pyle *et al.* (1996) propose that the characteristic stratospheric e-folding residence time of sulphate aerosols following a volcanic injection of up to tens of millions of tons of sulphur is ~9–12 months. For VEI 8 super-eruptions that emplace several thousand megatons (Mt) of sulphate aerosol into the stratosphere, however, residence times may be as long as 5 to 8 years (Rose and Chesner, 1990). Generally speaking, stratospheric

TABLE 5.2 Sulphur aerosol production and Northern Hemisphere temperature falls resulting from a range of volcanic eruptions of different sizes (Source: updated from McGuire, 1999)

Year	Eruption	Volcanic Explosivity Index (VEI)	Eruption column height (km)	Volume of eruptive products (km^3)	Sulphuric acid aerosols (Mt)	Northern Hemisphere temperature fall (°C)
73,500 y BP	Toba	8	>40	~3000	1000–5000	3–5
1783	Laki	4	n.a.	14–15	100	~1
1815	Tambora	7	>40	150	200	0.7
1883	Krakatoa	6	>40	>10	55	0.3
1902	Santa Maria	6	>30	~9	25	0.4
1912	Katmai	6	>27	15	25	0.2
1963	Agung	4	18	0.3–0.6	20	0.3
1980	Mount St Helens	5	22	0.35	1.5	0–0.1
1982	El Chichón	4	26	0.3–0.35	20	0.4–0.6

perturbations due to volcanic eruptions rarely last for more than 3 years. Because, however, the characteristic interval between eruptions that reach the stratosphere is only about 1.8 years (Pyle *et al.*, 1996) a volcanic contribution will always be present. The global temperature drop following an eruption has been shown by Palais and Sigurdsson (1989) to be broadly related to the total sulphur load, so in simple terms, the more sulphur-rich an eruption, the greater will be its impact on the climate (Fig. 5.3).

Bluth *et al.* (1993) estimate that generally non-explosive, continuously degassing volcanoes such as Mount Etna (Sicily) contribute, at a relatively constant annual rate, around 9 Mt of SO_2 to the atmosphere. In contrast, the SO_2 flux from periodic explosive eruptions is much more variable, and although Bluth and his co-authors proposed a time-averaged figure of ~4 Mt – less than half of the annual non-explosive output – individual explosive eruptions can contribute tens of millions of tons to the annual volcanogenic SO_2 flux over periods as short as days or weeks. In actual fact, the total volcanic emission of SO_2 to the atmosphere is less than 10 per cent of the current anthropogenic flux from coal-burning power stations and industry. The more noticeable impact of the volcanic output on weather and climate arises, however, from two factors. First, volcanogenic SO_2 is driven by explosion into the stratosphere, where residence times are much longer than in the troposphere and where it can be rapidly dispersed around the globe. Second, a single large explosive eruption has the potential dramatically and rapidly to increase the total volcanic flux to the stratosphere. Contrastingly, the contribution arising from human activity remains largely confined to the troposphere and demonstrates little interannual variability.

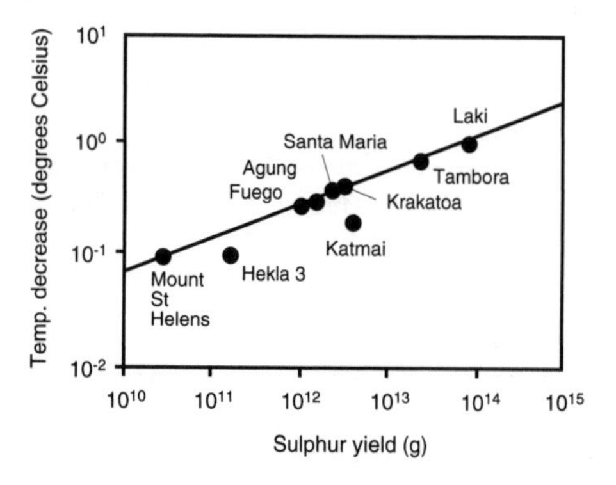

FIGURE 5.3 The total sulphur yield from volcanic eruptions is directly correlated with observed falls in temperature at the Earth's surface. (After Palais and Sigurdsson, 1989)

5.4 Physical and chemical effects of volcanogenic aerosols on the atmosphere

Once they reach the stratosphere in sufficiently large volumes, volcanic aerosols are capable of modifying both the chemistry and the radiative characteristics of the atmosphere (Fig. 5.2). The stratospheric aerosol clouds produced by large volcanic eruptions are associated with a range of radiative processes that can bring about both heating and cooling of the atmosphere. Because sulphate aerosol particles are about the same size (effective radius ~0.5 μm) as a wavelength of visible light, they interact strongly with incoming solar radiation, causing it to scatter (*see* Fig. 5.2). Some of the radiation is back-scattered into space, thereby increasing the net planetary albedo and reducing the amount of solar radiation reaching the Earth's surface, where a net cooling results. The sulphate aerosol particles also cause forward (downward) scattering of much of the incoming solar radiation, but although this leads to enhanced downward diffuse radiation, it does not compensate for the reduction in the direct solar flux due to the back-scattering component.

The degree of stratospheric perturbation due to volcanic eruptions is measured using determinations of *optical depth*, which is the attenuation of solar irradiance by the atmosphere. It is defined as the ratio of incoming to transmitted solar energy for a particular wavelength (e.g. 0.55 μm ~ visible light) (Bluth *et al.*, 1997). The degree of attenuation is determined by the sum of absorption and scattering by air molecules together with other particles such as volcanic aerosols. The effect of volcanic aerosol loading of the stratosphere is to increase the optical depth and thereby reduce the atmospheric transmission of solar radiation. The impact of individual eruptions can clearly be seen in a continuous atmospheric transmission log recorded at the Mauna Loa Observatory (Hawaii) (Fig. 5.4). Explosive eruptions at Agung (Indonesia) in 1963, El Chichón (Mexico) in 1982 and Pinatubo (Philippines) in 1991 can all be identified within the record, with the latter two events resulting in a 20 per cent or more fall in *apparent* (i.e. with solar constant variation dependence and instrument calibration eliminated) *transmission*. Although sulphate aerosol loading of the stratosphere leads to a net cooling in the lower atmosphere and at the surface, it also causes net heating in the stratosphere itself (Pollack *et al.*, 1976). This is due to absorption by the aerosol cloud of incoming solar radiation (in the near infra-red) and also infra-red radiation from the troposphere and the Earth's surface.

Through a complicated series of reactions, volcanic aerosols are also implicated in the depletion of stratospheric ozone (O_3), which is crucial in providing an effective barrier to the solar UV flux. Hydrochloric acid (HCl), formed through combination of volcanogenic chlorine (Cl_2) with atmospheric water, is released in enormous quantities in some eruptions (7 Mt in the 1783 Laki eruption; Thordarson *et al.*, 1996) and is capable of causing drastic reductions in ozone levels should it come into contact with ozone. Fortunately, most volcanogenic HCl appears to remain confined to the troposphere, where it is rapidly washed out by rain. Perhaps only 3 per cent of stratospheric chlorine – derived from HCl – is volcanic in origin, compared with over 80 per cent from anthropogenic chlorofluorocarbons (CFCs). Although volcanic HCl plays no direct role in destroying stratospheric ozone, it may still make a significant indirect contribution to ozone depletion (e.g. Solomon, 1999). This is because volcanogenic sulphate aerosol particles provide countless additional surfaces upon which reactions occur that remove chlorine-scavenging nitrogen species, thereby favouring the survival of anthropogenic chlorine and the more effective destruction of ozone (Prather, 1992). Following the 1991 Pinatubo eruption, for example, ozone depletion in the aerosol cloud was measured at around 20 per cent. Interestingly, this effect of volcanic aerosols on the environment is new, dependent as it is on available anthropogenic chlorine in the stratosphere, and is unlikely to have occurred even as recently as the 1963 Agung eruption (Robock, 2000). Furthermore, as a result of the Montreal Protocol and associated agreements, anthropogenic chlorine in the stratosphere is now decreasing, so the effect is likely to be lost in due course. The corollary of this, however, is that future large volcanic eruptions are likely to have a more noticeable cooling effect at the surface. This is because the destruction of stratospheric ozone increases the transmission of UV radiation and therefore compensates to some extent for the back-scattering effect due to the presence of the sulphate aerosols.

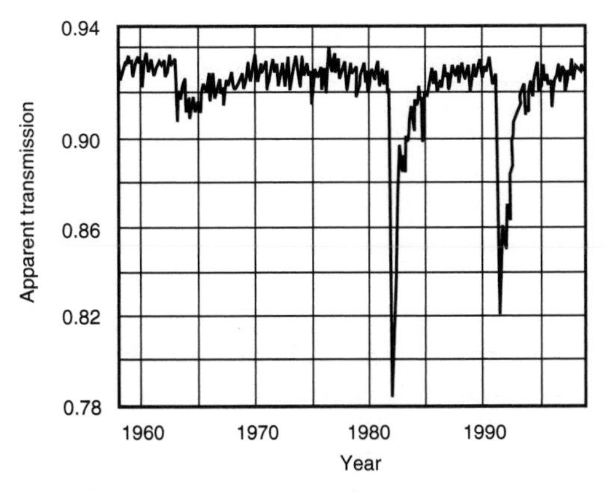

FIGURE 5.4 A log of atmospheric transmission as measured at the Mauna Loa Observatory (Hawaii) shows clearly how transmission of solar radiation was significantly reduced as a result of sulphur aerosols generated during the eruptions of Agung (1963), El Chichón (1982), and Pinatubo (1991). (Simplified from Robock, 2000)

5.5 Volcanoes and climate

In the context of global warming, it is crucial that we are able to evaluate the role of volcanic activity in the operation of the global climate system, and to understand the processes and mechanisms involved. This is of particular importance as the surface cooling associated with single large eruptions or multiple, temporally adjacent, smaller volcanic events can be sufficient to mask – at least temporarily – the impact of human activity on the climate due to greenhouse gas emissions. There is now abundant evidence in support of volcanic eruptions affecting the global climate on a range of timescales ranging from a few years to decades and centuries. In respect of the last super-eruption, at Toba in Indonesia, it has even been suggested that the severe cooling that followed the eruption might have been sufficient to accelerate the last major glaciation, which occurred at about the same time (Rampino and Self, 1992, 1993a; see also Sections 5.7 and 5.8).

Although Pyle *et al.* (1996) have drawn attention to a contribution made by moderate explosive eruptions that register 3 on the VEI, the long-term stratospheric sulphate budget is dominated by volcanic eruptions of VEI 4 and larger. Furthermore, individual climatologically significant eruptions register at 5 and above on the index. During the past 200 years, there have been 16 volcanic eruptions in this size range (Simkin and Siebert, 1994; Table 5.3), three-quarters of which constituted the first historical eruption at the volcano concerned. Assuming that all eruptions of VEI 5 and greater have the potential to lower global temperatures measurably in the short term, the time-averaged recurrence interval for climatically significant eruptions may be as low as 12 years. If events larger than or equal to VEI 6 are considered alone, however, this rises to over 30 years. VEI 7 eruptions, such as that of Tambora (Indonesia) in 1815 are far less common, and Decker (1990) proposes around 22 eruptions on this scale every 10,000 years, giving an average recurrence interval of around 450 years. Figure 5.5 (from Simkin and Siebert, 1996) demonstrates graphically the relationship between the numbers of Holocene eruptions and their size, the latter shown in terms of both tephra volume and VEI number. On the basis of statistical analysis of Northern Hemisphere eruptions over the past 600 years, Hyde and Crowley (2000) calculate that in any given decade there is a 25 per cent chance of a VEI 5 (El Chichón-scale) eruption and a 20 per

TABLE 5.3 The largest explosive eruptions of the nineteenth and twentieth centuries (Source: Simkin and Siebert, 1994)

Year	Volcano	First historical eruption?	Deaths
1991	Cerro Hudson (Chile)	No	0
1991	Pinatubo (Philippines)	Yes	800
1982	El Chichón (Mexico)	Yes	2000
1980	Mount St Helens (US)	No	57
1956	Bezymianny (Russia)	Yes	0
1932	Cerro Azul (Chile)	No	0
1912	Novarupta (US)	Yes	2
1907	Ksudach (Russia)	Yes	0
1902	Santa Maria (Guatemala)	Yes	>5000
1886	Tarawere (NZ)	Yes	>150
1883	Krakatoa (Indonesia)	No	36,417
1875	Askja (Iceland)	Yes	0
1854	Sheveluch (Russia)	Yes	0
1835	Cosiguina (Nicaragua)	Yes	5–10
1822	Galunggung (Indonesia)	Yes	4011
1815	Tambora (Indonesia)	Yes	92,000

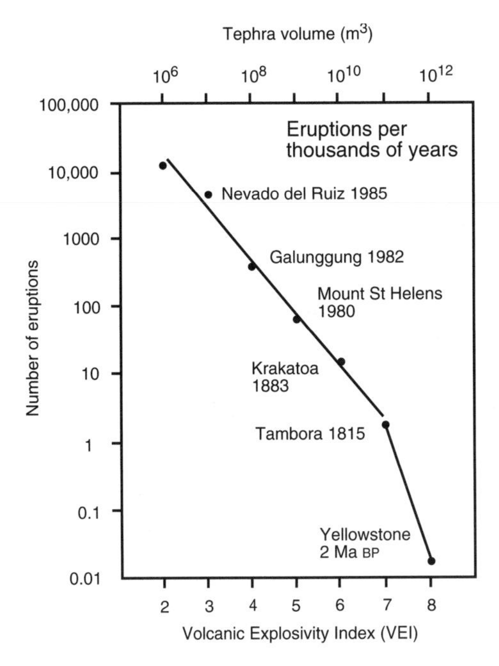

FIGURE 5.5 The frequency of volcanic eruptions during the Holocene plotted against eruption magnitude as shown by both tephra volume and Volcanic Explosivity Index. The total numbers of eruptions for each point on the VEI are shown, normalised to 'per 1000 years'. Best-fit line is determined by an exponential regression model for VEI 2-7 data. Examples of well-known eruptions illustrate each point on the VEI. (Simplified from Simkin and Siebert, 1994)

cent probability of a VEI 6 (Pinatubo-scale) event, both capable of having a measurable impact on global temperatures.

5.5.1 Linking historical eruptions and climate change

Supporting evidence for a volcanic influence on weather and climate comes from a number of sources. These include comparison of instrumental temperature records with the timing of volcanic eruptions in order to identify and quantify decreases in surface air temperatures expected to follow such eruptions (e.g. Kelly and Sear, 1984; Sear *et al.*, 1987). Prior to the past 100–200 years, however, the quality and completeness of both temperature records and volcanic eruption records becomes too poor to allow valid comparisons to be made between the two. In order to overcome this problem, indirect or *proxy* records are sought to provide information on the timing of temperature changes and eruptions and to look for correlations between the two. Owing to their annual resolution and absolute dating accuracy, tree rings provide an important source of proxy climate data (e.g. Schweingruber, 1988), and tree-ring records have been interpreted in the light of climatically significant volcanic eruptions and resulting lower surface temperatures (e.g. Scuderi, 1990; Grudd *et al.*, 2000). Problematically, however, individual tree rings and tree ring-derived temperature series are not only influenced by volcanic eruptions, but also subject to other forms of natural variability that can either mask or reinforce volcano-related effects. Notwithstanding this, Jones *et al.* (1995) have convincingly used tree-ring records to link some major eruptions of the past 400 years with falls in Northern Hemisphere summer temperatures. The important fact that some eruptions are recorded in tree-ring records while others of comparable size are not is addressed by Sadler and Grattan (1999). They propose that only trees that are environmentally stressed will respond negatively to volcanically induced impacts such as an increase in atmospheric optical depth. Trees that are less sensitive and/or unstressed will fail to record eruption years as significant.

In recent decades, ice cores have become vital in providing a high-resolution, long-term proxy record of volcanic activity (e.g. Hammer *et al.*, 1980). Although, unlike tree rings, ice cores do not record the climatic effect of an eruption, they do provide a time series of eruptive activity through the preservation of layers of volcanic products, and in particular sulphate concentrations due to deposition from aerosol clouds. In the GISP2 (Greenland) ice core, Zielinski *et al.* (1996) identified ~850 volcanic signals over the past 110,000 years characterised by sulphate spikes greater than those associated with historical eruptions from either equatorial or mid-latitude regions, which are known to have been climatically significant – either globally or in the Northern Hemisphere.

Over the past 30 years a number of indices have been devised with the purpose of providing a measure of volcanic impact on the climate, or *volcanic forcing* of the climate, as it is more commonly referred to. The earliest of these was devised by Lamb (1970), whose *Dust Veil Index* (DVI) was designed to analyse the impact of volcanoes on surface weather, lower and upper atmospheric temperatures, and large-scale wind circulation. The DVI was based upon a range of observations and measurements, including historical reports of eruptions, estimates of the volume of ejecta, optical phenomena (such as anomalous sunsets, blue moons and red suns), radiation measurements (subsequent to the 1883 Krakatoa blast) and temperature information. The index has been criticised because it includes climatic information in its derivation and therefore involves circular reasoning. Nevertheless, even if the climatic information is removed, the DVI still proves reliable as a measure of the volcanic contribution to the atmosphere over the past few centuries. Using Lamb's 1970 data, Mitchell (1970) compiled a more detailed time series of Northern Hemisphere volcanic eruptions for the period 1850–1968 (The *Mitchell Index*). In basing this on the total mass ejected during each eruption, the Mitchell Index is in some ways comparable with the VEI. More recently, Sato *et al.* (1993) have used a combination of ejecta volume and optical data (the latter since 1882), supplemented by satellite data (since 1979), to produce an index (The *Sato Index*) expressed in terms of optical depth at wavelength 0.55 μm (visible light). This provides a particularly good measure of the climatic impact of recent eruptions as it includes actual observations of the spatial and temporal distributions of aerosol clouds. In 1995, Robock and Free devised an *Ice Core Volcanic Index* (IVI) for the period 1850 to 1995, based upon acidity peaks and sulphate spikes, in order to attempt to identify the volcanic signal common to all records. Despite some problems and drawbacks, the IVI correlated well with existing non-ice core volcanic indices and with high-frequency records. An attempt to extend the IVI back to 2000 years BP (Robock and Free, 1996) suggested, however, that the ice core record currently available is inadequate for delineating climate forcing due to explosive eruptions before 1200 in the Northern Hemisphere and 1850 in the Southern Hemisphere.

5.5.2 The climate response to volcanic forcing

The response of the climate to volcanic forcing is broadly regarded as being one of cooling. The real situation is, however, considerably more complicated, and the impact of an eruption on global weather patterns may result in spatially heterogeneous variations in temperature and other meteorological parameters (Table 5.4). Groisman (1992), for example, noted that following nine large eruptions during the past two centuries, central Europe had warmer than usual winters while the north-eastern United States experienced colder than average winters. The important corollary of this is that volcanic modification of weather patterns can result in regionally diverse conditions involving the simultaneous prevalence of both poor and fine weather. This contrasts with the still common perception that large volcanic eruptions invariably result in the global or hemispherical deterioration of the climate. The environmental impacts of volcanic eruptions also vary from season to season, as well as geographically, and can be complicated or masked by other strong climatic signals such as the El Niño–Southern Oscillation (ENSO).

5.5.2.1 Cooling versus warming

A fall in the global average temperature following large explosive eruptions is now well established, with surface cooling resulting from clouds of sulphate aerosols reducing the total solar radiation reaching the surface. As might be expected, the cooling effect is most pronounced in the tropics and during the mid-latitude summer, as there is more solar radiation to block. Extracting the volcanic cooling effect from other simultaneous climatic variations is not easy, however, particularly as the volcanic signal is of approximately the same size as the powerful ENSO climatic perturbation. Angell

TABLE 5.4 The effects of large explosive volcanic eruptions on weather and climate (Source: Robock, 2000)

Effect	Mechanism	Begins	Duration
Reduction of diurnal cycle	Blockage of short-wave and emission of long-wave radiation	Immediately	1–4 days
Reduced tropical precipitation	Blockage of short-wave radiation; reduced evaporation	1–3 months	3–6 months
Summer cooling of Northern Hemisphere tropics and subtropics	Blockage of short-wave radiation	1–3 months	1–2 years
Stratospheric warming	Stratospheric absorption of short-wave and long-wave radiation	1–3 months	1–2 years
Winter warming of Northern Hemisphere continents	Stratospheric absorption of short-wave and long-wave radiation; dynamics	~ 6 months	1 or 2 winters
Global cooling	Blockage of short-wave radiation	Immediately	1–3 years
Global cooling from multiple eruptions	Blockage of short-wave radiation	Immediately	10–100 years
Ozone depletion; enhanced UV radiation	Dilution; heterogeneous chemistry on aerosols	1 day	1–2 years

(1988) showed that the ENSO signal partly obscured the impact of a number of recent large eruptions in terms of their effect on hemispheric annual average surface temperatures, making it difficult for the true level of volcanic cooling to be determined. More recently, however, Robock and Mao (1995) succeeded in removing the ENSO signal from the surface temperature record to reveal characteristics of the volcanic cooling effect. By superposing the signals of Krakatoa (1883), Santa Maria (Guatemala, 1902), Katmai (Alaska, 1912), Agung (1963), El Chichón and Pinatubo, the authors discovered that the maximum cooling occurred about 1 year after the eruptions. They also found that the cooling – of the order of 0.1–0.2 degrees Celsius – was displaced towards the Northern Hemisphere, with the maximum cooling in the Northern Hemisphere winter taking place at about 10º N and during the summer at around 40º N. The greater impact on the Northern Hemisphere reflects the higher percentage of landmass compared to the Southern Hemisphere. Because land surfaces respond more rapidly to radiative perturbations, the Northern Hemisphere is more sensitive to the reduction in solar radiation due to volcanic aerosols. Large volcanic eruptions may also have an impact on the hydrological cycle as the resulting cooling can be expected to reduce evaporation, and Robock and Liu (1994) report reduced tropical precipitation for 1–2 years after a large eruption. The same authors also tentatively suggest that the El Chichón eruption may have been implicated in a reduction in precipitation over the Sahel (Africa), contributing to the drought problem.

As the surface cools following large volcanic eruptions, so the stratosphere warms. After the 1982 El Chichón eruption, for example, the globally averaged stratospheric temperature rose by around 1 degree C for about 2 years. A similar duration stratospheric warming followed the 1991 Pinatubo eruption, although the peak warming following this event reached almost 1.5 degree C (Robock, 2000) (Fig. 5.6). The volcanic warming signals are superimposed upon a general stratospheric cooling trend, which is widely believed to represent anthropogenically related ozone depletion and CO_2 rise (e.g. Vinnikov *et al.*, 1996).

In contrast to summer cooling, winters in the Northern Hemisphere continents are characterised by tropospheric warming. This was first noted by Groisman (1985), who observed that large volcanic eruptions were followed by warm winters over central Russia. In a further study in 1992, Groisman identified similar warmings, following the El Chichón and Pinatubo eruptions, over both Russia and north-eastern North America. In another study, Robock and Mao (1992) examined the Northern Hemisphere winter surface air temperature patterns after the 12 largest eruptions of the past 100 years. They found a consistent pattern of warming over the continents, with cooling over the oceans and the Middle East, which in a later paper (Robock and Mao, 1995) they attributed to the impact of tropical rather than high-latitude eruptions. Following the Pinatubo eruption in 1991, satellite data show that the lower tropospheric temperature over North America, Europe and Siberia was much warmer than usual, while that over Alaska, Greenland, the Middle East and China was cold. In Jerusalem, snow fell (a very rare occurrence) while cooling led to coral die-back in the Red Sea – an event shown by Genin *et al.* (1995) to occur only following large explosive eruptions. More detailed discussion of volcanically forced Northern Hemisphere warming is beyond the scope of this book, but a useful summary is provided by Robock (2000), who also provides a reason for the Laki eruption of 1783 not fitting the pattern, causing a cold winter in Europe (Franklin, 1784) rather than a warm one. Most importantly, however, post-eruption tropospheric warming reinforces the point that the climatic impact of volcanic eruptions is not purely one of cooling, and also that different effects resulting from a large eruption will be experienced in different places.

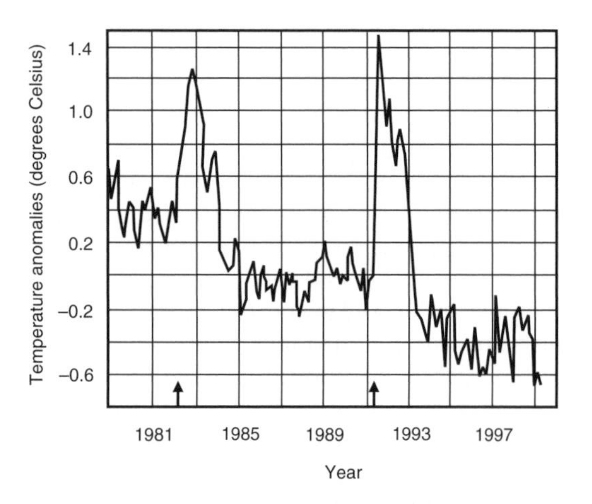

FIGURE 5.6 At the same time as cooling the troposphere, volcanic aerosols have a warming effect on the stratosphere. Satellite microwave sounding of the stratosphere clearly shows significant warming after both the El Chichón (1982) and the Pinatubo (1991) (arrowed) eruptions. (Simplified from Robock, 2000)

5.5.2.2 Volcanic eruptions and El Niño

Two major geophysical events occurred in April 1982: the unexpected start of the strongest El Niño of the twentieth century and the eruption of El Chichón in Mexico, the third largest eruption of the past 100 years. The strong El Niño resulted in one of the warmest ENSO events of the last century, with record high temperatures in the equatorial Pacific Ocean and serious perturbations to the weather across the planet. Owing to the close temporal relationship between the two events, it was inevitable that a causal link would be sought, and a number of researchers suggested, as a consequence, that large volcanic eruptions might be capable of triggering El Niños (e.g. Handler, 1986; Hirono, 1988). Hirono proposed that tropospheric aerosols from the eruption caused collapse of the trade winds and a resulting accumulation of warmer waters in the eastern Pacific. Although such a mechanism might be feasible, it cannot have been applicable in this case, as there was insufficient time for the aerosol cloud to have had any climatic impact. A similar situation occurred in 1991, when the Pinatubo eruption coincided with another strong El Niño. Yet again, however, the El Niño was already developing prior to the climactic explosions of Pinatubo in June. Perhaps we should consider the idea, therefore, that El Niño events might actually trigger volcanic eruptions. This is not as improbable as it might seem, and Rampino *et al.* (1979) proposed some time ago that climate change might trigger volcanic activity through causing stress changes in the crust. Admittedly, however, they were thinking of the much larger changes associated with glaciation–deglaciation cycles during the Quaternary. More recently, however, Marcus *et al.* (1998) have provided evidence of observable changes in the Earth's rotation rate following El Niño events, which might be capable of causing variations in lithospheric stress. Notwithstanding such speculation, however, Self *et al.* (1997) compared the timing of all large volcanic eruptions during the past 150 years with the timing of ENSO events. They noted that in many cases, the relative timing argued against volcanic triggering, and in fact five strong El Niños occurred between 1915 and 1960 when the atmosphere was largely free of volcanic aerosols. They also observed that of the three strongest tropical aerosol perturbations that coincided with or were followed by strong El Niños (Krakatoa in 1883, El Chichón in 1982, and Pinatubo in 1991), the two most recent occurred after El Niño-related sea surface warming had already commenced. The conclusion of this study is quite clear: there is currently no evidence that large volcanic eruptions can produce El Niños.

5.5.2.3 Multiple eruptions and interdecadal volcanic forcing

As the volcanic aerosols generated by large volcanic eruptions (barring super-eruptions) have residence times in the stratosphere of only 2 or 3 years, their impact on weather and climate must be of the same order or less. Any model that attempts to explain interdecadal climate change due to volcanic forcing has, therefore, to incorporate multiple eruptions occurring over a short space of time in order to maintain a sufficiently high aerosol contribution to the stratosphere. A number of authors have suggested that the pooled effect of temporally closely spaced eruptions may have been overestimated (Wigley, 1991) and that temperatures do not appear to be depressed for longer following grouped eruptions than after single events (Self *et al.*, 1981). Notwithstanding this, however, Crowley and Kim (1999) and Free and Robock (1999) propose that the medieval cold period known as the *Little Ice Age* can be explained in terms of multiple volcanic eruptions significantly raising the mean optical depth over a period long enough to cause decadal-scale cooling. Using new volcano chronologies, new solar constant data and new reconstructions of climate change for the Little Ice Age, the authors simulate the climate of the past 600 years using a model incorporating volcanic, solar and anthropogenic forcings, and compare the results with palaeoclimatic reconstructions. They conclude that both volcanic activity and changes in solar output were important in promoting the reduced temperatures of the Little Ice Age, and also that the warming of the past century has an anthropogenic rather than a natural cause.

5.5.2.4 Cultural aspects of eruption-induced climate change

For subsistence farming communities living under difficult conditions in marginal environments, even small, short-term perturbations in weather and climate can have devastating impacts. Both the 1783 Laki and 1815 Tambora eruptions, for example, led to severe famine through loss of harvest and livestock. The effects on crops in Europe of the Laki eruption are described by Grattan and Pyatt (1999), who summarise contemporary accounts of damage to vegetation. These include the yellowing of barley and rye, and the withering of the leaf tips of pine and fir trees. Other reports talk of leaves becoming dry and shrivelled, and losing their green colour. The authors attribute these effects to a cloud of tephra, toxic volatile gases and acid aerosols that arrived at the surface in Europe as a consequence of the prevailing meteorological conditions and which was held there by a high-pressure weather system (Fig. 5.7). Similar descriptions of withering crops are found in the writings of Plutarch for the year 44 BC and ascribed to a large eruption of Mount Etna, and in a number of contemporary accounts (see Stothers and Rampino, 1983) describing the effects of a so-called 'mystery cloud' in AD 536.

The impact on farming communities of such tropospheric volcanic clouds is debated by Dodgshon *et al.* (2000), who examine possible links between pre-Laki Icelandic eruptions and changes to the environment and traditional farming systems in the Highlands and Islands of Scotland. Over 20 Holocene tephra layers from Icelandic eruptions have been identified in north-west Europe, and one of these, known as Hekla 3 and associated with the Hekla eruption of 1159 BC, has been implicated in major cultural upheaval in Scotland. Burgess (1989) believes that the eruption drastically affected the climate, resulting in an apparent widespread abandonment of Late Bronze Age settlement on marginal moorland and upland sites in Scotland and elsewhere (Burgess 1989). A second eruption of Hekla, sometime around 2395–2279 BC, has also been linked with damage to trees in north-east Scotland (Blackford *et al.*, 1992). Given its distance from the source of the eruptions, it is unlikely that tephra fall could be the cause of the abandonment of settlements in Scotland, and in fact Thórarinsson (1971) showed that in Iceland, even for tephra depths of 10 cm – which would be impossible in Scotland – abandoned settlements were re-occupied a year later. Any serious cultural impact in Scotland leading to abandonment could, therefore, result only from noxious acid clouds similar to that generated over Europe by the Laki event. While the BC abandonment remains unsubstantiated, Dodgshon *et al.* (2000) do suggest that some abandonments in the Scottish Highlands prior to the 'clearances' of the eighteenth and nineteenth centuries may be attributable to volcanic activity in Iceland. The authors point out that at the time some communities in the Highlands and Islands would have been susceptible, for a variety of political or social reasons, to a subsistence crisis every four or five years. If such a crisis coincided with a deterioration in the environment resulting from a Laki-type acid fog, then the collapse and abandonment of a community could be triggered. Although, therefore, Icelandic volcanic activity may – from time to time and from place to place – have caused the break-up of communities in Scotland, region-wide abandonment of settlements appears unlikely.

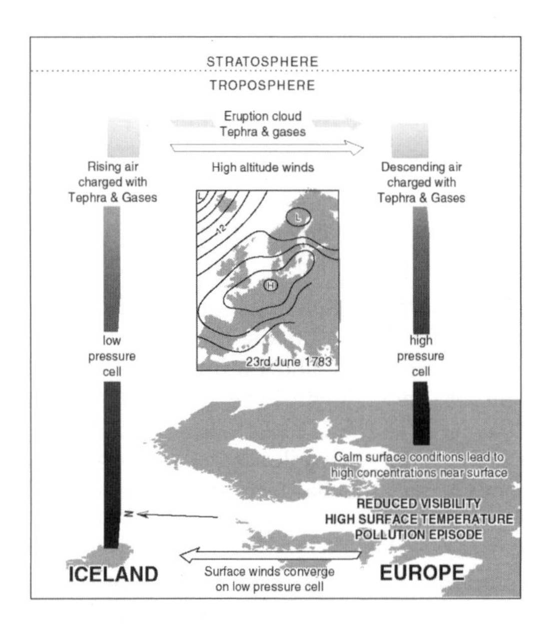

FIGURE 5.7 Dry volcanic fogs that hovered over Europe following the Laki eruption of 1783 resulted from a tropospheric cloud of volcanic gas that arrived at the surface in Europe due to fortuitous meteorological conditions, and which was held in place by anticyclonic conditions. (From Grattan and Pyatt, 1999)

5.6 Volcanoes as initiators of past environmental change: notes of caution

Within the current intellectual climate of neo-catastrophism and punctuated equilibrium (see Chapter 7.1), it has become fashionable to explain any dramatic change, either in the environment or in human history, in terms of anomalous physical events, of which asteroid and comet impacts and volcanic eruptions are most popularly invoked. Eruptions clearly have an impact on both weather and climate, but Sadler and Grattan (1999) urge more critical appraisal of the evidence for the link between volcanic events and environmental change, particularly for historical eruptions such as Tambora. The authors rightly point out that many studies linking past eruptions with climate change do so within a poorly constrained temporal framework, and stress that an inherent danger lies in linking disparate and geographically dispersed palaeo-environmental data with specific eruptions. They note that in trying to determine the environmental impact of an eruption the role of precursive environmental conditions, for example an ongoing cold snap, is often ignored, making it difficult to establish whether any relationship between an eruption and environmental change is dependent or fortuitous. In this context, a study of the climatic impact of nine eruptions between 1883 and 1982, including Krakatoa (1883), Katmai, Alaska (1912) and Agung (Indonesia) (1963), undertaken by Mass and Portman (1989) revealed that surface temperature declines were already under way prior to several of the eruptions.

Sadler and Grattan (1999) address questions arising about the cause, scale and patterns of post-eruption cooling by examining the regional responses to the Tambora eruption. For the United States they show that summer frosts and snowfalls are by no means uncommon and cannot, in themselves, be used as evidence of anomalously reduced surface temperatures. They also point out that although June 1816 was the coldest ever recorded in the city of New Haven, Connecticut, the temperatures recorded fall within the range of statistical normality. Furthermore, other North American temperature records show that 1816 was not a particularly cold year, but was reasonably typical of the decade. In Europe, eastern and central regions were cold and wet, but the weather in Northern Europe was mild and clement. In Scotland and England the winter of 1815–16 was not exceptional, and many colder winters are indicated for which there are no obvious forcing mechanisms. The summer of 1815–16 was certainly cool, but – as Sadler and Grattan note – this should be interpreted within the context of a cold decade. Furthermore, the summers of 1823 and 1830 were both colder.

In a critical evaluation of the usefulness and applicability of dendrochronological studies in identifying climatologically significant eruptions in the historical and pre-historical record, Sadler and Grattan propose that any association between eruptions and tree-ring minima (thin rings representative of poor growth and therefore a poorer climate) must specify the mechanism by which the eruption was able to trigger a response in the growing tree – which they attribute to acid deposition. They also draw attention to a problem of attribution first highlighted by Baillie (1991). If an anomalous dendrochronological horizon is identified, then it will tend to 'suck in' loosely dated archaeological and environmental and even apocryphal information, all of which seek to explain the dated event. Sadler and Grattan (1999, p. 191) illustrate the nature of the problem admirably in the form of a table (Table 5.5) that shows the range of associated phenomena and cultural events that have been linked to three ring-width minima in Irish oaks.

While recognising that a climatic signal may be detectable following a volcanic eruption, Sadler and Grattan observe that the scale of resulting temperature fluctuations falls within the normal range and is not, at least for 'normal' (i.e. up to VEI 7) eruptions, extreme. They discourage attempts to link all and every environmental change in the months and years following an eruption to that event, and rightly call for more stringent interpretation of the environmental impact of both individual and clustered eruptions.

5.7 Volcanic super-eruptions and environmental change

Alongside asteroid and comet impacts, volcanic super-eruptions – registering 8 on the Volcanic Explosivity Index – are capable of triggering rapid and dramatic changes to the Earth's physical environment. Over the past 2

Table 5.5 Phenomena associated with Irish tree-ring width minima (Source: Sadler and Grattan, 1999)

Irish oak ring width minima	Associated phenomena
1159 BC	Hekla 3 eruption (Iceland); Greenland ice core acidity peak, 1120 ± 50 BC Abandonment of the Strath of Kildonan, late twelfth century BC Movement of the 'Sea People' c. 1200 BC Fall of Mycenae, c. 1200 BC Plague in Ireland, 1180–1131 BC Hittites leave Anatolian Plateau, thirteenth century BC Libyan people migrate towards Egypt, c. 1200 BC Catastrophic flooding in Hungary, c. 1200 BC Rise in the level of the Caspian Sea, c. 1200 BC Wide tree rings in Turkey, 1159 BC Deepening of lakes in Turkey, 1200–1100 BC Lowered snowline in Norway, c. 1200 BC Crop failure in China
207 BC	Unknown eruption. Greenland ice core acidity peak, 210 ± 30 BC Irish murrain of cattle, 210–200 BC Stars invisible in China, 208 BC Chinese famines, 207–204 BC Chinese dynastic change, 202 BC Frost ring event, 206 BC
AD 540	Unknown eruption; Greenland ice core acidity peak, AD 516 ± 4 European mystery cloud, 536 AD Chinese atmospheric disturbance, 540 AD

million years, it is estimated that there have been around 40 such events, yielding an average return period of around 50,000 years (Decker, 1990). Little is known about the majority of these events, and the most closely studied are those that occurred at Yellowstone (Wyoming, USA) 2 million, 1.3 million and 0.65 million years ago (Smith and Braile, 1994), and the most recent super-eruption at Toba (Sumatra, Indonesia) around 73,500 BP (Chesner *et al.*, 1991) (Fig. 5.8). Invariably, explosive eruptions on this scale involve siliceous, high-viscosity rhyolitic magma and result in the formation of large calderas. At both Yellowstone and Toba these caldera systems remain active and *restless*, and are characterised by continuing hydrothermal activity, seismicity and surface deformation. The formation of the large volumes of rhyolitic magma required to feed a super-eruption necessitates the involvement of continental crust in magma formation and limits such volcanic systems, therefore, to ocean–continent destructive plate margins (e.g. Toba) or continental mantle plume settings (e.g. Yellowstone). During super-eruptions, magma is commonly ejected from ring fractures that during the later stages of the eruptions act as faults along which a central crustal block subsides to form a caldera. Magma is expelled in the form of curtain-like eruption columns that may last for up to two weeks (Ledbetter and Sparks, 1979), and which can loft tephra to altitudes of 40–50 km. Column collapse typically results in the formation of extensive pyroclastic flows that deposit *ignimbrite* (pumice-rich pyroclastic flow material) over thousands of square kilometres. While the coarser tephra component falls to earth locally, progressively finer fractions are deposited over an area the size of a continent, with the finest material being distributed globally by stratospheric winds. On the VEI, super-eruptions are

FIGURE 5.8 The Toba super-eruption 73,500 years ago was the largest volcanic event of the past 100 millennia. The eruption ejected at least 3000 km^3 of debris and left a crater (now lake-filled) 50 km long. Elevated lake sediments and the central island testify to continued uplift related to continued magma movement at depth. (Courtesy of Landsat Pathfinder Project)

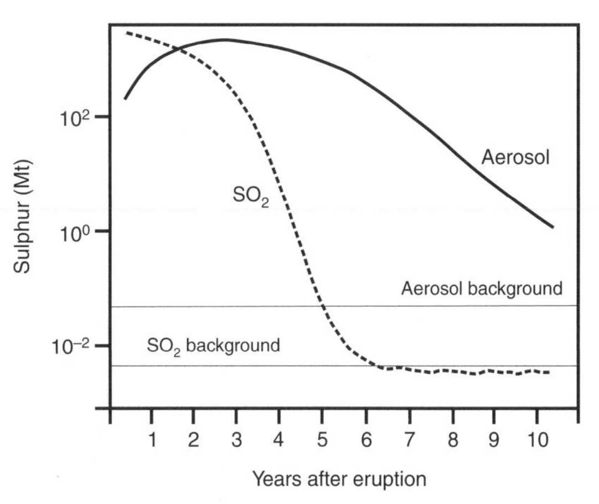

FIGURE 5.9 Modelling of the total mass of stratospheric SO_2 and sulphate aerosol as a function of time shows that stratospheric aerosol levels after the Toba eruption are likely to have remained well above background (non-volcanic) level for several years following the event. (From Bekki *et al.*, 1996)

classified as those that eject volumes of debris of the order of 1000 km^3. In actual fact, however, volumes may be considerably greater, with the 2.2-million-year eruption at Yellowstone ejecting around 2500 km^3 of debris and the 73,500-y Toba expelling at least 2800 km^3 (Rose and Chesner, 1990) and possibly nearly 6000 km^3 (Bühring *et al.*, 2000) of material.

In terms of its environmental impact, the most recent eruption of Toba is the most closely studied (e.g Rampino and Self, 1992, 1993a; Bekki *et al.*, 1996; Yang *et al.*, 1996; Zielinski *et al.*, 1996). According to Rampino and Self (1992), the 73.5-ka eruption lofted 10^{15} g of fine ash and 10^{16} g of sulphur gases to altitudes of between 27 and 37 km, creating dense stratospheric clouds of dust and aerosols. Zielinski *et al.* (1996) estimate, on the basis of volcanic sulphate recorded in the GISP2 Greenland ice core, that the total stratospheric sulphate aerosol loading due to the combination of SO_2 and atmospheric water may have been as high as 4400 Mt. This is perhaps 20 times greater than that caused by the 1815 eruption of Tambora, which resulted in a Northern Hemisphere temperature decrease of 0.7 degrees C. The resulting global aerosol optical depth following the Toba eruption is estimated to have been 10, in comparison to 1.3 following the Tambora blast, and is sufficient to have caused a Northern Hemisphere temperature fall of 3–5 degrees C (Rampino and Self, 1992, 1993a).

The length of the *volcanic winter* triggered by the Toba event is not well constrained, but assuming an *e-folding* stratospheric residence time for the Toba aerosols of about 1 year, Rampino and Self suggest that it could have lasted for several years. In support of this, an approximately 6-year-long period of volcanic sulphate recorded in the GISP2 ice core at the time of the Toba eruption suggests that the residence time of the Toba aerosols may have been of this order (Zielinski *et al.*, 1996). This is supported by modelling undertaken by Bekki *et al.* (1996), which suggests that SO_2 aerosol levels in the stratosphere would have been above background for nearly a decade (Fig. 5.9). Zielinski and his colleagues also recognise a 1000-year-long cooling event immediately following the deposition in the ice of the Toba sulphate, and suggest that the longevity of the Toba stratospheric loading may account at least for the first two centuries of this cooling episode.

The impact on our human ancestors of such an extended period of volcanogenic cooling remains largely a matter for speculation, although Rampino and Self (1993b) have made a link with a late-Pleistocene 'bottleneck' in human evolution that appears to have reduced the race to a few thousand individuals. More recently, Ambrose (1998) has invoked the Toba eruption to explain a severe culling of the human population, from the survivors of which the modern human races differentiated around 70,000 years ago. Clearly, there is little doubt that even for our technologically advanced civilisation the consequences of the next super-eruption will be devastating, and the chances of a consequent breakdown of global society must be high (McGuire, 1999).

5.8 Volcanoes and Ice Age

'Fire and ice' have long been linked in some way or other, and as far back as 1975 Kennett and Thunell proposed, on the basis of the tephra content of deep-sea cores, that a dramatic and global increase in explosive volcanism coincided with the glacially dominated Quaternary. The precise nature of the link between volcanism and glaciation remains problematical, however, and while some authors (e.g. Bray, 1976, 1977; Rampino and Self, 1992) suggest that volcanic activity is able to initiate or accelerate Quaternary glaciation, others (e.g. Matthews, 1969; Rampino *et al.*, 1979; Hall, 1982; Wallmann *et al.*, 1988; Nakada and Yokose, 1992; Sigvaldason *et al.*, 1992; McGuire *et al.*, 1997) propose that the dramatic changes in climate over the past 2 million years may have actively triggered volcanism. Closer to the present, Nesje and Johannessen (1992) invoked volcanic aerosol loading of the stratosphere as a major factor influencing climate and glacier advance throughout the Holocene, while Porter (1981, 1986) has suggested that decadal-scale glacier variations over the past thousand years may be linked to the eruption chronology over the same period. Of major importance from a hazard viewpoint, Day *et al.* (1999; 2000), have also recently proposed that climate change may trigger the collapse of ocean island volcanoes, resulting in the formation of giant, ocean-wide tsunami, and these ideas are addressed in more detail in the next chapter (Section 6.4).

In glaciated terrain, a causative relationship between volcanism and changes in ice cover is now well established, and as early as 1953, Thórarinsson suggested that the pressure drop resulting from the removal of a 100-m-deep glacier meltwater lake from the ice-dammed Grímsvötn caldera (Iceland) triggered eruptions. More recently, Sigvaldason *et al.* (1992) were able to show that increases in the rate of lava production in the Dyngjufjöll area of central Iceland coincided with the disappearance of glaciers at the end of the last glaciation. The authors' explanation was that the reduction in ice mass caused a fall in lithostatic pressure and vigorous isostatic crustal rebound, which triggered increased volcanism until a new pressure equilibrium was established. A similar anti-correlation of glaciation and volcanism has also been reported for the Quaternary of eastern California (Glazner *et al.*, 1999), where the maxima of episodic volcanic activity coincide with interglacial periods, supporting modulation of volcanism by changes in climate. The authors suggest a range of possible explanations including ice and meltwater unloading and changes in the groundwater regime.

Environmental change may also be capable of triggering volcanism in unglaciated terrains through the global redistribution of planetary water that accompanied Quaternary glaciation–deglaciation cycles. In 1979 Rampino and others addressed the possibility that volcanoes remote from glaciated areas might be induced to erupt by the large, dramatic changes in sea level that characterised much of the Quaternary. Wallmann *et al.* (1988) put forward a mechanism to explain such a link when they proposed that late-Quaternary eruptions at the Mediterranean island of Pantelleria were triggered by reduced pressures on the underlying magma chamber that occurred during the large (~100-m) falls in sea level that accompanied glaciation at high latitudes. Nakada and Yokose (1992) also implicated large sea-level changes in triggering Quaternary volcanism, pointing out that a 130-m sea-level rise would cause a 13-megapascal (MPa) stress change in regions of thin lithosphere such as volcanic island arcs. This, they suggest, could have been sufficient to trigger increased levels of volcanism.

More recently, by analysing statistically the ages of tephra layers in deep-sea cores, McGuire *et al.* (1997) have been able to show that explosive volcanism in the Mediterranean region over the past 80,000 years is positively correlated with the rate of sea-level change. Here, the intensity of volcanism is greatest during periods of rapid sea-level change, and in particular during the period from 17 to 6 thousand years ago, when sea levels rose by over 100 m. The authors feel that a single causal link between the rate of sea-level change and the level of explosive activity is unlikely and propose a range of local and regional mechanisms that might be responsible. These result from the dynamic responses of volcanoes to an array of stress-related influences linked to sea-level change, and are discussed in more detail in Section 6.3 of the following chapter. The episodes of enhanced volcanism recognised by McGuire *et al.* (1997) in the Mediterranean region also coincide well with periods showing the greatest number of volcanic sulphate signals in the GISP2 and GRIP Greenland Summit Ice Cores (Zielinski *et al.*, 1997), suggesting that the link between volcanism and rate of sea-level change is not regionally constrained, but representative of a hemispherical, and perhaps global, effect.

Compelling evidence now exists in support both of volcanic eruptions causing environmental change and for such change triggering volcanic activity. The circular nature of the relationship makes it difficult in specific circumstances, therefore, to determine cause and effect. This is particularly well illustrated by the relative timing of the Toba eruption and the onset of the last glaciation. Rampino and Self (1993a) suggest that severe global cooling caused by the Toba eruption might have been sufficient to increase snow cover and sea-ice extent at northern latitudes sufficiently to accelerate the deterioration of the climate into ice age conditions. They also note, however, that the planet was already cooling prior to the Toba eruption, and suggest that falling sea levels related to ice accumulation might have triggered the eruption through stressing the crust. The solution of the authors to this apparent dilemma is to propose a climate–volcanism feedback at times of climate transition during the Quaternary, whereby climate change triggers volcanism, which then accelerates the onset of glaciation through positive feedback. This hypothesis has been taken further by the European SEAVOLC team (1995), who suggest that at glacial terminations, a higher intensity of volcanic explosions might hinder planetary warming by means of negative feedback, perhaps offering an explanation for short, sudden cold 'snaps' during post-glacial times such as the Younger Dryas.

In conclusion, it is worth returning here to some of the ideas promulgated in the thought-provoking 1979 paper 'Can rapid climate change cause volcanic eruptions?' by Mike Rampino and his colleagues, many of which have not been taken much further. Rampino and his co-authors put forward a number of suggestions for the manner in which climate change might trigger volcanic activity, including hydro-isostatic and glacio-isostatic readjustments and global spin axis adjustment due to asymmetric mass loading. More speculatively, they also address the possibility that, as the timing of major glaciations is correlated with orbital variations, the Earth tides caused by planetary motions may at the same time cause increased levels of volcanic activity. On shorter timescales, they explore the possibility that changes in solar output that may be linked to climatic fluctuations over periods of the order of decades might be related to solar tide variations that might induce comparable Earth tide changes capable of triggering eruptions. The main point of both these speculative scenarios is that climate change and increased levels of volcanism may be coincident but that this reflects a common underlying cause rather than a direct causative link.

5.9 Volcanoes and mass extinctions

The largest eruptions of all are not in fact of the explosive variety, but involve instead the relatively quiet effusion of gigantic volumes of basaltic lava. Such *continental flood basalt* (CFB) eruptions occur – as the name suggests – on continental crust and are typically associated with mantle plumes or 'hot spots'. The volumes involved are often in excess of 1,000,000 km^3, or over 300 times the volume of Toba ejecta, and the areas of the flows are typically of the order of hundreds of thousands of square kilometres. Unlike large explosive eruptions, which are generally over in a few days, or weeks for the largest, CFBs are emplaced over periods of hundreds of thousands of years, although single flows of the order of 1000 km^3 may have been erupted in a matter of weeks. This is an emplacement rate 100 times greater than that of the 1783 Laki lavas, suggesting that the environmental impact resulting from associated degassing is likely to have been many orders of magnitudes more severe. Exactly how CFBs might have caused mass extinctions remains to be established, however, although one model involves severe greenhouse warming associated with the release of vast amounts of CO_2 triggering perturbation of the carbon cycle, ecological instability and consequent extinctions (McLean, 1985).

In recent decades, CFB eruptions have been implicated, as an alternative to asteroid and comet impacts, in mass extinctions within the geological record, and argument continues between advocates of a comet impact and supporters of a CFB cause for the Cretaceous/Tertiary (K/T) extinction event 65 million years ago. The case for CFBs as initiators of mass extinctions is succinctly and clearly put in a review paper by Courtillot *et al.* (1996). With regard to the K–T mass extinction, Courtillot and his co-authors point out that major CFB volcanism was already under way in the Deccan region of western India when a 10- to 15-km object impacted at Chicxulub off the east coast of Mexico. On this basis they propose that the extinction event was already under way at the time of the impact, which, they suggest, caused the extinction of fewer than half of the species that disappeared in the crisis.

Just as impact events are held responsible by their supporters as the source of all mass extinctions, some advocates of CFB-triggered mass extinctions propose that most, if not all, mass extinction events are a result of this style of volcanic activity. A causal link between CFBs and mass extinctions was first proposed by Vogt (1972), but has been reiterated more recently by others researchers (e.g. McLean, 1982; Morgan, 1986). Stothers (1993) noted that the correlation between the start of bursts of CFB volcanism and major extinctions is significant to a high level of confidence, and Courtillot *et al.* (1996) (Fig. 5.10) claim that over

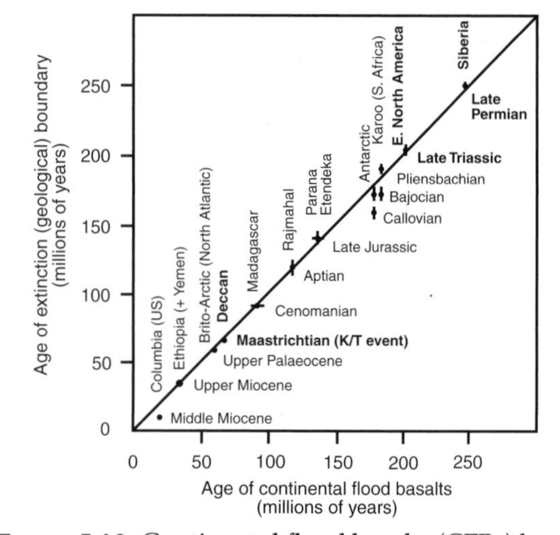

FIGURE 5.10 Continental flood basalts (CFBs) have been correlated by a number of authors with mass extinction events. This plot (simplified from Courtillot *et al.*, 1996) shows both CFBs having erupted during the past 300 million years and significant extinctions. Major events are shown in bold.

three-quarters of all CFBs are contemporaneous with a mass extinction event.

In an interesting twist that might have been expected to unite advocates of impact with supporters of the CFB model, Rampino and Stothers (1988) suggested that CFBs might actually be generated by impact events. To date, however, this idea has failed to gain substantial support, largely owing to problems with the relative timing of events. As previously mentioned, CFB volcanism had, for example, already started prior to the Chicxulub impact at the K–T boundary, a situation difficult to explain in terms of the Rampino and Stothers model without invoking an earlier impact event, for which there is as yet no evidence. The CFB versus impact cause of mass extinctions is an issue that is likely to run for some time, and it is addressed once more in Section 7.5.3.

6

Sea-level change as a trigger of natural hazards

6.0 Chapter summary

Significantly elevated sea levels are now widely recognised as being an inevitable consequence of the current period of global warming. In this chapter we address the causes and scale of past sea-level fluctuations, particularly during the late Quaternary and, most recently, in the twentieth century. We look forward through examining sources of future sea-level rise and projections for elevated sea levels over the next few thousand years. Uncertainties in such forecasts are considered, arising from poor constraints on the response to warming of the Greenland and Antarctic Ice Sheets, and the degree of greenhouse gas mitigation in coming decades. Coastal flooding, the most obvious consequence of future sea-level rise, is discussed in Chapter 3, and although coastal erosion can be hazardous, we leave treatment of this to texts in environmental management and geomorphology. Here, we focus on the less obvious hazard implications of sea-level change, including the triggering of earthquakes, volcanic eruptions and submarine landslides. The chapter ends with an evaluation of the intriguing link between climate change, fluctuating sea levels, the collapse of large volcanic ocean islands, and the threat from giant ocean-wide tsunami.

6.1 Changing sea levels and natural hazards

Sea levels have changed throughout geological time (e.g. Haq *et al.*, 1987) in response to a range of different and sometimes interacting isostatic, eustatic and tectonic processes (Dawson, 1992; Box 6.1). Natural hazards related to such sea-level change are surprisingly many and varied, and the relationship between the two is often far from clear. Broadly speaking, rising sea levels can be expected to increase the threat to coastal zones, primarily owing to the inundation or flooding of low-lying terrain (see Chapter 3) but also through increased erosion, destabilisation and collapse of elevated coastlines. Higher sea levels will also exacerbate the impact and destructive potential of storm surges and tsunami, partly because of the elevated level of the sea surface but also through increasing the exposure of many coastlines as a result of inundation of wetlands and other protective environments. The hazard implications of falling sea levels are less obvious, although it has been suggested that rapid drops in sea level may trigger submarine landslides (e.g. Maslin *et al.*, 1998; Rothwell *et al.*, 1998). On a much broader scale, a number of authors have proposed that large sea-level changes – either up or down – may trigger

Box 6.1 Causes of sea-level change in the geological record

Eustatic sea-level changes

Eustatic changes in sea level represent changes in the form and level of the surface of the Earth's oceans that exist in equilibrium with the planet's gravitational field. Eustatic sea-level changes occur as a result of three geophysical mechanisms. *Glacio-eustatic* changes reflect vertical changes to the global ocean surface as a result of variations in water volume due to the growth and melting of ice caps and glaciers. During the Quaternary, major fluctuations in sea level were largely glacio-eustatic, occurring in response to changes in global ice volume during glaciation–deglaciation cycles. *Tectono-eustatic* changes arise from modifications to the shapes of the ocean basins as a result of plate tectonic processes such as the widening of the ocean basins due to sea-floor spreading or their shrinkage through the consumption of ocean floor at subduction zones. For example, Bloom (1971) proposed that owing to sea-floor spreading during the late Quaternary, the ocean basins are capable of holding 6 per cent more water by volume now than during the last interglacial. *Geoidal-eustatic* changes involve modifications to the shape of the ocean surface. Like the land, the surface of the ocean possesses a 'topography' with swells and depressions that can nowadays be contoured and monitored using precise satellite altimetry. This watery 'topography' is defined as the *geoid*, which is the equipotential surface of the Earth's gravity field. Gravitational attraction is affected by, for example, density variations within the crust and mantle and the sizes of ice masses at high latitudes. Variations in geoidal sea-surface altitudes are today measured using satellite instruments (such as TOPEX/POSEIDON and Jason-1), which have revealed vertical differences in the sea surface of the order of 200 m. In contrast to glacio- and tectono-eustatic sea-level changes, those due to variations in the geoid are not global in their effects. The importance of the nature of geoidal sea-level variations lies in the fact that many published curves of Quaternary sea-level change may be accurate on only a regional rather than a global scale.

Isostatic sea-level changes

Isostatically related sea-level changes are a consequence of accommodations in the equilibrium that exists between the lithosphere and the underlying asthenosphere on which it 'floats'. During the Quaternary, most isostatic changes in sea level reflected loading and unloading effects associated with the redistribution of ice and water during glaciation–deglaciation cycles. *Glacio-isostatic* changes relate to the depression and rebound of glaciated continental and adjacent areas, while *hydro-isostatic* changes are related to the subsidence (and, arguably, the uplift) of continental margins that are respectively loaded and unloaded by changing ocean volumes. Owing to the elastic behaviour of the lithosphere, depression and succeeding uplift in response to ice sheet growth and decay can exceed 1 km. Over the past 7000 years alone, Chappell (1974) has proposed that sea-level rise has resulted in an overall depression of the ocean floor by 8 m, accompanied by an average 16 m rise of the continents, although these numbers have been questioned.

Tectonic sea-level changes

In addition to large-scale tectono-eustatic changes related to modifications to the ocean basins, local sea-level changes may also result from vertical crustal movements, particularly in tectonically active settings such as island arcs and collisional plate margins. Such behaviour has drastically modified the global or regional Quaternary sea-level change records for a number of tectonically active areas including the Mediterranean region, Japan and the Aleutian Islands.

increased volcanism (e.g. Wallmann *et al.*, 1988; Nakada and Yokose, 1992; McGuire *et al.*, 1997) and seismicity (Nakada and Yokose, 1992) along continental margins. Proposed mechanisms range from global spin (Matthews, 1969) or hydro-isostatic crustal (Anderson, 1974; Chappell, 1975; Rampino *et al.*, 1979; Nakada and Yokose, 1992) readjustments related to the redistribution of planetary water, to responses by individual coastal and island volcanoes to large sea-level changes (Wallmann *et al.*, 1988; McGuire *et al.*, 1997).

6.1.1 Sea-level changes in the Quaternary

Most of our knowledge of sea-level change and its impact derives from the rapid and dramatic sea-level changes that occurred during the Quaternary – the past 1.65 million years – when successive glaciation–deglaciation cycles led to changes in global sea level of over 130 m (Shackleton, 1987). These large Quaternary sea-level fluctuations were primarily glacio-eustatic and can be broadly attributed to changes in global ice volume. At a regional scale, however, isostatic processes were also probably important, with the wholesale redistribution of planetary water resulting in loading and unloading of the lithosphere, particularly along continental margins and on glaciated continental terrain, sufficient to cause significant spatial variations in relative sea level. The rate of global eustatic Quaternary sea-level change, particularly at glacial terminations, was very rapid, and at times during the Holocene (the past ~10,000 years) may have reached several metres per millennium. Global eustatic sea-level changes over the past 140,000 years are relatively well constrained from the oxygen isotope record determined primarily from planktonic foraminifera (*see* Dawson, 1992, for more detailed discussion of oxygen isotopes and their application to the late Quaternary). At the same time, analysis of foraminifera in deep-sea sediment cores provides information on the history of global continental ice volume, thereby permitting the glacio-eustatic component of sea-level change to be determined (e.g. Shackleton and Opdyke, 1973). These high-resolution records of ocean volume changes represent, however, only a first approximation to variations in global sea level as they are not consistent with sea-level records derived from dated marine terraces in areas undergoing uniform crustal uplift (e.g. Bloom, *et al.*, 1974; Dodge *et al.*, 1983). Shackleton (1987; Fig. 6.1) attempted to resolve this situation by producing a general eustatic sea-level curve that presents a more detailed record of ocean volume changes than that from raised terraces. Discrepancies remain between the two records, but both do show the same general trends in sea level (Chappell and Shackleton, 1986). It is worth reiterating here, however, that although a global eustatic sea-level curve is a valid concept, owing to changes in the shape of the geoid (see Box 6.1), sea-level changes over the entire planet, during glaciation–deglaciation cycles, were neither of the same magnitude nor even the same sense. This was a consequence of the large-scale redistribution of planetary

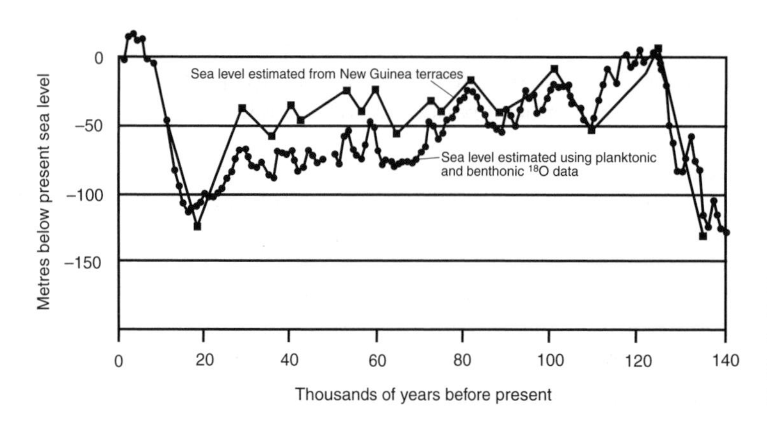

FIGURE 6.1 The oxygen isotope global sea-level history compared with sea-level data estimated from marine terraces in New Guinea. (After Shackleton, 1987)

water, which had a profound effect on the topography of the geoid through the massive transfers of load from the land to ocean basin (and vice versa), loading and unloading effects on the upper mantle, and the growth and decay of ice masses (Tooley and Turner, 1995). The large variations in sea-level change that occurred during the Holocene have important implications for future global warming-related sea-level rise, which may not affect all parts of the globe equally.

Sea-level changes during the past 140,000 years (approximately corresponding to the late Quaternary, which started 130,000 years ago) are important because they provide a backdrop against which we can view the current period of rising sea levels. Over this time, sea-level behaviour can be subdivided into three broad categories (Box 6.2), each linked to the prevailing global climate and therefore also to ice-cap growth and decay. As might be expected, reconstructions of the timing and magnitude of sea-level changes are better constrained for the late Pleistocene and Holocene, a period covering the past 30,000 years. The early part of this time-span, from 30 ka to 18 ka BP, is characterised by a large fall in sea level from between 45 and 80 m to a base level of around –120 m. The former figure is based upon oxygen isotope data (Shackleton, 1987) and the latter on the ages of marine terraces (Bloom *et al.*, 1974). Depending on which starting level is accepted, a time-averaged rate in sea-level fall of between 3.75 and 6.55 mm/y is obtained.

Subsequent rises in sea level after 18 ka BP coincide with progressive warming at the end of the last glaciation, and are dated by incremental drowning of coral reefs in the Caribbean–Atlantic region (Fig. 6.2). If we look more closely at time-averaged sea-level curves (e.g. Fairbanks, 1989), we find that they obscure the real nature of sea-level rise, which is actually episodic and step-like. Some of these rises are in fact revealed to be near-instantaneous, and Blanchon and Shaw (1995) report three metre-scale catastrophic rise events (CREs) in the post-glacial record: a 13.5 m rise dated at 14.2 ka BP, a 7.5-m rise at 11.5 ka BP, and a 6.5-m rise as recently as 7.6 ka BP. Although sea level rose, on average, by around 10 m per millennium during the period of rapid post-glacial ice sheet melting, Blanchon and Shaw's reef-drowning studies from the Caribbean and the Atlantic suggest that the CREs may have involved an increase in sea level by as much as 13.5 m in less than 290 ± 50 years. These rates – of up to 46 mm a year – are overprinted on an annual general post-glacial rise of between 4 and 13 mm, and reflect major and rapid changes in global climate and geography. The 7.5-m rise at 11.5 ka BP, for example, coincided with the end of the Younger Dryas cold phase, when changing dust concentrations and snow accumulation rates suggest that global climate change might have

Box 6.2 Sea-level changes over the past 140,000 years

Dates are in thousands of years BP; sea-level changes are compared to current sea level.

Rapid rises in sea level at the end of major glacial stages (glacial terminations): The principal rises occurred from 140 to 128 (–130 m to 0 m) and from 18 to 7 (–125 m to +5 m)

Stable sea levels during interglacial periods (warm phases between distinct glacial episodes): from 128 to 115, during interstadials (short periods of climate amelioration within otherwise cold phases) from 105 to 100, 95.5 to 94, 82 to 80, 60 to 58 and 40 to 36, and during one stadial (glacial period) from 70 to 60.

Step-like falls in sea level during the last (Weichselian/Devensian) glacial period: with major fall between 115 and 110 (–8 m to –50 m), 80 and 74 (–27 m and –60 m), 72 and 70 (–50 m and –80 m) and 18 and 29 (–75 m to –120 m).

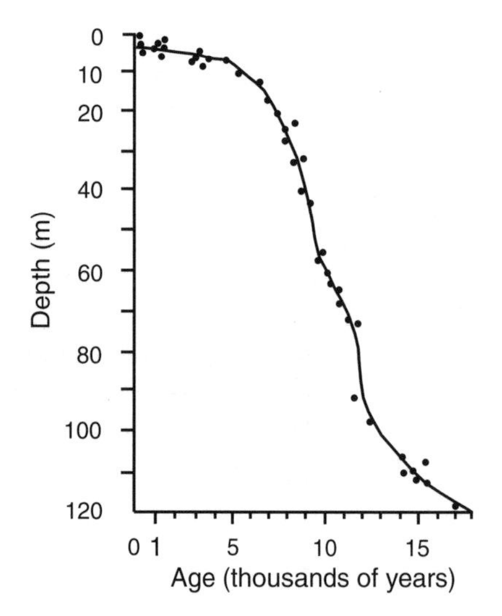

FIGURE 6.2 Sea-level curve for the past 18,000 years based upon the dating of submerged corals from the Barbados region of the Caribbean (after Fairbanks, 1989). Such time-averaged curves obscure the real nature of post-glacial sea-level rise, which is actually episodic and step-like.

occurred within 20–50 years (Mayewski *et al.*, 1993; Alley *et al.*, 1993). Others may reflect the breaching of continental meltwater lakes that then catastrophically emptied into the ocean. Blanchon and Shaw (1992) do, in fact, explain their CREs in terms of the spectacular release of meltwater megafloods from glacial and proglacial reservoirs, which led to unprecedented sea-level rises over very short periods of time. Discharge from Lake Agassiz in North America around 8000 years ago is estimated (Hillaire-Marcel *et al.*, 1981), for example, to have caused a eustatic sea-level rise of between 20 and 42 cm. Such dramatic rises were succeeded in the late Holocene by a more steadily consistent rise in sea level, the rate of which also began to slow, first to 5.45 mm per year between 7000 and 4800 years ago, and since then to a time-averaged annual rate of 1.25 mm.

6.1.2 Recent changes in sea level

How then do recent changes in global sea level compare with the major variations observed in the post-glacial record? First, it must be remembered that sea-level change measured at a point on a coastline can reflect either vertical displacement of the land or the changing volume of ocean, or a combination of the two. Consequently, great care must be taken when interpreting the raw data in terms of rising sea levels on a global scale, and effects that are not a result of environmental change must be subtracted before a true record of absolute sea-level rise is achieved. These are many and varied, and occur on both a local and a regional scale. Examples of the former include the compaction of coastal sediments and the abstraction of water from coastal aquifers, while the latter include tectonic uplift and continuing post-glacial rebound of the crust. In order to obtain a meaningful global average for absolute sea-level change, the effects of such local or regional vertical movements of the crust must be removed from the observed record, and many such records from numerous geographical locations across the planet combined. The situation is further complicated if one wishes to extract that part of the sea-level rise signal that reflects anthropogenic global warming. This is because sea levels have been slowly rising since long before human activities could have had any effect on global temperatures.

Despite these difficulties, a number of different approaches have been utilised in order to constrain better estimates of sea-level rise over the past 150 years. These include (a) averaging all tide gauge observations from stable coastal locations, (b) determining the long-term late-Holocene rise from dated shorelines and subtracting this from the local tide gauge record, and (c) generating a geophysical model that simulates local tectonic and isostatic effects and subtracting this from the local tide gauge data. A number of estimates for recent sea-level rise have been obtained using such approaches, ranging from 1.8 ± 0.1 mm/y (Douglas, 1997) for the period from 1880 to 1980 to 2.4 ± 0.9 mm/y (Peltier and Tushingham, 1989) for the twentieth century up to that date. These rates are significantly higher than those prior to the middle of the nineteenth century, for which some estimates (e.g. Douglas, 1995) suggest that the rise was either negligible or even indistinguishable from zero. Although it seems clear, then, that twentieth-century sea level rose probably more rapidly than at any time during the past 1000 years or more, so complex are the interdecadal and other variations in the sea-level record that a statistically significant acceleration remains to be identified (Douglas, 1992). This is a major part of the problem that has resulted in such disparate estimates of just how much sea level will have risen by the end of this century.

6.2 Perspectives on future sea-level change

The worrying element of anthropogenic sea-level change is that it involves a considerable time lag between the forcing agent, for example greenhouse gas (GHG)-induced planetary warming, and the response of the system. The corollary of this is that even if we reduced GHG emissions to zero today, the warming resulting from previous emissions would still be sufficient to cause a substantial rise in global sea level over the coming century. To illustrate this, the UK Meteorological Office (1999) modelled an atmospheric CO_2 increase of 1 per cent per year for 70 years, after which the level was held constant for 700 years (Fig. 6.3). The result showed that melting land ice (glaciers) continued to contribute to sea-level rise for about 400 years after stabilisation of CO_2 levels. However, thermal expansion of the ocean, due to the slow penetration of heat to greater depths, contributed to rising sea level for perhaps a thousand years or more.

Sea-level rise during the past century is generally agreed to reflect contributions from a number of sources, the most important of which are melting of small glaciers and the Greenland Ice Sheet (4–7 cm; Zuo and Oerlemans, 1997), thermal expansion of the seawater (2–5 cm; de Wolde *et al.*, 1995) and changes in the mass balance of the Antarctic Ice Sheet (~4 cm; Huybrechts and de Wolde, 1999). Continued sea-level rise will reflect contributions from the same sources (Box 6.3), although the relative contributions from the Greenland and Antarctic ice sheets are very uncertain.

6.2.1 Projections of future sea-level rise

Inevitably, given the current level of research, new projections of future sea-level rise are constantly being published, and that which follows may well have been overtaken by others – either more or less optimistic – by the time this book is published. Nevertheless, the 1999 forecast of the UK Meteorological Office

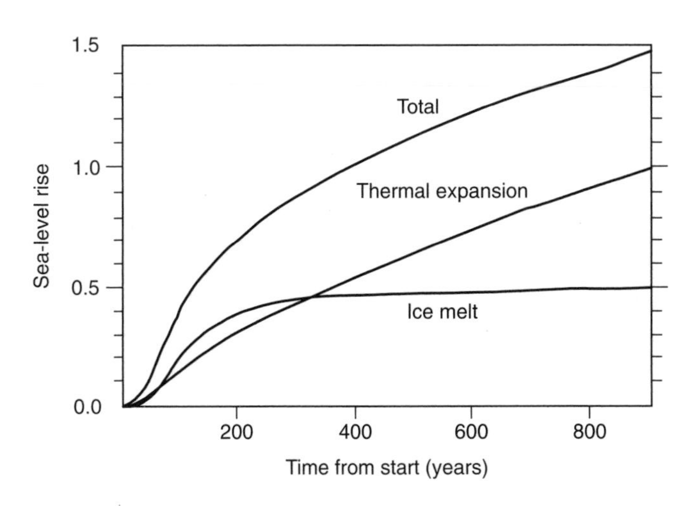

FIGURE 6.3 Sea-level rise will continue long after GHG emissions are stabilised. This is illustrated in an idealised model experiment run by the UK Meteorological Office. Following a 1 per cent increase in CO_2 for the first 70 years and no further change for the next 700 years, sea level continues to rise for many hundreds of years more. The long-term component of this rise is due to continued thermal expansion of the oceans after all land ice (glaciers) has melted.

Box 6.3 Sources of future sea-level rise

Thermal expansion: The thermal expansion of sea water is based upon a simple premise, which is that as it warms, its density decreases and therefore it occupies more volume. The sea-level rise predicted from ocean warming and expansion is difficult to constrain, however, as it is strongly dependent on the vertical profile of warming within the deep ocean. Furthermore, for a given average ocean warming, sea-level rise depends upon where warming occurs. Because the expansion coefficient increases with water temperature, the greatest rise in sea level will occur if warming is concentrated in regions where the ocean waters are already the warmest, in other words the upper few hundred metres of the sea at low latitudes (Harvey, 2000).

Growth and decay of glaciers and small ice caps: Growth and decay of glaciers and small ice caps is dependent on the relative rate of ice accumulation versus melting. Mountain glaciers and ice caps can be simply divided into two zones: an upper zone of accumulation, where ice formed from new snowfall annually exceeds that lost due to melting; and a lower zone of ablation, in which annual melting (*ablation*) exceeds annual accumulation. For a glacier to maintain its mass, the net mass loss from the ablation zone must be exactly balanced by a net mass gain in the accumulation zone. The two zones are separated by the *equilibrium-level altitude* (ELA), the position of which determines whether a glacier will grow or decay. A rise in temperature, for example, will tend to raise the ELA, which, if it rises above the highest point on the glacier, will eventually cause it to vanish entirely. Some glaciers are more sensitive to changes in the ELA than others, and, broadly speaking, smaller glaciers or those with a greater mass throughput (higher rates of both accumulation and melting) will respond more rapidly to variations in the altitude of the ELA. The mountain glaciers and small ice caps that exist today are tiny remnants of those that covered upland areas between 15,000 and 20,000 years ago, and most are now confined to the Himalayas, Andes, Western Cordillera of North America and glaciated terrain in the Arctic and Antarctic. Recent work by Warrick *et al.* (1996) suggests that even should all the mountain glaciers and small ice caps melt as a result of global warming, the total sea-level rise would still be only of the order of 50 cm ± 10 cm.

Melting of the Greenland Ice Sheet: One of the big unknowns in trying to forecast future sea-level rise, particularly in the long term, lies in the contributions from melting of the polar ice caps. Melting of the Greenland Ice Cap would cause a 7.4-m rise in sea level, resulting in major inundation of low-lying coasts worldwide. The UK Meteorological Office (1999) proposes that over the next 100 years the contribution from melting of the Greenland Ice Cap will probably be offset by increased levels of precipitation over Antarctica. Beyond this, however, the Earth may well have warmed sufficiently for melting of the Greenland ice to double the rate of sea-level rise. Recently, Huybrechts and de Wolde (1999) calculated that on a multiple-century timescale, melting of the Greenland Ice Cap will make the biggest contribution to global sea-level rise. This contribution, they predict, will probably be only around 10 cm by 2100, but could amount to several metres by the year 3000 if greenhouse gas emissions are sustained. Letréguilly *et al.* (1991) think that a regional warming of as little as 6ºC might be enough to melt the Greenland Ice Cap eventually. Assuming unmitigated emissions of greenhouse gases, this is a figure that would be exceeded globally within the next 200 years, and probably much sooner at high latitudes. Similarly worryingly, in the longer term, Crowley and Baum (1995) show that once the Greenland Ice Cap melts it will not regrow, even if greenhouse gas emissions are reduced to pre-industrial values, so the post-twenty-second-century world might have to live permanently with elevated global sea levels. On the basis of their calculations, Huybrechts and de Wolde (1999) also forecast that even if greenhouse gas concentrations are stabilised by the early twenty-second century, inertia in the system would ensure that melting of the Greenland Ice Cap was irreversible for equivalent CO_2 concentrations more than double the present value.

Mass balance changes in the Antarctic Ice Sheet: Of even greater concern than the Greenland scenario is melting of the Antarctic Ice Sheet. This consists of Eastern (EAIS) and Western (WAIS) Antarctic Ice Sheets, with most of the former sitting on bedrock above sea level, but much of the latter sitting on bedrock below sea level or comprising gigantic floating ice shelves. Although it is unlikely that the EAIS would start to melt substantially in the absence of a catastrophic rise in global temperatures, the WAIS is much more susceptible to rising global temperatures. Over 20 years ago, Mercer (1978) suggested that melting of the huge floating ice shelves, such as the Ross Ice Shelf, might debuttress the remainder of the WAIS currently grounded on bedrock, causing it to slide into the sea and causing a global sea-level rise of 5–6 m. The UK Meteorological Office (1999) warns that in the longer term, continued warming might threaten the existence of the WAIS, although it admits that the timing of this is highly uncertain. Recently, Huybrechts and de Wolde (1999) calculated that up to the end of the twentieth century, the Antarctic Ice Cap would just about balance the sea-level rise contribution of the Greenland Ice Cap, because increased ice accumulation rates would dominate over melting. In the longer term, however, they predict a positive contribution from the Antarctic to sea-level rise, while finding no evidence for any catastrophic collapse of the WAIS.

does provide a flavour of the speed and scale of sea-level rise expected over the next century or so. The forecast predicts that by 2080 sea level will, on average, be 41 cm higher than it is at present if GHG emissions are not reduced. Because of the inertia in the system, however, substantial rises can be expected even if emissions fall significantly, and the main effect of any reduction is really just to slow the rate of sea-level rise (Fig. 6.4). For example, the 22-cm rise expected by 2050, assuming unmitigated emissions, would be delayed by around 15 years if emissions were stabilised at 750 ppm CO_2 and by 20 years if stabilisation were achieved at the 550 ppm level. Similarly, the 50-cm rise predicted by 2100, under the unmitigated scenario, would be delayed by around 35 and 55 years respectively. The Met Office study also predicts that – even assuming a 550-ppm CO_2 stabilisation level – sea levels by the 2230s will be 75-cm higher than they are now and still rising. With unmitigated emissions the situation is forecast to be far more extreme, with sea level rising at a rate of 60 cm/century by 2080. It should be noted, however, that, as mentioned earlier in this chapter, the ocean surface varies in height, and even today there are marked regional differences in sea level. This situation is likely to persist in a warmer world, with some regions experiencing a higher rise than others, and regional variations approaching the magnitude of the global sea-level rise itself (Bryan, 1996).

Attempting to forecast longer-term sea-level rise is extremely difficult, mainly because there are so many variables, interactions and potential feedback mechanisms that we do not as yet fully understand. For example, a model developed by Harvey (1994) predicted that a 3 degree Celsius global mean surface warming – as forecast to occur by 2080 with unmitigated emissions (UK Meteorological Office, 1999) – could produce a sea-level rise anywhere between 0.5 to 3 m, depending on how the

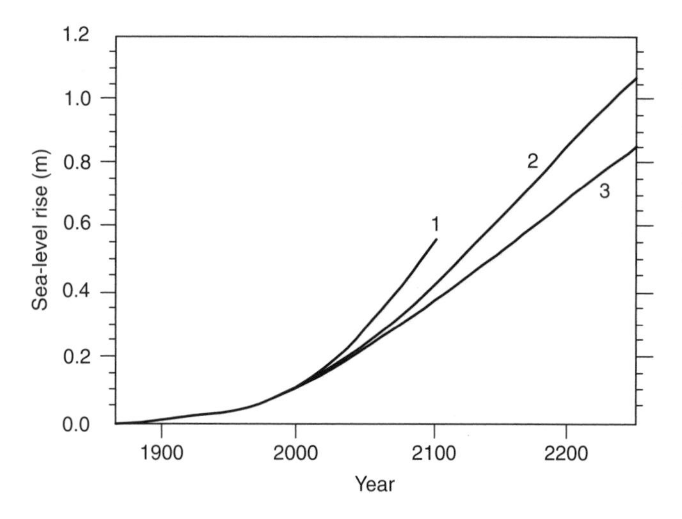

FIGURE 6.4 A reduction in GHG emissions would slow sea-level rise but not stop it. For example, the 22-cm rise expected by 2050 (1) would be delayed by 15 years if emissions were stabilised at 750 ppm CO_2 (2) and 20 years if stabilisation was achieved at 550 ppm (3). (After UK Meteorological Office, 1999)

temperature change varies with depth in the oceans. Similarly, the model showed that the length of time required for sea level to rise varied considerably, with the time needed to reach 75 per cent of the final rise ranging from 400 to 2400 years. These large uncertainties derive primarily from a lack of knowledge about how exactly the oceans will warm at depth and therefore exactly what the contribution will be from thermal expansion. Until these uncertainties are ironed out, forecasts of the magnitude and rate of future sea-level change are likely to remain poorly constrained.

Looking further ahead, and assuming complete melting of the Greenland Ice Cap and substantial melting of the West Antarctic Ice Sheet, sea-level rise could be between 14 and 17 m. As pointed out by Harvey (2000), however, a progressive sinking of the oceanic crust due to increased water loading would accommodate some of this eustatic rise. As this is likely to occur slowly, and at roughly the same rate as melting of the polar ice caps, the final peak sea-level rise might be expected to be of the order of 10–13 m. With the huge East Antarctic Ice Sheet likely to remain reasonably stable up to a regional temperature increase of 20ºC, which is unlikely to happen without very high GHG emissions, this probably represents the maximum extent of the sea-level rise threat. How rapidly this will be achieved, however (assuming unmitigated emissions), is unknown. A conservative estimate is 'several thousand' years, but others suspect that the rise could be much rapid. In their 1992 paper, Blanchon and Shaw demonstrated just how rapidly the climate flipped during the last glacial to interglacial transition – perhaps switching between glacial and interglacial conditions in less than a decade – and how this led to rapid ice-sheet melting and sea-level rise. They also issued a warning for the future, suggesting that sea-level rise resulting from global warming might not be slow and continuous, but involve periodic catastrophic rises driven by sudden, dramatic climate change and rapid melting of the Greenland Ice Cap and the West Antarctic Ice Sheet.

As discussed earlier in this chapter, it is unlikely, owing to changes in the geoid, that the coming rise in sea level will be equally severe at all points on the globe. Clarke and Primus (1987), for example, modelled instantaneous partial melting of the Greenland Ice Sheet sufficient to raise the global sea level by 100 cm. As a result of the substantial mass loss, however, and the consequent reduction in its gravity field, Greenland would be less effective at 'attracting' the ocean, and sea level here and in Iceland would actually fall by up to 2 m. The effect of the Greenland mass loss on the geoid would persist as far as the UK, where sea level

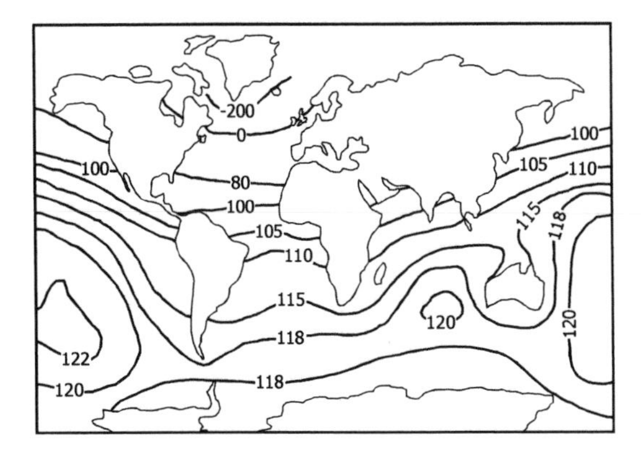

FIGURE 6.5 Due to the resulting modification of the geoid, instantaneous melting of the Greenland Ice Sheet sufficient to raise global sea level by 100 cm can be expected to produce different changes in ocean topography in different parts of the world. Adjacent to Greenland, sea level would actually fall by around 200 cm, while around the United Kingdom it would remain largely unchanged. The Indian Ocean, in contrast, would experience a sea-level rise of around 120 cm. (After Clarke and Primus, 1987)

would barely change. Contrastingly, however, sea levels in the southern Indian, Pacific and Atlantic Oceans would rise by up to 122 cm (Fig. 6.5). Clearly, such variations mean that the sea-level rise curves published by the UK Meteorological Office (1999) must be treated with caution in terms of extrapolation across the planet as a whole. Similarly, geoidal-related variations will dramatically influence geographical susceptibility to future coastal flooding, which is likely to depend strongly on the relative contributions to the oceans from the Greenland and Antarctic ice sheets.

6.3 Sea-level change as an initiator of seismicity and volcanism

During the Quaternary, the continental margins were subjected to major stress changes as the enormous variations in ocean volume repeatedly loaded and unloaded the crust. These stress changes have been implicated in triggering a range of potentially hazardous phenomena, including increased tectonism and seismicity (e.g. Nakada and Yokose, 1992), enhanced volcanic activity (e.g. Rampino *et al.*, 1979; Wallmann *et al.*, 1988; McGuire *et al.*, 1997), and landslide formation (e.g. Weaver and Kuijpers, 1983; Prior *et al.*, 1986; Roberts and Cramp, 1997; Maslin *et al.*, 1998; Rothwell *et al.*, 1998).

6.3.1 Sea-level change as a trigger of active faulting and seismicity

It has been known for some time that post-glacial ice unloading of depressed continental crust caused an isostatic rebound effect accompanied by fault reactivation and seismic activity, and Mörner (1980) has argued that active faulting and large earthquakes were characteristic of the last episode of deglaciation in Fennoscandia. Similar evidence for widespread palaeoseismicity related to deglaciation has more recently been reported from both Scotland (Ringrose, 1989) and North America (Wu and Johnston, 2000). In the more recent work, the authors even question whether or not stresses related to post-glacial isostatic rebound of the crust might have triggered the major intraplate earthquakes at New Madrid (Missouri) in 1811–12. They conclude, however, that New Madrid was probably too far from the former ice margin for this to be the case.

Loading and unloading due to water has also been known for some to time to trigger seismicity, and this effect is both common and expected as reservoirs accumulate behind new dams and progressively load the crust (e.g. Simpson *et al.*, 1988; Kilburn and Petley, 2001). Similar effects were associated with the drainage of post-glacial ice-dammed lakes such as that at Glen Roy in the Scottish Highlands. Although one of the smallest, with an estimated water volume of 5 km^3, the catastrophic drainage of the lake during the Younger Dryas stadial triggered widespread landslide formation and faulting (Sissons and Cornish, 1982) and was probably accompanied by earthquake activity (Ringrose, 1989). Given the huge loading and unloading effects caused by ~100 m changes in sea level, fault reactivation and seismicity at continental margins can be expected to be orders of magnitude greater at glacial terminations and periods of rapid sea-level fall. In an attempt to quantify this effect, Nakada and Yokose (1992) numerically evaluated stress accumulation in the crust caused by the surface mass redistribution associated with Quaternary glaciation–deglaciation cycles. Focusing on island arc environments around the Pacific Rim, where the lithosphere is relatively thin, the authors showed that for a lithospheric thickness of 20–30 km the changes in stress difference during a deglaciation event involving a 130-m sea-level rise and lasting 10,000 years totalled 13 megapascals (MPa). For the Japanese island arc, Nakada and Yokose proposed that this value represents about half that of the 'normal' tectonic compressive stress (~30 MPa), and that its addition is sufficiently large, therefore, to trigger or accelerate active tectonism in this and other arcs with similar characteristics.

One of the most bizarre links between environmental change and seismicity – at least at first glance – is that proposed by Daniel Walker (1988, 1995), who correlates the timing of El Niño events with increased levels of seismicity along the East Pacific Rise spreading ridge system from 20º S to 40º N. Walker observes that between 1964 and 1994, El Niños and East Pacific Rise earthquake swarms seem to have occurred almost simultaneously, with the longest El Niño events corresponding to the longest earthquake swarms. Although the idea is highly speculative, Walker suggests that perhaps the earthquake swarms reflect massive submarine volcanic activity along portions of the East Pacific Rise (for which there is currently no evidence), which may trigger El Niños through the generation of magmatic heat that warms the surface waters. An equally plausible alternative explanation, however, is that increased seismic activity is being triggered by loading effects resulting from the changing sea levels in the eastern Pacific (of up to half a metre) that accompany El Niño events. Should this ever be established, it would mean that very small changes in seawater load pressures are capable of triggering seismic activity – presumably provided that the system is in a critical state – with serious implications for the impact of contemporary sea-level rise on seismically active coastlines.

6.3.2 Sea-level change and increased volcanic activity

In Chapter 5 we examine the link between volcanism and ice ages, and briefly introduce evidence for a correlation between changes in ocean volumes and increased volcanic activity. Here we return to the issue to look in more detail at the potential mechanisms that might permit large and dramatic changes in sea level to trigger or accelerate volcanism in island and coastal settings. The earliest quantitative work in this area was published by Wallmann and co-workers in 1988, who modelled the effects of sea-level drawdown on late-Quaternary eruptive activity at the Mediterranean island of Pantelleria. They showed that the timing of ring-fracture eruptions around the margins of a 114,000-year-old caldera, which occurred 67,000 years ago and again between 22,000 and 17,000 years ago, correlated with glacial maxima when sea levels were depressed by ~100 m. The authors calculated that such a fall would amount to a load pressure reduction of 10 bar (1 MPa) on the magma reservoir feeding the volcano, while their modelling showed that the removal of only around 5 bar should be sufficient to trigger an eruption. The precise mechanism advocated by Wallmann *et al.* (1988) is illustrated in Fig. 6.6. Within the model, sea-level drawdown results in the development of tensional stress conditions above the magma reservoir and above its margins, which coincide with zones of weakness associated with ring fractures around the periphery of the caldera. Dykes propagate into the weak zones, either cooling beneath the surface or feeding new ring-fracture eruptions.

A problem with this model lies in the calculated load pressure reduction on the magma reservoir. If the state of the magmatic system was such that half the reduction attributed to a 100-m sea-level fall was sufficient to trigger an eruption, why did eruptions not occur when sea level had fallen by half this amount?

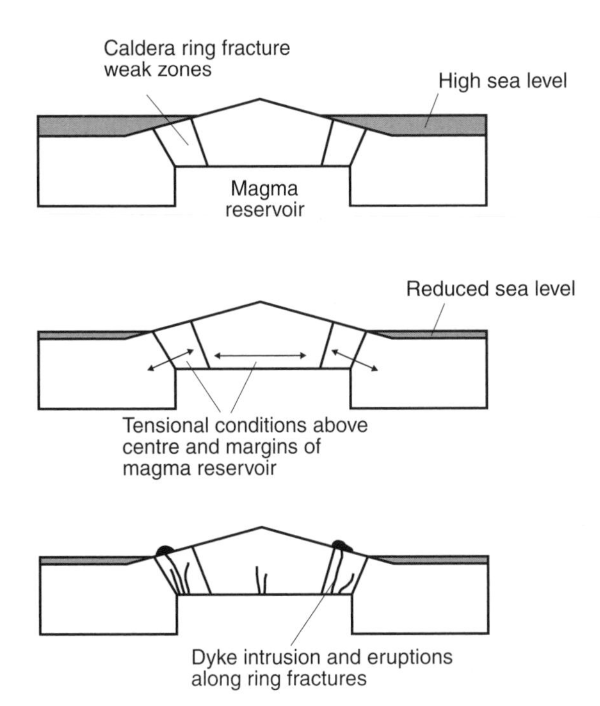

FIGURE 6.6 Sea-level drawdown has been implicated in eruptions from the Pantelleria caldera during the late Quaternary. At times of low sea level, prevailing stress conditions may permit magma to erupt through the caldera ring fractures. (After Wallmann *et al.*, 1988)

Intuitively, this would seem to be more likely, otherwise how does the magma reservoir 'know' when the sea-level fall has reached its maximum value and why would it wait until then to erupt if pressure conditions were sufficient to promote eruption at an earlier stage in sea-level drawdown? This is an issue addressed by McGuire *et al.* (1997), who argue for eruption triggering at coastal and island volcanoes during the dynamic condition of falling or rising sea level rather than during periods of stable sea-level depression or elevation.

McGuire and his colleagues investigated evidence for a link between sea-level change during the late Quaternary and explosive volcanic activity at Mediterranean volcanoes by comparing the published ages of tephra layers in Mediterranean deep-sea cores (e.g. Keller *et al.*, 1978; Paterne *et al.*, 1988) with established global sea-level curves for the past 80,000 years. A cumulative plot of ordered event times (representing tephra-layer occurrence and by assumption notable eruptions in the region) versus time, produced by the authors (Fig. 6.7) shows that over the past 80,000 years, tephra layer-producing events occurred – on average – every 1.05 ka. Three anomalous periods are also identified, however, from 55–61, 34–38 and 8–15 ka BP, within which the median repose periods (time to next tephra-producing event, or effectively the time between eruptions) are much smaller, respectively 0.80, 0.45 and 0.35 ka. In addition, a period centred on 22 ka BP is characterised by a much longer median repose period of 1.40 ka. These anomalous periods of tephra production are suggested by the authors to reflect episodes of either increased or reduced levels of explosive volcanic activity in the region, which are in turn linked – respectively – to periods of rapid sea-level change or sea-level stability. McGuire and

FIGURE 6.7 A cumulative plot of ordered event times (representing tephra-layer occurrence and, by assumption, notable eruptions in the region) versus time reveals systematic variations in the timing of eruptions, which correlate with changes in the rate of sea-level change during the late Quaternary. Note the anomalous periods within which the time to next tephra-producing event (median repose period) is noticeably reduced. (After McGuire *et al.*, 1997)

his colleagues also point out that although the rate of sea-level change appears to be modulating the record of explosive volcanism at late-Quaternary Mediterranean volcanoes, there is no evidence for an increase in the overall output of the volcanoes involved.

Following further statistical treatment of the data (see McGuire *et al.*, 1997), the authors conclude that there is a clear correlation between a decrease in repose period (time between tephra-producing eruptions) and the rate of change of mean sea level, either upwards or downwards. In other words, there is a strong positive relationship between the absolute rate of change of late-Quaternary sea level and the number of explosive eruptions in the Mediterranean region as represented by tephra layers in the marine record. McGuire and his colleagues also report (SEAVOLC, 1995) that the maximum correlation occurs at lags of 1.5 and 6.5–7 ka, suggesting the existence of both a relatively rapid response of volcanic systems and a slower-acting effect. As reported in Chapter 5, a temporal correlation between increased periods of volcanism and sea-level change has also been identified, on the basis of the ages of volcanic sulphate signals in ice cores from Greenland (Zielinski *et al.*, 1997), suggesting that the relationship may be valid on a hemispherical or even global scale.

Determining the causative mechanism(s) whereby explosive volcanism can be triggered by changing sea level is problematical, given that there are numerous candidates. The volcanic response may be a global one occurring, as mentioned earlier, in response to hydro-isostatic crustal (e.g. Anderson, 1974; Rampino *et al.*, 1979; Nakada and Yokose, 1992) or global spin (Matthews, 1969) adjustments related to the redistribution of planetary water. McGuire *et al.* (1997) propose, on the other hand, that the existence of a single causal link between the rate of sea-level change and the level of explosive volcanic activity is unlikely. While not ruling out possible global effects, the authors suggest that a range of local or regional mechanisms may be responsible, resulting from the dynamic responses of coastal and island volcanoes to an array of stress-related influences. At the individual volcano level, these might include changing magma-confining pressures, water-table variations and edifice destabilisation (Fig. 6.8). Using finite-element modelling, McGuire *et al.* (1997) demonstrated that large changes in sea level have contrasting effects on the internal stress regimes of coastal and island volcanoes. In the former case (for example, Mount Etna in Sicily), hydrostatic loading due to a 100-m sea-level rise adjacent to the volcanic edifice reduces

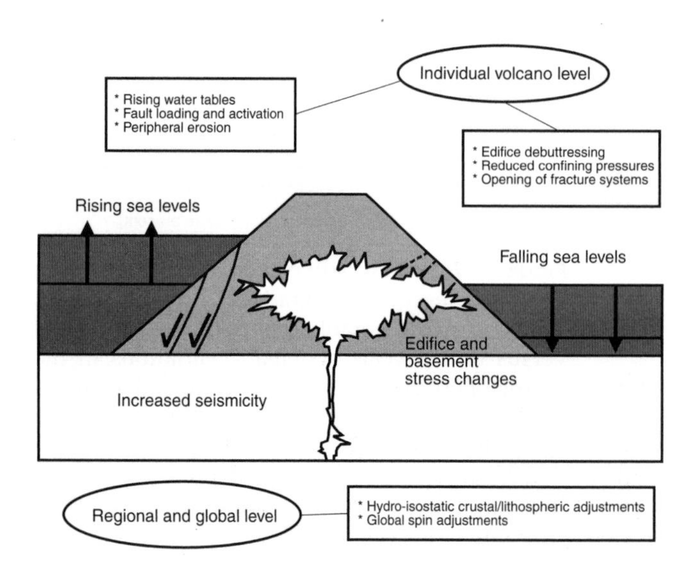

FIGURE 6.8 Changes in sea level/ocean volume may trigger volcanic eruptions in a number of ways. Some may act globally or regionally while others operate at an individual volcano level.

compressive stresses by around 0.1 MPa (1 bar), a pressure variation shown by Jaupart and Allégre (1991) to be sufficient, in magma chambers, to lead to a transition in eruption regime. Such a reduction in stress may be sufficient to trigger explosive eruptions at 'charged' volcanoes in which bodies of relatively differentiated magma reside at depths of 5 km or less. In contrast, the expulsion of magma at island volcanoes (such as Stromboli, Aeolian Islands) seems to be favoured by sea-level drawdown, which reduces radial compressive stresses by as much as 1 MPa (10 bar) (as also determined by Wallmann *et al.*, 1988), easily sufficient to promote the eruption of high-level stored magma. McGuire *et al.* (1997) also note that a 100-m sea-level change additionally results in an increase in shear stress at the land–sea interface, thereby increasing the potential for slope instability, lateral collapse, and consequent decompression-related explosive eruption (*à la* Mount St Helens, 1980). The collapse of coastal and island volcanoes constitutes a major hazard, particularly at times of significant environmental change, and the phenomenon is discussed in more detail in the following section.

To augment the impact of sea-level changes on the internal stress regimes of individual volcanoes, McGuire and his colleagues also advocate broader-scale influences, such as the slower-acting stresses that Nakada and Yokose (1992) have proposed result in increased tectonism and volcanism at island arcs. The latter believe that during periods of planetary cooling, when sea levels are falling, water unloading causes tensional stresses to operate at the base of the island arc lithosphere, allowing cracks to open that provide conduits for magma to rise into the lower part of the lithosphere (Fig. 6.9). Contrastingly, when sea levels start to rise at glacial terminations, water loading causes tensional conditions to switch to the upper half of the lithosphere, permitting the magma-filled fractures to feed eruptions.

On a global scale, the number of volcanoes that were susceptible to large, rapid, late-Quaternary sea-level changes is large. Current spatial distributions of active volcanoes show

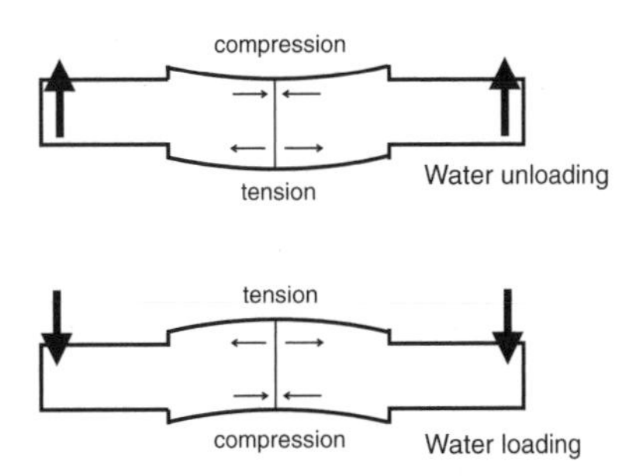

FIGURE 6.9 A schematic model illustrating the proposed relationship between changing ocean volumes and island-arc volcanism during the Quaternary. During glacial stages, water unloading creates tensional conditions at the base of the lithosphere, allowing magma to rise from below. Water loading caused by sea-level rise during interglacials creates tensional conditions in the upper lithosphere, permitting stored magma to reach the surface. (After Nakada and Yokose, 1992)

that 57 per cent form islands or occupy coastal sites while a further 38 per cent are located less than 250 km from a coastline. Assuming a similar distribution for around 1500 volcanoes known to have been active during the Holocene, then over 1400 volcanoes are likely to have been susceptible to the more direct effects of rapid sea-level change during the latter part of the Quaternary. Given current forecasts of sea-level rise in the next 100 years, it is fortunate that accompanying stress changes acting on coastal and island volcanoes are unlikely to be sufficient to trigger any noticeable increase in eruptive activity. There is, however, a possibility that the climatic changes that accompany global warming may initiate a volcanic response from marine volcanoes, and this new and fascinating hypothesis is addressed in the following section.

6.4 Changing sea levels, submarine landslides and collapsing ocean islands

It would perhaps be surprising if large and rapid changes in sea level did not have a destabilising influence on continental margins and ocean islands, and both sea-level falls and rises have been implicated in catastrophic submarine landslides and sediment failures (e.g. Weaver and Kuijpers, 1983; Prior *et al.*, 1986; Masson and Weaver, 1992; Roberts and Cramp, 1996; Maslin *et al.*, 1998; Rothwell *et al.*, 1998; Alibés *et al.*, 1999) and collapses of the flanks of ocean islands (e.g. McGuire *et al.*, 1997).

6.4.1 Submarine landslides

Continental margins are notoriously unstable environments that frequently undergo structural failure and collapse. Instability may arise as a result of a number of factors, including sea-level change and sediment accumulation, while structural failure and landslide formation could be promoted by changes in pore-water pressures, gas hydrate activity or earthquakes. In the last of these circumstances a recent submarine landslide triggered by an offshore earthquake contributed to the formation of a devastating tsunami that took around 3000 lives at Aitape on the north coast of Papua New Guinea in 1998 (Tappin *et al.*, 1999). Similarly, the submarine collapse of part of the Grand Banks (Newfoundland, Canada) due to a magnitude 7.2 earthquake in 1929 contributed to a tsunami that killed 51 people on the south coast of Newfoundland (Murty and Wigen, 1976). These events pale into insignificance, however, when compared with some of the gigantic landslides and sediment failures that occurred during the late Quaternary.

One of the largest, and probably the most recent, of these giant continental margin slides is located off western Norway. Although the structure had long appeared on maritime charts, the *Storegga Slide* began to attract attention in academic circles really only during the 1980s, following the publication of a number of papers by Tom Bugge and his colleagues (e.g. Bugge, 1983; Bugge *et al.*, 1987, 1988). These present statistics that provide a clear impression of the staggering scale of the slide. In fact, the 'slide' was formed in three separate events which, together, removed around 5600 km^3 of sediment from the Norwegian continental margin. In total the slide deposit covers an area of about 34,000 km^2, almost equal to the land area of mainland Scotland. The first of the mass movements is not temporally well constrained and might have occurred at any time between 30,000 and 50,000 years ago. The second and third movements occurred one after the other around 7200 years ago (Bondevik and Svendsen, 1995), only a few hundred years after one of the catastrophic sea-level rise events identified by Blanchon and Shaw (1995). Bugge *et al.* (1987) propose three possible trigger mechanisms for the Storegga Slide: rapid sedimentation, gas hydrate decomposition and earthquakes. The last two triggers at least can be convincingly linked, either directly or indirectly, to rising sea levels.

The accumulation of gases from the decomposition of organic detritus leads to the formation of *gas hydrates* (or *clathrates*) in marine sediments. These are solids, with an appearance rather similar to that of ice, whose physical state is highly sensitive to changes in either pressure or temperature. As little as a 1 degree Celsius warming may cause rapid dissociation of the solid into a gaseous state, thereby increasing sediment pore pressures, and a similar response accompanies a fall in pressure acting on the solid hydrate – as might be experienced at times of sea-level drawdown. Such changes in the physical state of clathrates contained in sediment can lead to destabilisation and eventual failure of the sediment pile. Driscoll *et al.* (2000) propose, for example, that gas hydrates – particularly in the upper several metres of sub-seafloor – could have been undergoing progressive melting (changing from the solid to the gaseous state) ever since the end of the last glaciation, in response to slow warming of bottom waters. At the same time, progressive loading of continental margins during post-glacial times could reasonably lead to fault activation and increased levels of seismicity. Either the increased pressures exerted by clathrate dissociation, or earthquakes, or a combination of the two, might have contributed to collapse of at least stages 2 and 3 of the Storegga Slide.

The primary hazard arising from such large submarine landslides involves the generation of destructive tsunami, which today have the capability of bringing devastation to densely populated coastlines thousands of kilometres from the source. Evidence for such tsunami, generated by the second-stage Storegga Slide collapse, has been recognised in Scotland by Alastair Dawson and colleagues (1988), who identified a prominent sand layer of comparable age in Holocene deposits along the coast of eastern Scotland. The authors estimate that the tsunami run-up was at least 4 m and propose that an earthquake of at least magnitude 7 would have been required to generate such a run-up at a distance from source of around 500 km. This figure is in close agreement with the results of numerical tsunami modelling undertaken by Harbitz (1992), which suggest that run-up heights of 3–5 m would have been experienced by exposed areas along the east coasts of Greenland, Iceland and Scotland, and the west coast of Norway.

Submarine landslide formation has also been linked to falling sea levels and low sea-level stands, when conditions are perhaps even more conducive to structural failure and collapse than during periods of sea-level rise. The deep-water Currituck landslide on the continental slope north-east of Cape Hatteras (eastern United States), for example, was initiated in the late Pleistocene between 48,000 and 16,000 years ago, and its formation is attributed to rapid sedimentation at the edge of the continental slope during a period of low sea level (Prior *et al.*, 1986). Similarly, the emplacement in the western Mediterranean Sea of a huge late-Pleistocene 'mega-turbidite' with a volume of around 500 km^3 has been shown to coincide closely with the last low sea-level stand (Rothwell *et al.*, 1998). The authors propose that sediment movement was triggered by the destabilisation of gas hydrate deposits as a consequence of the reduced pressures caused by sea-level drawdown (in contrast to gas hydrate disassociation due to higher bottom waters during periods of sea-level rise), but do not rule out seismic triggering resulting from water unloading effects. In the Atlantic too, sea-level falls are linked with the destabilisation and catastrophic mass movement of huge volumes of sediment, and Alibés *et al.* (1999) advocate global cooling and falling sea levels as one of the factors contributing to turbidite formation and emplacement in the Madeira Abyssal Plain (north-east Atlantic).

6.4.2 Collapsing volcanic islands: a sea-level change or climate change trigger?

Landslides from volcanic ocean islands (Keating and McGuire, 2000) are among the biggest catastrophic mass movements on the planet. Around 70 major landslides have been identified around the Hawaiian Island archipelago, the largest having volumes in excess of 5000 km^3 and lengths of over 200 km (e.g. Moore *et al.*, 1994). Such volcanic landslides are now proving to be widespread in the marine environment (Holcomb and Searle, 1991; McGuire, 1996) and have been identified around other island groups, such as the Canary (e.g. Weaver *et al.*, 1994; Watts and Masson, 1995) and Cape Verde (Jacobi and Hayes, 1982) islands, and around individual island volcanoes, including Stromboli (Kokelaar and Romagnoli, 1995), Piton des Neiges and Piton de la Fournaise (Réunion Island) (e.g. Labazuy, 1996), Tristan de Cunha (Holcomb and Searle, 1991), the Galápagos Islands (Chadwick *et al.*, 1992), Augustine Island (Alaska) (Begét and Kienle, 1992) and Ritter Island (Papua New Guinea) (Johnson, 1987).

Serious attention became focused on the unstable nature of volcanic edifices, and their tendency to experience structural failure, following the spectacular landslide that triggered the climactic eruption of Mount St Helens during May 1980 (Lipman and Mullineaux, 1981). Such behaviour is now recognised as ubiquitous, and evidence for collapsing volcanoes has been recognised both within the geological record and at many of the world's currently active volcanoes (e.g. Ui, 1983; Siebert, 1984). Francis (1994) observed that 75 per cent of large volcanic cones in the Andes had experienced collapse during their lifetimes, while Inokuchi (1988) reported that over a hundred landslides had been identified around

Quaternary volcanoes in Japan. Although Siebert (1992) estimated that structural failure of volcanic edifices had occurred roughly four times a century over the past 500 years, this may be an underestimate. Belousov (1994), for example, points out that there were three major collapses in the last century occurring in the Kurile–Kamchatka region of Russia alone.

Active volcanoes are dynamically evolving structures, the growth and development of which are typically punctuated by episodes of edifice instability, structural failure and, ultimately, collapse (McGuire, 1996). Growing volcanoes may become unstable and experience collapse at any scale, ranging from minor rockfalls with volumes of the order of a few hundred to a few thousand cubic kilometres to the giant 'Hawaiian-type' megaslides involving in excess of 1000 km^3. Low-volume collapses occur at some volcano or another on an almost daily basis, whereas the largest events have frequencies of tens to hundreds of thousands of years. The causes of volcano instability are manifold, and some volcanoes clearly have a greater propensity to become destabilised than others. Broadly speaking, small *monogenetic* (single-eruption) cones and shields are rarely characterised by anything greater than minor slumping and sliding. Structural instability on a massive scale is limited to major, *polygenetic* volcanoes that have grown over long periods of time (ranging from tens of thousands to millions of years) as a result of countless eruptions. Despite low slope angles and an essentially homogeneous structure, instability development is common at large basaltic volcanoes, where persistent dyke-related rifting is implicated in large-scale failure of the flanks. In the marine environment, instability at large basaltic volcanoes may be increased by edifice spreading along weak horizons of oceanic sediment (e.g. Nakamura, 1980). Strato-volcanoes composed of a mixture of lavas and pyroclastic materials are also easily destabilised, partly owing to their unsound mechanical structure and partly owing to their characteristic steep slopes and the high precipitation rates that often accompany their elevation. The development of instability and the potential for failure are enhanced at all types of volcano by the fact that actively growing edifices experience continuous changes in morphology, with the *endogenetic* (by intrusion) and *exogenetic* (by extrusion) addition of material often leading to oversteepening and overloading at the surface. Once a volcanic edifice has become sufficiently destabilised, structural failure and collapse may be induced by a number of triggers. These include earthquakes, mechanical stress or the pore-water pressurisation resulting from the emplacement of fresh magma, or environmental factors such as changes in sea level or variations in the prevailing climate.

At volcanic oceanic islands and coastal volcanoes, large, rapid changes in sea level are likely to play a significant role in contributing to edifice destabilisation and consequent collapse. In fact, while linking sea-level change and the incidence of explosive volcanism in the Mediterranean, McGuire *et al.* (1997) proposed that some at least of the eruptions might be triggered by structural failure and collapse. The seaward-facing flank(s) of any volcano located at the land–sea interface is inevitably the least buttressed. This applies both to coastal volcanoes such as Mount Etna (Sicily), where the topography becomes increasingly elevated inland, and to island volcanoes such as Hawaii, where younger centres (such as Kilauea) are buttressed on the landward side by older edifices (such as Mauna Loa). This morphological asymmetry leads to a preferential release of accumulated intra-edifice stresses due, for example, to surface overloading or to persistent dyke emplacement, in a seaward direction. Stress release may take the form of co-seismic downfaulting towards the sea, the slow displacement of large sectors of the edifice in the form of giant slumps, or the episodic generation of catastrophic landslides, or a combination of all three. Inevitably, the relatively unstable nature of the seaward-facing flanks of any marine volcano is further enforced by the dynamic nature of the land–sea contact. As mentioned earlier in this chapter and in Chapter 5, McGuire *et al.* (1997) demonstrated that large sea-level changes are implicated in significant internal stress variations within coastal and island volcanoes, which may contribute towards eruption, collapse or both.

More directly, peripheral erosion associated with rapid sea-level rise, and the removal of lateral buttressing forces due to a large sea-level fall might also be expected to promote collapse of the flanks of island and coastal volcanoes (McGuire, 1996).

One of the problems with attempting to link collapses at marine volcanoes with changes in sea level lies in the fact that connection might not be directly causal in nature, and both the collapsing volcanoes and the sea-level change may have a common cause. This is an avenue that has recently been explored by Simon Day and others (Day *et al.*, 1999; 2000), who propose that the common cause might be climate change. They point to the fact that most mid- to low-latitude volcanic ocean islands experienced arid conditions during glacial periods and propose that contemporary water tables are therefore likely to be depressed. This, the authors suggest, would minimise opportunities for new magma to induce pore-fluid pressurisation – a dominant mechanism for volcano destabilisation during the rift-zone eruptions common at ocean island volcanoes such as those of the Hawaiian, Cape Verde and Canary Island archipelagos (e.g. Elsworth and Voight, 1995; Elsworth and Day, 1999). At the same time, the penetration of salt water into the cores of the volcanoes might promote alteration and weaken the structure. During interglacials, in contrast, Day and his co-workers note that – for low latitudes at least – oceanic climates are humid, and ocean island volcanoes at such times are likely to have elevated water tables that are capable of being pressurised by magma intrusion, thereby increasing opportunities for structural failure. The corollary of this, as the authors propose, is that such collapses are likely to occur during the warming following glacial terminations, when the destabilising effects of pore-pressure increases are greatest and reinforce the effects of earlier coastal erosion and volcanic core alteration.

More recently, Day and co-workers (2000) have updated their model, which now advocates a correlation between the timing of prehistoric giant lateral collapses on oceanic islands and the precession-forced sea surface temperature (SST) (Fig. 6.10). Day and his colleagues note that as sea levels rise following glacial terminations, so does the low-latitude SST. This sea surface warming is in turn accompanied by changes in the pattern and characteristics of the trade winds so that they bring increased humidity to low-latitude volcanic islands and increased precipitation on their mid-flanks and summit regions. This, they propose, leads to a rise in the water table of the order of several hundred metres and increased opportunity for collapse as a result of magma intrusion-related groundwater pressurisation in the core of the volcano (Fig. 6.11). Day and his co-workers also point out that, at least over

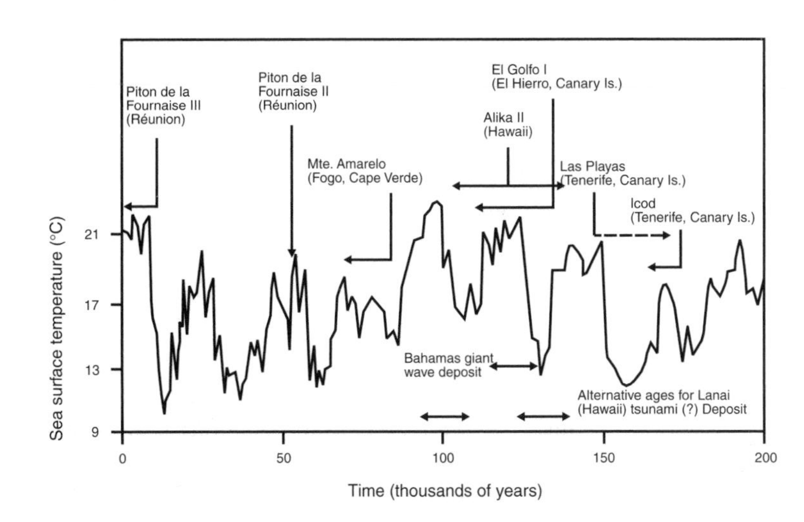

FIGURE 6.10 Ages of oceanic island collapses over the past 200,000 years plotted against the global sea-level curve and sea surface temperature change. (After S.J. Day, personal communication)

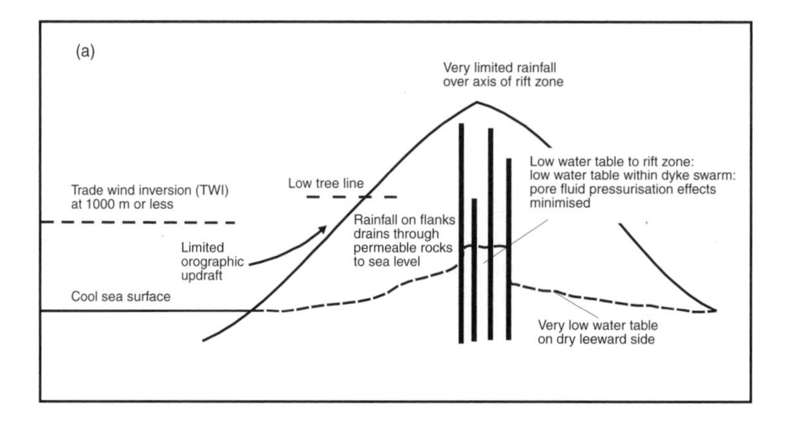

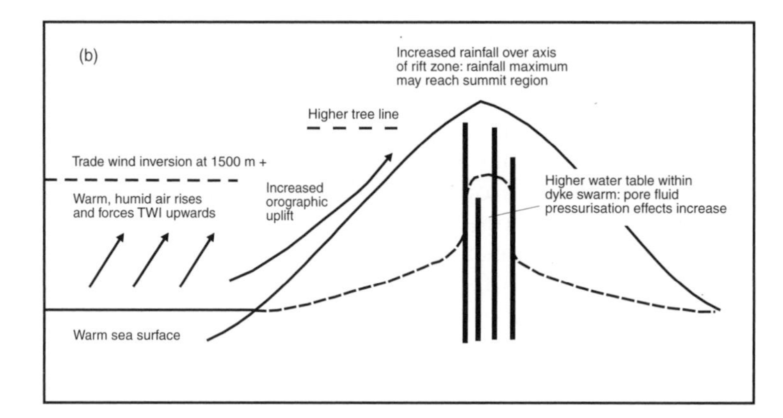

FIGURE 6.11 Increased precipitation on the mid-flanks and summit regions of low-latitude volcanic islands, accompanied by a large rise in water tables, has been advocated as contributing towards a clustering of ocean island collapses during periods when the climate is warm. (After S.J. Day, personal communication)

the past 200,000 years, giant volcanic ocean island collapses appear to be clustered, with the clusters having periodicities that reflect the *c.* 20-ka Milankovitch precessional forcing of sea surface temperature maxima at low latitudes (Ruddiman and McIntyre, 1984; Berger and Loutre, 1991). In proposing that the ocean volcano collapse hazard is greatest during warm periods such as the present, they also tentatively suggest that global warming might further exacerbate the situation.

6.4.3 Tsunami generation from ocean island collapses

From a hazard viewpoint, the major threat from large collapses at oceanic islands arises from the generation of giant tsunami, for which there is increasing evidence in the geological and geomorphological records. Around 5 per cent of all tsunami are related to volcanic activity, and at least a fifth of these result from volcanic landslides entering the ocean (Smith and Shepherd, 1996). The tsunami-producing potential of a large body of debris entering the sea is much greater than that of a similar-sized submarine landslide, and even small sub-aerial volcanic landslides can generate highly destructive waves if they enter a large body of water. In 1792 at Mount Unzen (Japan), for example, a landslide with a volume of only a third of a cubic kilometre – which was not connected with an eruption – entered Ariake Bay and triggered a series of tsunami that caused 14,500 deaths. More recently, many deaths are thought to have been caused by the collapse of the Ritter Island volcano (Papua New Guinea) in 1888, which generated tsunami with wave run-up heights of 12–15 m (Johnson, 1987).

Tsunami associated with giant collapses at oceanic islands can, however, be an order of magnitude greater. For example, a wave train associated with collapse of part of Mauna Loa (Hawaii) – the so-called Alika Slide – around 105,000 years ago has been implicated in the deposition of coral and other debris at an altitude of 326 m above current sea level on the neighbouring island of Lanai. The giant waves generated by collapses in the Hawaiian Islands appear to have been of Pacific-wide extent, and Young and Bryant (1992) explain signs of catastrophic wave erosion up to 15 m above current sea level along the New South Wales coast of Australia – 14,000 km distant – in terms of impact by tsunami associated with the Mauna Loa collapse. Although such run-up heights testify to unprecedented tsunami that would be completely devastating should they occur today, some remain unconvinced. In particular, Jones (1992) disputes the Hawaiian source of the Lanai wave event, citing problems with wave attenuation and evidence of uplift of the Hawaiian Islands to explain the Lanai deposits. Furthermore, Jones and Mader (1996) claim to have demonstrated that the Alika Slide could not have produced the catastrophic erosion observed on the south-east coast of Australia, preferring instead to invoke an asteroid impact as the most likely cause.

Nevertheless, potential giant-tsunami deposits are being identified at increasing numbers of locations. On Gran Canaria and Fuerteventura in the Canary Islands, for example, deposits consisting of rounded cobbles and marine shell debris have been recognised at elevations of up to 100 m above present sea level (Simon Day, personal communication) and may have been emplaced by waves associated with ancient collapses in the archipelago. Similarly, large coral boulders weighing up to 2000 tonnes on the Rangiroa reef (French Polynesia) have been linked by Talandier and Bourrouilh-le-Jan (1988) with giant tsunami formed by the early-nineteenth-century collapse of the Fatu Hiva volcano (Marquesas Islands, south-east Pacific) (Filmer *et al.*, 1994).

Most spectacularly of all, boulders and geomorphological features in the Bahamas may provide evidence of the impact of giant tsunami from a major collapse event at the Canary Island of El Hierro that occurred around 120,000 years ago (Simon Day, personal communication). Seven limestone boulders measuring up to 1000 m^3 are scattered along the north-eastern coast of Eleuthera Island, some 20 m above current sea level (Hearty, 1997). These are accompanied by landward-pointing V-shaped sand ridges – *chevron ridges* – several kilometres long, which are found at several locations along the east-facing coasts of the Bahamas (Hearty *et al.*, 1998) with other deposits and erosional features interpreted as resulting from giant waves with run-up heights of up to 40 m (Fig. 6.12). Hearty (1997) is uncertain as to whether or not the giant boulders were emplaced by tsunami or unprecedented late-Quaternary storms. In his later, co-authored paper (Hearty *et al.*, 1998),

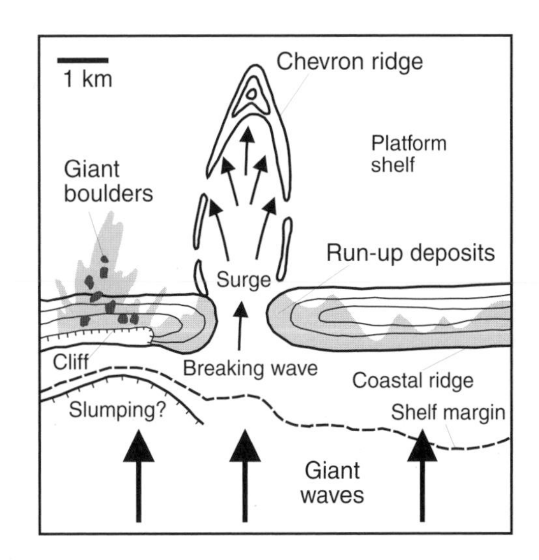

FIGURE 6.12 Schematic of giant boulders, run-up deposits and a chevron ridge emplaced along the eastern coasts of the Bahamas by giant waves around 120,000 years ago. (After Hearty *et al.*, 1998)

however, he comes down on the side of greatly amplified storms. From a future hazard point of view this is an equally worrying conclusion, as Hearty and his colleagues implicate abrupt climate change at the end of oxygen isotope stage 5e in the amplification of North Atlantic storms. They also hint that perhaps global warming might lead to an abrupt and destabilising climate change similar to that which brought into existence such unprecedented storms around 120,000 years ago. Hearty *et al.* (1998) also point out, however, that the giant waves appear to have been generated as sea level fell rapidly from a high stand at the end of isotope substage 5e, which would fit nicely with the ideas of McGuire (1996) and McGuire *et al.* (1997). Furthermore, the age of the Bahamian deposits appears to be close to those proposed for the El Golfo collapse on the Canary Island of El Hierro (S.J. Day, personal communication), which lies on a great circle route with respect to the Bahamas and which therefore fits the bill of a north-eastern source for the Bahamas deposits.

7

Asteroid and comet impacts as initiators of environmental change

7.0 Chapter summary

In this chapter we examine the threat to the Earth from asteroids and comets and assess the role of impact events in causing environmental change. Impact studies represent one of the most dynamic of all research areas, and the mass of papers, articles and books published on the Cretaceous/Tertiary (K/T) impact event, for example, is probably only marginally less than the mass of the impactor itself. Similarly, research into the contemporary threat from near-Earth objects is rapidly moving, and even during the writing of this chapter, estimates of the numbers of large, Earth-threatening asteroids have been at first halved and then raised again. Following a brief scene-setting and introduction to the history of impact studies, the threat from asteroids and comets is addressed through consideration both of their orbital populations and of the terrestrial cratering flux. Environmental effects of impacts are related to the energy of collision and are described on scales ranging from local to global. Short-term effects, including blast, re-entering ejecta, wildfires and tsunami, and longer-term effects, such as impact winter and post-impact warming, are discussed and evaluated, drawing both on direct evidence from known impacts and from the results of modelling studies. The chapter concludes with an examination of the role of impact events in mass extinctions and a brief introduction to competing theories.

7.1 Setting the scene: the new catastrophism

Before Charles Darwin proposed his revolutionary model to explain the diversity of life on the Earth, the general consensus, both scientific and populist, held that the planet and all life upon it had been created – exactly as it appeared – near-instantaneously and at the whim of a deity. Any subsequent changes, such as Noah's flood, were interpreted as sudden, devastating events sent by the aforementioned deity to chastise humankind for one misdemeanour or another. In short, *catastrophism* was very much the order of the day. Slowly but surely, the work and writings of Charles Darwin (e.g. Darwin, 1859), and his Scottish geologist contemporary Charles Lyell, contributed to the birth of a new scientific paradigm, tagged *uniformitarianism*, which demeaned the importance of sudden and drastic change to life and the environment and embraced the merits of gradual, incremental change (Lyell, 1830–33). Central to this principle was the idea that change was accomplished by the very same processes – both biological and geological – that operate today; simply explained by Lyell's less celebrated predecessor, the eighteenth-century Scottish geologist James Hutton, in terms of 'the present is the key to the past' (Hutton, 1788). In the past few decades, however, the pendulum has started to swing back – at least some of the way – towards catastrophism, as recognition of the existence of infrequent but globally devastating natural hazards has revealed that the present is not necessarily always the key to the past. A new working model for the Earth as a combined geological and biological system now envisages a *steady state* situation, within which natural processes operate slowly and incrementally, but which is periodically interrupted by catastrophes that dramatically change the environment of the planet. This concept of *punctuated equilibrium* is much better suited to explaining a world that we now know is periodically subjected to geological upheavals of sufficient size to cause dramatic and rapid environmental change on a global scale and perhaps cause, or at least contribute towards, mass extinctions. Within this new paradigm, both volcanic 'super-eruptions' (see Chapter 5) and asteroid and comet impacts on the Earth adopt important pivotal roles.

7.2 Recognising the impact threat

As at November 2000, 278 potentially hazardous asteroids (PHAs) had been identified in orbits close enough to threaten collision with the Earth at some time in the future (NASA NEO Program: http://neo.jpl.nasa.gov/neo/pha.html), and this number is increasing almost weekly. The Earth is constantly being bombarded by debris from space, most of which burns up in the atmosphere. Once or twice a century, however, objects in the 50-m size range penetrate the atmospheric shield and either impact on the surface or explode sufficiently violently at low enough altitudes to cause severe damage. In the twentieth century four such events were recorded: in Siberia in 1908 (Tunguska) and 1947 (the so-called Sikhote-Alin meteorite fall), in Brazil (1931) and in Greenland (1997). Although capable of destroying a major city, such objects are not large enough to have a significant effect on the Earth's environment. Of much greater concern, in this respect, are the rare collisions with objects 1 km across and above. This diameter is a broadly favoured threshold (Chapman and Morrison, 1994) above which an impactor has the potential to trigger major changes to the global environment, primarily through the release of large quantities of debris into the atmosphere. These changes in turn are capable of causing a dramatic reduction in the level of solar radiation reaching the surface (e.g. Pope *et al.*, 1994; Toon *et al.*, 1994) and the consequent initiation of a so-called *impact winter* (Turco *et al.*, 1991; Pope *et al.*, 1994). Collisions with still larger objects, of the order of 5 km and above, are incriminated in mass extinction events (e.g. Rampino *et al.*, 1997) and charged with instigating dramatic environmental changes lasting as long as half a million years.

It may seem surprising in retrospect, particularly given the accretionary origin of the Earth and other planetary bodies, but only since the 1960s has impact cratering become widely accepted as a normal 'geological' process. Not until after World War II did it slowly dawn on the scientific establishment that the countless lunar craters were the result of collisions with asteroids, comets and smaller fragments of space debris, and even in the late 1970s some still argued for a volcanic origin. Similarly, lunar crater-like structures on Earth were interpreted as having been formed by volcanic activity or other unknown geological process, even though the physicist Benjamin Tilghman wrote as early as 1905 of the Barringer (Meteor) Crater (Arizona): 'the formation at this locality is due to the impact of a meteor of enormous and hitherto unprecendented size'. Although Tilghman and his mining engineer colleague D.M. Barringer drilled into the crater for a quarter of a century, hoping to find and exploit traces of a metallic impacting body, they were unsuccessful. This, as is now known, was the result of the wholesale disintegration and vaporisation of the object as the kinetic energy associated with a body travelling with a velocity on the order of 20 km/s was transformed instantaneously into heat.

Despite the failure of Tilghman and Barringer to find evidence for an extra-terrestrial origin for the Barringer Crater, this is now accepted, along with the idea that the Earth periodically suffers collisions with other bodies in space. Until some 20 years ago, however, little consideration was given to the possible consequences of such impacts in terms of environmental change and the threat to life on Earth. Then in 1980, Alvarez *et al.* (1980) and Smit and Hertogen (1980) proposed, in groundbreaking papers, that the Cretaceous/Tertiary (K/T) mass extinction that wiped out perhaps up to a half of all existing species was the result of a major impact event at the K/T boundary. Their evidence was based upon the existence of an anomalous layer of iridium in sediments crossing the boundary at a location in Italy, which – they proposed – could only have an extraterrestrial origin. The contentious work of Alvarez and his colleagues triggered a search for alternative mechanisms to explain the source of the iridium anomaly without recourse to an extraterrestrial cause, and continental flood basalt (CFB) volcanism was championed by some (e.g. Officer and Drake, 1985; Chapter 5). The widespread occurrence of impact-melted ejecta in the form of crystalline spherules (e.g. Smit and Klaver, 1981; Kyte and Smit, 1986) and glassy tektites (Izett, 1991; Sigurdsson *et al.*, 1991; Smit *et al.*, 1992), 'shocked' quartz and other minerals (Bohor *et al.*, 1984; Bohor, 1990) and soot (Wolbach *et al.*, 1988), in association with iridium in K/T boundary clay sediments, led in the decade following the Alvarez paper, however, to general acceptance that a major impact event occurred at this time. A search for the 'smoking gun' followed, and by the early 1990s attention had focused on the Chicxulub Basin in the Yucatán platform (Gulf of Mexico) (Fig. 7.1; Box 7.1), for which evidence that this was indeed the site of the K/T impact is now overwhelming (e.g. Sharpton and Marin, 1997).

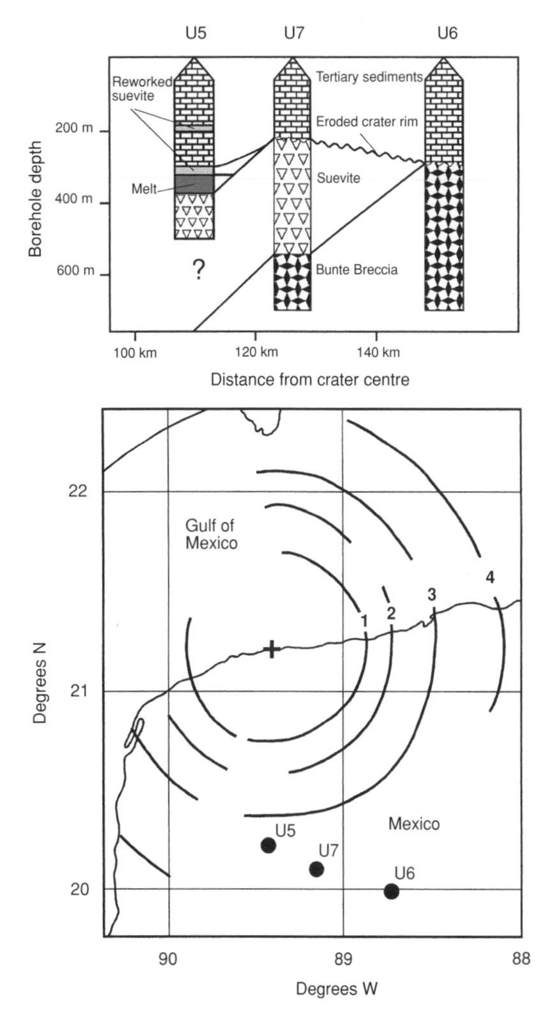

FIGURE 7.1 The Chicxulub multi-ring basin in cross section (*top*) and plan view (*bottom*). The stratigraphy is dominated by impact-related lithologies including shocked and melted basement rocks, breccias and suevites (*see text for details*). Plan view shows ring locations determined from gravity data, and borehole locations. (Adapted from Sharpton and Marín, 1997)

A second event in the early 1990s contributed to an increased awareness of the impact threat as a contemporary issue. On 16 July 1994 the fragments of the disrupted Comet Shoemaker–Levy 9 collided with planet Jupiter (Levy, 1998), providing the world – via the *Galileo* spacecraft (then approaching Jupiter) and the Internet – with spectacular images of the event (Fig. 7.2). Over a seven-day period, 21 separate major fragments of the comet collided with the planet, the largest approaching or exceeding 1 km across. The total energy equivalent of the collisions is estimated by Crawford (1997) to have approached 300 gigatons of TNT. Inevitably, immediately following the event there was much speculation about what effect such a series of impacts would have on the Earth and life thereon (e.g. Boslough and Crawford, 1997).

Box 7.1 The Chicxulub Basin: the anatomy of a major impact site

The impact that formed the Chicxulub Basin occurred 65 million years ago on the Yucatán platform, which at the time was a shallow carbonate-depositing sea. The resulting crater was buried under post-impact sediments and now lies beneath a 1-km-thick sequence of Caenozoic limestones and marls. The structure first came under scrutiny in the 1950s when Petróleos Mexicanos (Pemex) sank test wells to examine its petroleum-generating potential. Unexpectedly, the wells revealed no sign of hydrocarbons, but instead a sequence of unusual breccias and silicate crystalline rocks. Almost 30 years later, Penfield and Camargo (1981) proposed, on the basis of aeromagnetic and gravity data, an impact origin for what they described as 'a major igneous zone' in the central Yucatán platform. Following a hiatus of another 10 years, analysis in the early 1990s of samples from the original drilling campaigns revealed diagnostic evidence of shock metamorphism (Hildebrand *et al.*, 1991; Sharpton *et al.*, 1992), demonstrating beyond reasonable doubt that the crystalline igneous complex had an impact rather than a volcanic origin.

Confirmation of a link with the iridium anomaly at the K/T boundary followed soon after, based upon a number of lines of evidence. These included a 65-Ma Chicxulub ^{40}Ar–^{39}Ar date coincident with the age of the boundary (e.g. Swisher *et al.*, 1992), and the compositional similarity of melt rocks (in terms of major and trace element and isotope properties) within the Chicxulub structure with impact glass spherules recognised at K/T boundary sites in Haiti and north-eastern Mexico and shocked lithic fragments found at boundary sites around the world (e.g. Bohor, 1990; Izett, 1991). Shallow drilling undertaken in 1994 generated nearly 3000 m of cores from seven wells and provided more detailed information about the size, shape and stratigraphy (Fig. 7.1) of the Chicxulub structure (Sharpton and Marín, 1997). A well 112 km from the centre encountered the upper boundary of the impact sequence at a depth of 330 m. The sequence itself comprises a 30-m-thick altered melt-rock formation containing cobble-sized clasts of shocked and partially melted basement rocks. This rests conformably on *'crater suevite'*, a polymict breccia containing clasts of impact-generated glass. Further out, at 125 km from the centre, no impact melt-rock was encountered, and post-crater Tertiary sediments were observed to lie unconformably on top of the suevite deposit. At 150 km from the centre, post-Tertiary sediments lie unconformably on an anhydrite–dolomite breccia (the Bunte Breccia), similar to that found in association with the Ries impact crater in Germany. The breccia represents a mixture of primary ejecta expelled from the transient crater (the ephemeral, bowl-shaped depression produced by excavation and downward displacement of the target on impact) and local materials incorporated into the ejecta during emplacement. In terms of crater size, the borehole stratigraphy is in agreement with gravity data that locate the crater rim at a distance of 150 km from the centre of the structure (Sharpton *et al.*, 1996). Both the gravity survey and seismic reflection data (Camargo and Suárez, 1994) support the existence of a 1-km-deep central basin 130 km across that marks the transient crater. Between this central basin and the outer crater rim are two concentric gravity highs that are interpreted by Sharpton and Marín (1997) as the buried terrestrial equivalents of the circular topographic highs that characterise multi-ring impact basins on the Moon and other rocky bodies in the solar system.

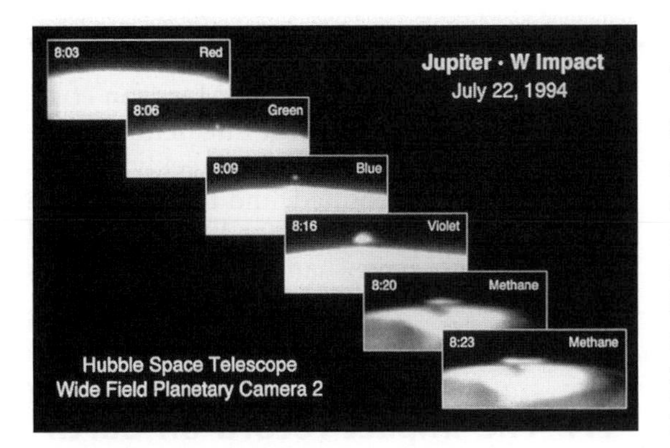

FIGURE 7.2 Impact of Comet Shoemaker–Levy fragment W on Jupiter. The series of images cycles through several filters while observing the sequence. The impact plume is clearly visible, first rising, then 'pancaking' back down into Jupiter's cloud tops. The last two methane images show bright features in Jupiter's cloud deck, which mark the remnants of the fragment K impact. (Courtesy, NASA Hubble Space Telescope Comet Team)

There is no doubt that the collision of Comet Shoemaker–Levy 9 with Jupiter, together with the widespread recognition that a large impactor was at least implicated in the K/T mass extinction, has led directly to increased interest in the impact threat to human society and the environment in general. This is reflected in the media by countless documentaries and two major feature films on the theme, and in scientific and political circles by a number of initiatives. These include the launch of an international programme (Spaceguard) to locate and map the orbits of near-Earth asteroids, the establishment of a NASA Near Earth Asteroid Tracking (NEAT) Program and a European Near Earth Object (NEO) Search Project, and the formation of a UK Government Task Force to evaluate the impact threat (UK Task Force on Potentially Hazardous Near Earth Objects, 2000). (Go to http://www.nearearthobjects.co.uk for a copy of the Task Force report.)

7.3 Asteroids and comets

Comets and asteroids represent debris left over after the accretion of the planets and their attendant moons, although some of the latter are themselves likely to be captured asteroids or comets. The majority of asteroids circle the Sun in the *Asteroid Belt*, a region between the orbits of Mars and Jupiter between two and four times further from the Sun than the Earth. Many thousands of asteroids and some comets have, however, had their orbits perturbed – usually by the strong gravitational field of Jupiter – and have entered orbits closer to the Sun that cross or at least approach the Earth's orbit. Asteroids are solid, rocky bodies of generally low volatile content that range in size from large *meteoroids* 5–10 m across up to bodies such as Ceres, which has a diameter of around 1000 km. To date, tens of thousands of asteroids have been observed, but detailed orbital parameters have been determined for fewer than 7000. In contrast to asteroids, comets generally have lower densities and are relatively volatile rich, with perhaps between 50 and 80 per cent of their mass consisting of water ice and other frozen gases such as carbon dioxide, methane and ammonia, with the rest being rocky material. Owing to the partial evaporation of the volatile component as it enters the inner solar system, resulting in the development of a 'tail' that may be tens of millions of kilometres long, a comet is relatively easy to recognise. Comets increase in brightness as they approach the Sun, owing to the accelerated vaporisation of frozen gases. This contributes to an envelope or *coma* of vapour around the *nucleus* that makes it difficult to study. It is generally accepted that most observed comets have nuclei around 1 km across, although observations of Halley and Swift–Tuttle revealed discrete bodies over ten times this size.

Three groups of objects threaten the Earth (Fig. 7.3): *near-Earth asteroids* (NEAs) and short-period comets, which together constitute *near-Earth objects* (NEOs), and long-period comets (Morrison *et al.*, 1994). NEOs are confined to the inner solar system and are relatively slow-

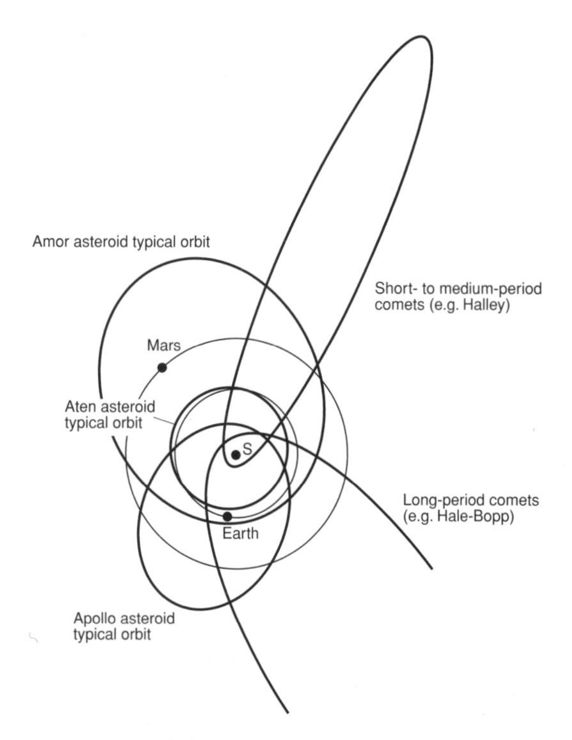

FIGURE 7.3 Orbital characteristics of Earth-threatening objects (not to scale).

moving with respect to the Earth. Consequently they have impact velocities that are typically in the range 15–22 km/s. In contrast, long-period comets, which have orbits that can take them perhaps a third of the distance (>1 light year) to the nearest star, travel much more rapidly when they enter the inner solar system and are capable of striking Earth at speeds of up to 55 km/s. Much remains unknown about properties such as the structure and strength of Earth-threatening objects, which are critical to the manner in which they will interact with the Earth's atmospheric shield. This is really only relevant to objects up to a few hundred metres across, however. Kilometre-scale objects capable of global environmental impact carry sufficient mass and kinetic energy to ensure that the atmosphere has little effect during their passage through it. The devastating effects of a large impactor are primarily a reflection of the kinetic energy of the object, which is – broadly speaking – almost two orders of magnitude greater than an equivalent mass of TNT. Energy release predicted for impactors of various sizes is typically expressed in megatons (Mt) of TNT, where 1 Mt = 4.2×10^{15} joules (Morrison *et al.*, 1994). The inventory of Earth-threatening space debris is changing on an almost weekly basis as more objects are identified, and the statistics in Box 7.2 – although correct for late 2000 – will almost certainly have changed by the time this book is published.

7.4 The impact record on Earth

The dynamic nature of the Earth provides conditions that are far from ideal for the preservation of impact structures. Seventy per cent of the planet's surface is covered with water, and perhaps 50 per cent of the land surface has been altered sufficiently over geological time to eradicate evidence of all but the very largest collisions. This leaves only 15 per cent of the Earth's surface suitable for preserving the evidence of most past impacts. Around 165 impact sites have now been identified, and they range in age from 2.4 billion (Suavjavi Crater, Russia) to 50,000 years (Barringer Crater, Arizona, USA) (Fig. 7.4; Table 7.1). The youngest crater to be formed by an impactor with the potential to cause global environmental change is 900,000 years old and

FIGURE 7.4 Spatial distribution of confirmed terrestrial impact structures. Note the apparent concentrations in Scandinavia, Australia and North America. (Modified from Grieve, 1998)

Box 7.2 Near-Earth asteroids and potentially Earth-threatening comets

Prior to the year 2000, the number of large (≥1 km) near-Earth asteroids, was estimated at between 1000 and 2000. This was revised downwards substantially in January 2000 (Rabinowitz *et al.* 2000) to 700 ± 230, using the results of the NASA Near Earth Asteroid Tracking (NEAT) Program, which has located and determined the orbital parameters for 322 large NEAs. Numbers of smaller objects, capable of local to hemispherical damage, are less well constrained, but best estimates from Rabinowitz and his co-authors predict at least 100,000 100-m NEAs and over 20 million objects in the 10-m size range. In October 2000, however, the downward revision of large NEAs was challenged in a paper given at a meeting of the American Astronomical Society by Stuart Scott (a researcher on the MIT NEA research project – LINEAR), who believes that the real number of large NEAs is closer to 1100.

Near-Earth asteroids are defined as objects having a perihelion distance (*q*; closest distance from the Sun) of less than 1.3 AU (astronomical unit = the mean orbital distance of the Earth from the Sun = ~150,000,000 km). Three groups of NEA are recognised, based upon their orbital parameters:

1. *Atens*. Earth-crossing NEAs named after asteroid *2062 Aten* that have orbital semi-major axes smaller than the Earth's. These objects have orbits that are generally closer to the Sun than the Earth's and orbital periods that are less than 1 year.
2. *Apollos*. Earth-crossing NEAs named after asteroid *1862 Apollo* that have semi-major axes larger than the Earth's. Objects in the Apollo group have orbits that are generally further from the Sun than the Earth and orbital periods that are greater than 1 year.
3. *Amors*. Earth-approaching NEAs named after *1221 Amor* that have orbits beyond the Earth's but within that of Mars.

Of the NEA population, 278 (as of November 2000) are catalogued as *potentially hazardous*. This definition is quite specific, and based upon an asteroid's potential to make threateningly close approaches to the Earth. All PHAs have an Earth minimum orbit intersection distance (MOID; effectively the estimated closest approach) of 0.05 AUs or less, and an absolute V-magnitude (*H*; a measure of brightness assuming an albedo of 13 per cent and therefore an estimate of size) of 22.0 or less. In other words, to be catalogued as a PHA, an asteroid must be larger than 150 m in diameter and have the potential to approach to or within ~7,480,000 km. Objects smaller than the cut-off size have the potential to be locally or regionally destructive, but not to have major or lasting impact on the global environment. Furthermore, any detection programme focusing on objects in the 50–150 m size range, the number of which may be up to a million, would prove both difficult to undertake and expensive. The largest NEA whose orbit crosses that of the Earth is *1627 Ivar*, which has a diameter of around 8 km and an estimated (Morrison *et al.*, 1994) mass of about 10^{15} kg. Larger objects, such as *1036 Ganymed* – around a hundred times more massive than Ivar – have orbits that approach the Earth but which do not come sufficiently close for them to be defined as PHAs. Of the ten closest approaches by NEAs within the next 40 years, it is interesting to note that all bar one of these objects was discovered during the past three years, raising the question of how the 'top ten' might look in another three years. Clearly, there are likely to be undiscovered NEAs with orbits that will bring them closer to the Earth in the next half-century than those currently catalogued. In fact, initiatives such as the NASA NEAT Program have increased the rate of discovery of new NEAs enormously in recent years, with 47 new objects being logged in 1999 and a further 39 so far in 2000. Of these, 2000SG344, an object in the 100 m size range, looks as if it will just miss the Earth in September 2030, when it is predicted to approach to 0.003 AU, offering perhaps a 1 in 500 chance of collision.

On the basis of their orbital periods, three classes of comet are also recognised:

1. short-period comets that have orbital periods (P) of less than 20 years;
2. intermediate-period comets for which P = 20–200 y;
3. Long-period or *parabolic* comets for which P >200 y.

The short- and intermediate-period comets make up the cometary constituent of the near-Earth object population, and are together known as *near-Earth comets* or NECs. Most of these objects originate in the *Edgeworth–Kuiper Belt,* a comet 'cloud' beyond Neptune orbiting the Sun at distances of between 30 and 1000 AUs. Comets from the inner region of the belt are sometimes perturbed by the gravitational fields of the giant outer planets, and enter new orbits that bring them into the inner solar system. Like NEAs, near-Earth comets are all characterised by a perihelion distance (q) of less than 1.3 AU. NECs include Comet Halley (with an orbital period of 76 years) and other well-observed comets such as Encke and Swift–Tuttle. Long-period comets originate in the *Oort Cloud,* a spherical 'shell' of cometary debris left over from the formation of the solar system. The Oort cloud comets have orbits that take them far beyond the outer planets, to between 40 and 50 thousand AU or around 1 light year (around a quarter of the distance to the nearest star). Many long-period comets have orbital periods of thousands or even millions of years, leading to their having made only one entry into the inner solar system in modern times, while countless others have yet to be observed.

For a number of reasons the contribution from comets to the terrestrial impact threat is poorly constrained, and estimates range from 2 to 30 per cent (Steel, 1995). Of some concern is the fact that owing to their close approaches to Jupiter, the orbits of short-period comets are potentially unstable, and can change significantly over centuries or millennia. At the other end of the spectrum, long-period comets tend to be large (perhaps up to 100 km across), particularly on first approach to the Sun, when they will have suffered no significant devolatilisation previously, and travel at up to three times the velocity of their short-period equivalents. Owing to the spherical nature of their Oort Cloud source, they can also approach the Earth from any direction. Furthermore, less than six months' warning is likely of a long-period comet on a collision course with the Earth. Steel (1995) suggests that the contribution from short- to intermediate-period comets to the 1-km-plus impactor flux to the Earth is likely to be of the order of 5 per cent, while owing to their speed and size, long-period comets may comprise 25 per cent of the impact hazard.

located in Kazakhstan. Five craters dated at 300,000 y or less, formed by objects large enough to have a regional impact on the environment have also been identified, however: in Arizona (the Barringer object), Algeria, Argentina, South Africa and Western Australia. Inevitably, the terrestrial cratering record is biased towards younger and larger structures preserved within stable cratonic areas, and no craters with diameters of less than 5 km have been identified in rocks older than 15 million years. Approximately a third of currently identified impact structures are buried under sediments, and the discovery of both continental and marine impact sites has depended strongly on their anomalous appearances in geophysical (often petroleum-prospecting) surveys. Concentrations of known continental impact sites are found in Scandinavia and the Baltic States, Australia and North America. In part, this is a function of the age and stability of these shield areas. It also

TABLE 7.1 Representative impact structures (After Grieve, 1998)

Crater	Country	Position	Diameter (km)	Age (Ma)
Barringer	USA	35.2° N 111.1° W	1.1	0.049
Zhamanshin	Kazakhstan	48.2° N 60.5° E	13.5	0.9
Ries	Germany	48.5° N 10.3° E	24	15
Popigai	Russia	71.4° N 111.4° E	100	35.7
Chicxulub	Mexico	21.2° N 89.3° W	170	64.98
Gosses Bluff	Australia	23.4° S 132.1° E	22	142.5
East Clearwater	Canada	56.5° N 74.7° W	26	290
West Clearwater	Canada	56.1° N 74.3° W	36	290
Kelly West	Australia	19.5° S 133.5° E	10	>550
Acraman	Australia	32.1° S 135.2° E	90	>450
Sudbury	Canada	46.3° N 81.1° W	250	1850
Vredefort	South Africa	27.0° S 27.3° E	300	2023

reflects, however, the more intense impact site surveys that have been undertaken in these more fruitful regions. The limited sample size makes it difficult to use impact crater data to make sensible estimates about the size range and frequency of impactors over time. As previously mentioned, the size–frequency distribution is biased in favour of larger structures owing to their relatively rapid burial or erosion. The bias also reflects the fact that smaller objects – up to 500 m for icy bodies, and perhaps up to ~200 m for iron-rich bodies – are susceptible to break-up, dispersal and velocity reduction during passage through the atmosphere (Melosh, 1981). The cumulative size–frequency curve for the terrestrial crater record (Grieve, 1997; Fig. 7.5) clearly shows the lower than expected numbers of smaller craters, with a deficit starting to appear for crater diameters of less than ~25 km.

In theory it is possible to convert the observed size–frequency distribution of impact structures into a similar distribution for impactors. However, there are a number of problems in doing so, particularly in choosing the appropriate energy scaling relations that allow impact structure diameters to be translated into impactor sizes (e.g. Croft, 1985). In general terms, however, the size–frequency distribution for recent impacts appears to be of the order expected, assuming current estimates of the size distribution of the near-Earth asteroid population. On the 15 per cent of the Earth's surface capable of retaining the evidence, three impacts of objects in excess of 1 km are recognised over the past 4 million years, yielding a return period of 700,000 years for this proportion of the surface. This translates into a global 1-km-impactor frequency of around 100,000 years.

A terrestrial cratering rate of $5.6 \pm 2.8 \times 10^{-15}$ per square kilometre per year has been derived for impact structures ≥20 km across (Grieve and Shoemaker, 1994), which is in agreement

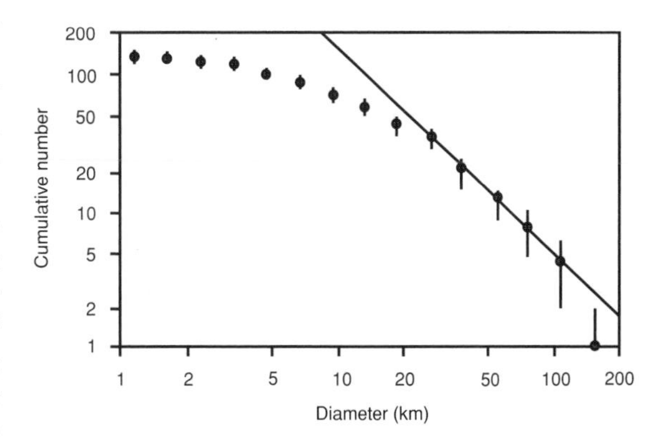

FIGURE 7.5 Cumulative size–frequency curve for terrestrial impact craters. Note the lower than expected number of smaller craters that starts to become apparent for diameters of less than around 25 km. (From Grieve, 1997)

with that determined from astronomical observations. Both estimates, however, contain uncertainties of the order of ±50 per cent. Some authors claim to recognise periodicities in the terrestrial cratering rate (e.g. Alvarez and Muller, 1984; Raup and Sepkoski, 1984), statistically linking periods of apparently higher impact rates with enhanced comet activity (due to a range of astronomical mechanisms) and the timing of mass extinctions (e.g. Davis *et al.*, 1984; Rampino and Stothers, 1984; Rampino *et al.*, 1997). This is challenged, however; by Grieve and Pesonen (1996), on the basis that uncertainties in the terrestrial cratering record are too great to permit their use in such calculations, and by Weissman (1990), who questions both the evidence for periodic comet swarms and the astrophysical mechanisms proposed to invoke them. Most recently, however, Napier (1998) has reported that the temporal distribution of well-dated terrestrial craters does indeed reveal periodic changes in the impact flux. This he interprets as a continuum upon which is superimposed a number of peaks with a period of 27 ± 1 million years. Surprisingly, the current terrestrial cratering rate is higher than that integrated over the past ~3.2 billion years, based on lunar cratering data, and it has been suggested (Shoemaker *et al.*, 1990) that this may reflect an increasing number of comet impacts. Clustering of impacts of similar age has also been suggested, and the evidence is in some cases compelling (e.g. the East and West Clearwater impacts in Canada, and the Kamensk and Gusev sites in Russia) for binary impactors. Currently, however, no hard evidence exists for temporally close multiple impacts that might reflect episodes of increased bombardment.

Well-preserved continental impact structures display a range of crater morphologies, from simple concave depressions to complex, multi-ring basins, the full spectrum of which can be observed on less dynamic worlds such as the Moon, Mercury and Mars. Simple impact structures on Earth have diameters of up to 4 km, and a characteristic bowl form bounded by a structurally elevated rim. Typically, a blanket of ejecta surrounds the depression, which is also partly filled by brecciated debris. Continental impact craters with diameters in excess of 4 km are structurally more complex, and may display a range of features – if well preserved – including faulted rims, annular troughs, central peaks, and ring basins (Grieve, 1997). Almost invariably, post-impact deformation and erosion later modify the original form of an impact crater. In particular, the original diameter may be difficult to determine, resulting in poor constraining of the impactor size. For example, published values for the diameter of the Chicxulub crater range from around ~180 to ~300 km.

Of the 165 terrestrial impact structures so far identified, only 13 per cent have been located in the marine environment (Table 7.2), and with a single exception all occur in shallow waters (mostly <200 m) where the diagnostic crater morphology is relatively well preserved (Gersonde and Deutsch, 2000). The exception is the deep-water (around 5 km) Eltanin impact, which occurred during the late Pliocene (2.15 Ma) in the Bellingshausen Sea to the west of the tip of South America. In contrast to shallower-water impacts, no cratering structure is associated with the Eltanin event, which is distinguished as an impact structure on the basis of a characteristic iridium anomaly (Kyte *et al.*, 1981) and seismic and sedimentological data (Gersonde *et al.*, 1997). With 70 per cent of the Earth's surface covered by water, it might be expected that the crater count in the marine environment would be perhaps twice as great as that on land. A number of factors contribute, however, to the poor preservation of impactor evidence in the oceans. Because of the formation of new lithosphere at constructive plate margins and its destruction in subduction zones, the ocean basins are floored by material that is, on average, much younger than that making up the continents. Oceanic crust older than the Jurassic is unknown, leading to the deep ocean floor having the potential to record only impact events confined to the last twentieth or so of the geological record. This is one of the main reasons for the location of all but one of marine impact sites so far identified in shallow water underlain by continental crust, and half of these sites occur in rocks that are in excess of 360 million years old. The

TABLE 7.2 Marine impact structures (Source: Gersonde and Deutsch, 2000).

Name	Impact site water depth (m)	Position	Diameter of structure (km)	Age (Ma)
Deep ocean impacts				
Eltanin	~5000	57S 91° W	No crater observed	2.15
Shallow water impacts				
Chesapeake Bay	200–500	37.16° N 76.07° W	85	35.5
Shiyli	~300	49.10° N 57.51° E	3.2	45
Montagnais	<600	42.53° N 64.13° W	45	50.5
Marquez	Shallow	31.17° N 98.18° W	22	58
Kamensk	100–200	48.20° N 40.15° E	38	65
Chicxulub	100–200	21.20° N 89.30° W	180	65
Mulkarra	Very shallow	27.51° S 138.55° E	17	105
Tookoonooka	Very shallow	27.07° S 142.50° E	55–65	128
Mjolnir	300–500	73.45° N 29.45° E	40	140
Flynn Creek	~10	36.17° N 85.40° W	3.8	360
Kaluga	?100–400	54.30° N 36.15° E	15	380
Ames	10–50	36.15° N 98.10° W	15	450–460
Brent	0–25	45.05° N 78.20° W	4	453
Kärdla	25–75	57.00° N 22.42° E	14	455
Lockne	>200	63.00° N 14.48° E	24	455
Tvären	100–150	58.46° N 17.25° E	2	457
Granby	50–100	58.25° N 14.56° E	3	465
Hummeln	25–75	57.22° N 16.15° E	1.2	465
Possible impact structures				
Toms Canyon	200–500	39.08° N 72.51° W	15	?35.5
Tore 'seamount'	Deep ocean	39.30° N 13° W	122 × 86	91

For some of the age data, precision is less than 10 per cent.

absence of a crater at the Eltanin site also suggests that even quite large impactors, in excess of 1 km, may suffer substantial deceleration and disintegration during their passage through a high water column. Consequently, the deep-water record for impactors with smaller diameters is liable to be poor. Even in shallow marginal or epicontinental seas, where crater formation is likely, thick sediment cover may disguise the evidence, which would be unlikely to be uncovered, as at the Chicxulub site, without recourse to dense geophysical surveys and borehole studies.

As it is the only documented impactor in the deep ocean, the characteristics of the Eltanin impact site are worth further consideration. Although evidence for an impact in the region was provided by Kyte *et al.* (1981) on the basis of the presence of anomalous concentrations of iridium in sediments, a comprehensive account of the site did not appear for another 16 years (Gersonde *et al.*, 1997). This delay highlights a general problem in recognising and studying marine impact sites, which is both costly and time-consuming. Gersonde and his colleagues combined seismic, bathymetric and borehole surveys to build a picture of the effects of the impact and to characterise a type of sedimentary sequence formed by an impact in the deep-ocean floor. Most distinctive is a 60-m-thick acoustically transparent sediment layer extending over 200 km from 'ground zero' and presumed to reflect impact shock-related turbulence in the deep-sea sedimentary cover. In the absence of a crater, it has proven difficult to constrain the diameter of the Eltanin impactor. Recent numerical simulation (Artemieva and Shuvalov, 1999) has, however, allowed an upper estimate to be made. They demonstrated that crater formation does not occur if the water depth/impactor ratio (*h*/*d*) exceeds 5, and no bottom effects occur at all if *h*/*d* is greater than 25. This argues for a 1 km or less diameter for the Eltanin object. The effects of an Eltanin impact in 5 km of water have been simulated by Ivanov (1999), who reports consequent expected flow velocities on the ocean floor of the order of 100 m/s at distances of up to 20 km from the centre of the impact. These high velocities are the probable cause of the widespread disturbance of ocean-floor sediments and the rapid degradation of any crateriform morphology (Ormo and Lindstrom, 2000).

Because of their greater number (currently 18 are confirmed) and accessibility, the characteristics of shallow-water impact sites are better known. At very shallow depths, of the order of tens of metres, a crater is present that is morphologically similar to those formed in land targets. At depths of a few hundred metres, however, the post-impact rush of water (known as *resurging*) typically reduces the profile of the crater by erosion.

Grieve (1997) reports only 15 known occurrences of impact-related materials in the stratigraphic record, including those produced by the Chicxulub event. These take the form of shocked minerals, tektites and microtektites, meteoritic spherules, or geochemical (especially iridium) anomalies. As pointed out by Rampino *et al.* (1997), however, many of the near-source ejecta products of large impacts are likely to be unrecognised as such in the geological record or misinterpreted as turbidites, tillites, tectonic megabreccias and other deposits. While physical and geochemical evidence for the Chicxulub impact has been identified at over a hundred sites across the planet, the spatial distribution of material associated with other impacts in the stratigraphic record currently appears too limited for any to have had a serious effect on the global environment. Worthy of mention here is the late-Eocene North American tektite field (Glass *et al.*, 1985), probably representing ejecta from an impact in Chesapeake Bay (which has been implicated by some in an Upper Eocene extinction event), and the Miocene moldavite tektite field (e.g. Engelhardt *et al.*, 1987) associated with the Ries impact structure in Germany.

7.5 Environmental effects of impact events

The scale of environmental damage due to an impact event is not simply a reflection of the impactor size. Morrison *et al.* (1994) observe that it is the kinetic energy of the impactor that is its most significant property, which is a function of both object size *and* velocity. All other parameters being equal, a large object will have a greater kinetic energy and be more destructive than one of smaller dimensions (Fig. 7.6; Table 7.3). Owing to their much higher velocities, however, long-period comets are more destructive than near-Earth asteroids or 'local' comets with orbital periods of less than

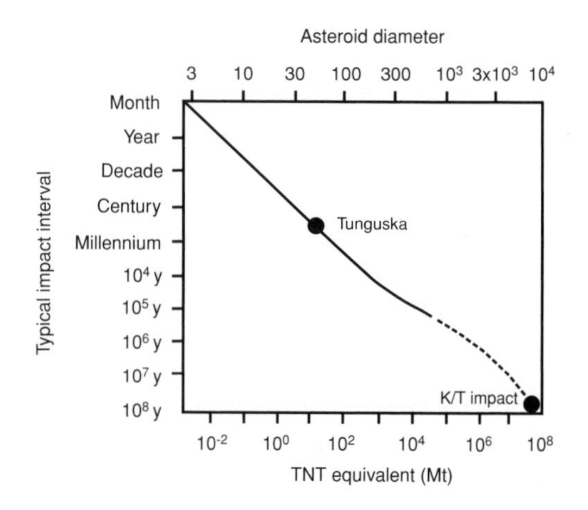

FIGURE 7.6 Cumulative energy–frequency curve for impacts on the Earth, also showing asteroid diameters. The continuous curve is the 'best estimate' from Shoemaker (1983) for the average interval between impacts equal to or greater than the indicated energy yield. The dashed line is an extension to a minimum estimate of the K/T impact energy. (From Chapman and Morrison, 1994)

Table 7.3 Impact scales, energies and predicted fatality rates for three different estimates of the global threshold (After Chapman and Morrison, 1994)

Type of event	Diameter of impactor	Energy (Mt)	Frequency (years)	Deaths
Tunguska-scale event	50–300 m	9–2000	250	5×10^3
Large sub-global events	300–600 m	2000–1.5×10^4	35×10^3	3×10^5
(3 estimates)	300 m to 1.5 km	2000–2.5×10^5	25×10^3	5×10^5
	300 m to 5 km	2000–10^7	25×10^3	1.2×10^6
Low global threshold	>600 m	1.5×10^4	7×10^4	1.5×10^9
Nominal global threshold	>1.5 km	2×10^5	5×10^5	1.5×10^9
High global threshold	>5 km	10^7	6×10^6	1.5×10^9
Rare K/T-scale events	>10 km	10^8	10^8	5×10^9

200 years. An impactor's effect on the environment will also depend upon whether it hits a land or ocean target. Although a deep-ocean impact is likely to release less debris into the atmosphere, it has the potential to generate devastating ocean-wide giant tsunami (e.g. Hills and Mader, 1997; Hills and Goda, 1998; Mader, 1998; Crawford and Mader, 1998; Ward and Asphaug, 2000) and may affect the climate through the injection of huge volumes of water and salt into the stratosphere.

Much of what we currently know about the environmental effects of impacts comes from study of two events on very different scales: the 10–20 Mt airburst of a 50-m object over Tunguska (Siberia) in 1908 (Chyba *et al.*, 1993), and the impact of a ~10-km asteroid or comet (energy equivalent = 10^8 Mt) in the Yucatán platform 65 Ma ago (e.g. Pope *et al.*, 1994). As a result, much extrapolation has been involved in trying to determine the consequences for the environment that might have arisen from impacts in the intermediate size (e.g. 100 m to 6.5 km) and comparable energy (10^2–10^7 Mt) range. Summaries of the environmental effects of impacts of various sizes and at various energies (Table 7.4) are given by Morrison *et al.* (1994) and Chapman and Morrison (1994) and in more detail by Toon *et al.* (1994). Although this is a rapidly changing research field, the scenarios remain essentially valid. Chapman and Morrison (1994) observed that the Earth's atmospheric shield prevents penetration of many potential impactors at the lower end of the size/energy range. For example, a stony object needs to be at least 50 m and perhaps 100 m across to have any chance of reaching the surface, or even the lower levels of the atmosphere (Melosh, 1997), although iron-rich objects an order of magnitude smaller might survive transit. Objects large enough to penetrate to the surface or close to it, and retaining velocities of tens of kilometres per second, will cause explosions comparable to those associated with the detonation of nuclear warheads, and the effects will be the same, although without a radiation signal. Depending on the energy released in the impact, Chapman and Morrison (1994) recognised both locally devastating and globally catastrophic impacts. The best-documented example of the former is the 1908 Tunguska event, when a stony impactor exploded in the lowest few kilometres of the atmosphere with a force sufficient to fell trees over an area of around 2000 km^2 (Toon *et al.*, 1994) (Fig. 7.7), igniting fires in the vicinity of 'ground zero' (Chyba *et al.*, 1993). Such locally devastating impacts occur a few times a century and have the potential to wipe out a major city, or a small US state or European country.

At the other end of the spectrum, the consequences of the impact of a ~10-km object, on a scale of the K/T event, are truly global in extent. For impacts of this order, Toon *et al.* (1997) expect, among others, the following environmental effects: global fires due to hot ballistic ejecta re-entering the atmosphere; light levels so

Table 7.4 Environmental effects of impacts (Adapted from Morrison *et al.*, 1994, and Toon *et al.*, 1994)

Energy	Impactor size	Crater (km)	Effects
10^1–10^2	75 m	1.5	Iron objects make craters, stone objects produce airbursts (Tunguska). Land impacts destroy area the size of a city (e.g. London, Moscow)
10^2–10^3	160 m	3	Iron and stone objects produce groundbursts; comets produce airbursts. Land impacts destroy area the size of a large urban area (e.g. New York, Tokyo)
10^3–10^4	350 m	6	Impacts on land produce craters; ocean tsunami become significant. Land impacts destroy area the size of a small state (e.g. Wales, Estonia)
10^4–10^5	700 m	12	Tsunami from marine impacts reach global scales and exceed damage from land impacts. Land impacts destroy area the size of a moderate state (e.g. Virginia, Taiwan)
10^5–10^6	1.7 km	30	Land impacts raise enough dust to affect climate and trigger impact winter. Ocean impacts generate hemispheric-scale tsunami. Global destruction of ozone shield. Land impacts destroy area the size of a large state (e.g. UK, Japan, California)
10^6–10^7	3 km	60	Both land and ocean impacts raise enough dust to trigger impact winter. Photosynthesis ceases. Impact ejects are global, triggering widespread fires. Land impact destroys area the size of a large state (e.g. India, Mexico)
10^7–10^8	7 km	125	Vision impossible due to reduced light levels. Global conflagration. Probable mass extinction. Direct destruction approaches continental scale (e.g. Brazil, Australia, United States)
10^8–10^9	16 km	250	Large mass extinction (e.g. K/T)
$> 10^9$			Threatens survival of all life

low that vision might not be possible; tsunami with heights of 100 m and extending 20 km inland in coastal zones (in the case of a deep-ocean impact); destruction of the ozone layer; and dramatically reduced temperatures leading to an *impact winter*. Both Morrison *et al.* (1994) and Chapman and Morrison (1994) address the critical problem of determining the threshold impact energy required to trigger a global natural catastrophe. In terms of its effect on human society, Chapman and Morrison (1994) define such an event as one 'that would disrupt global agricultural production and lead, directly or indirectly, to the deaths of more than a quarter of the world's population (>1.5 billion people)'. In this regard, Toon *et al* (1994) propose that a 10^6-Mt impact (corresponding to a 2-km asteroid with an impact velocity of 20 km/s) into either a land or an ocean site will inject sufficient sub-micrometre dust into the atmosphere to induce global freezing and large-scale crop loss. Other energy thresholds put forward for the triggering of a global catastrophe are 3×10^5 Mt (corresponding to an asteroid of 1.7 km with an impact velocity 20 km/s) (Morrison *et al.*, 1994), and 1.5×10^4 Mt to 10^7 Mt (0.6–5 km for an asteroid; Chapman and Morrison, 1994). Because of their higher velocities, and therefore kinetic energies, all estimates require a smaller diameter for a globally destructive comet than for an asteroid of similar destructiveness (Table 7.5). Although estimates

FIGURE 7.7 In 1908 a stony impactor around 50 m across exploded in the lowest few kilometres of the atmosphere over Tunguska, Siberia. The resulting blast flattened around 2000 km^2 of mature forest. This photograph was taken in 1928 by the first expedition to reach the region. (Reproduced with permission of The Tunguska Page of Bologna University at http://wwwth.bo.infn.it/tunguska/)

for the global catastrophe threshold vary by almost two orders of magnitude in energy terms and by more than a factor of 2 in terms of projectile size, there is a very general consensus that collision of the Earth with a 1-km object will have severe ramifications for both human society and the global environment.

7.5.1 Immediate to short-term effects

Immediate to short-term environmental effects of impacts can be categorised in terms of (a) those resulting directly from crater formation on land or in a shallow sea or from a deep-ocean impact, namely blast, re-entering ejecta and earthquakes; and (b) those secondary effects that may persist over several hours or days, such as tsunami, or are initiated within hours or days of the impact event, such as wildfires.

7.5.1.1 Blast effects of small impactors (energy yields up to 104 Mt)

The immediate effects of an impact of any size are associated with an explosive blast, the size and damage potential of which are related to the kinetic energy of the impacting body and the effective altitude of the explosion. Studies of the effects of nuclear detonations (e.g. Glasstone and Dolan, 1977) reveal that the area of devastation from the explosion is equal to the explosive yield to the power of ⅔. The explosion associated with the 50-m Tunguska object affected a larger area (≥2000 km^2) than predicted, owing to the more destructive nature

TABLE 7.5 Threshold energies and impactor sizes for global catastrophe (Source: Chapman and Morrison, 1994)

	Energy (Mt)	Asteroid diameter (km)	Comet diameter (km)	Typical interval (y)
Lower limit	1.5×10^4	0.6	0.4	7×10^4
Nominal	2×10^5	1.5	1.0	5×10^5
Upper limit	10^7	5	3	6×10^6

(up to a factor of 2) of a low airburst compared to a surface detonation. The blast wave generated by impacts consists of a sudden pressure pulse followed immediately by very strong winds (Toon *et al.*, 1994). The strength of the shock wave is dependent on the *peak overpressure*, in other words the difference between ambient pressure and the pressure of the blast front. A peak overpressure contour corresponding to 4 p.s.i. ($= 2.8 \times 10^5$ dyn/cm^3) is used by a number of authors (e.g. Chapman and Morrison, 1994; Toon *et al.*, 1994) to define the area of severe destruction due to blast. The peak overpressure is related to the maximum wind speed, which is approximately 70 m/s for 4 p.s.i. This is more than double the force of hurricane-force winds. For nuclear detonations, Hills and Goda (1993) derived an equation for the maximum distance at which a 4-p.s.i. overpressure occurs:

$$r = ah - bh^2E^{-1/3} + cE^{1/3} \tag{7.1}$$

where r is the maximum distance (km) of the overpressure contour from the point below the detonation, h is the height of the explosion (km), E is the energy of the explosion in megatons, $a = 2.09$, $b = 0.449$ and $c = 5.08$. Because only around 50 per cent of the energy in a nuclear explosion is partitioned into the shock waves, Toon *et al.* (1994) assumed, using eqn (7.1), that for impact events

$$E = 2\varepsilon Y \tag{7.2}$$

where Y is the kinetic energy of the impactor and ε is the fraction converted to shock energy.

Assuming that the 4-p.s.i. contour defines the limit of complete destruction of (poorly constructed) buildings and the deaths of their inhabitants, Chapman and Morrison (1994) defined an area of *lethal damage* given by

$$A = 100Y^{2/3} \tag{7.3}$$

where Y is the explosive yield in megatons and A is expressed in square kilometres.

For an impactor with an explosive yield of 100 Mt, for example a 100-m cometary object, Morrison *et al.* (1994) calculated a radius of destruction, due to blast, of 25 km. Similarly, Chapman and Morrison (1994) estimated that a stony or metallic impactor of 250 m diameter would, provided it struck land, excavate a crater 5 km across (crater diameter in a land target is typically 20 times the size of the impacting object). They calculated an energy yield of around 10^3 Mt and an area of devastation of the order of 10,000 km^2. A cometary object of similar size would be more likely to break up during its passage through the atmosphere, generating a number of airbursts that would substantially increase the area of damage. For impact energies of the order of 10^4 Mt (corresponding to a comet ~400 m across, or an asteroid with a diameter of 650 m), Toon *et al.* (1997) forecast a damage area (due to blast, ground shaking and fire) of up to 10^5 km^2.

7.5.1.2 Blast effects of medium to large impactors (105–108 Mt)

As the kinetic energy of the impacting object is increased, so the area affected by the direct blast also rises. For impactors with the potential to trigger global catastrophe (a 1-km comet or a 1.5-km asteroid), Morrison *et al.* (1994) calculated an explosive yield of the order of 10^5 Mt, which translates into a zone of complete destruction of radius 250 km and a potential death toll due to blast of up to 3 million people. For very large (~10^8 Mt) impacts of the order of the K/T event, the size of the explosion blows away the atmosphere above the impact site, limiting the amount of energy going into the shock waves. This causes a decline in the efficiency of blast wave generation, so that for a K/T-size impactor only about 3 per cent of the impact energy enters the shock waves (O'Keefe and Ahrens, 1982). This means that the blast generated by even K/T-size objects will be restricted to a small percentage of the Earth's surface and will not pose a threat on a global scale.

7.5.1.3 Re-entering ejecta and the ignition of global wildfires

For larger impacts into land or shallow water targets, the expulsion of huge volumes of debris and vapour poses an additional immediate hazard. At the point of impact, shock waves are driven into the target material and back into the impactor, which disintegrates to form a fireball containing molten debris and superheated vapour. The fireball expands as a powerful blast wave that propagates outward and upwards. For impacts with energies greater than 150 Mt (Boslough and Crawford, 1997) it breaks through the atmospheric envelope, and entrained melt particles follow ballistic trajectories that return to the surface as quenched glassy tektites, sometimes covering areas of hundreds of thousands of square kilometres. For impactors of Chicxulub size, the distribution of this ballistic debris can be global (Argyle, 1989), which explains why pulverised material from the K/T impact averages 3 mm thickness across the planet. Closer to ground zero, millimetre-size spherules associated with the Chicxulub impactor are found in an arc over 2500 km in length that stretches from Alabama to Haiti (e.g. Kring and Boynton, 1991). For K/T-size impactors, Melosh *et al.* (1990) estimated that the mass of ejected debris re-entering the atmosphere on ballistic trajectories is such that atmospheric temperatures are increased to levels high enough for downward thermal radiation to trigger wildfires. They propose that the global radiation flux may have increased by factors of 50 to 150 times the solar output for periods of up to several hours, posing a serious threat to unprotected animal life as well as causing wholesale destruction by burning of vegetation. The global ballistic dispersion of ejecta requires ejection velocities of the order of 7 km/s, comparable with that required for satellites to maintain a circular orbit, and an impactor energy of 10^6 Mt. Of the total energy deposited in the atmosphere by re-entering ejecta, Toon *et al.* (1994) report that perhaps 25 per cent will reach the surface, arriving within tens of minutes of the impact. The same authors show that for a 10^7 Mt impact, the energy flux would be comparable with tens of minutes of solar radiation, and levels of thermal radiation high enough to trigger fires over a fifth of the Earth's surface. For a K/T-size event, the amount of thermal radiation generated by re-entering ejecta could be sufficient to ignite fires anywhere on the planet.

The existence of global-scale wildfires at the K/T boundary is supported by a distinct soot layer, described in detail by Wolbach *et al.* (1988), which has been detected at a number of sites as far afield as New Zealand and Europe. Here, boundary clays are 10^2–10^4 times enriched in elemental carbon, which is isotopically uniform and attributed to a single global wildfire. The soot layer coincides with the iridium-rich layer, suggesting that the fire was triggered by a major impact and had started even before the ejecta had settled. A global soot volume of 7×10^{16} g is estimated by Wolbach *et al.* (1988), which is about 3 per cent of the maximum Cretaceous carbon level (assuming all land masses were covered with biomass comparable in density with today's tropical rain forest). This figure is somewhat higher than that obtained for present-day forest fires, which result in soot yields ranging from 0.1 to 2 per cent. Wolbach and his co-authors suggest, however, that the higher conversion rate from living carbon to soot might result from a combination of circumstances, including the higher partial pressures of oxygen during the late Cretaceous, and incessant lightning storms.

7.5.1.4 Earthquakes and volcanism

Medium to large impacts will inevitably generate severe ground-shaking over an area dependent on the energy of the collision and the local geology, but which may comprise millions of square kilometres for the largest events. For 10^6-Mt impacts such as caused the Manson impact structure in north-west Iowa, for example, the radius of severe seismic shocks (Richter magnitude 9) is estimated at about 1000 km (Adushkin and Nemchimov, 1994). Seismic waves diverge radially from the impact site, experiencing attenuation with generally increasing distance from the site, but are thought to focus again at its antipode. Using computational simulations, Boslough *et al.* (1996) demonstrated that seismic displacement and strain amplitudes at the Earth's surface are here larger than anywhere else except close to the impact itself. For a K/T-size impactor, peak displacement at the antipode approaches 10 m, and seismic energy remains focused all the way down to the core–mantle boundary. Boslough and his co-workers showed that focusing of seismic waves is greatest beneath the impact site and in the asthenosphere beneath the antipode. They also suggest that such focusing may contribute to the rapid triggering of basaltic volcanism at both the impact site and the antipode, and hypothesise about a potential link between seismic focusing, mantle plume development and flood basalt magmatism.

7.5.1.5 Tsunami

Because of their potential to near-simultaneously devastate the margins of a target ocean basin, a considerable amount of research has been undertaken into impactorgenic tsunami, resulting in estimates of tsunami height and run-up that vary significantly (e.g. Hills *et al.*, 1994; Hills and Mader, 1997; Hills and Goda, 1998; Crawford and Mader, 1998; Mader, 1998; Ward and Asphaug, 2000). The importance of this work lies in the fact that substantial tsunami can be generated, not only by impactors of 1 km and above, but also by smaller objects in the 100–200 m size range, which occur at frequencies on historical time scales (e.g. Hills *et al.*, 1994; Ward and Asphaug, 2000). Tsunami generated by submarine earthquakes (sometimes augmented by associated submarine landslides) rarely exceed 10–15 m in height, although, as shown by the Aitape (Papua New Guinea) tsunami, even waves on this modest scale can cause severe destruction and take many thousands of lives. In the Pacific Basin alone, around 400 tsunami took over 50,000 lives during the twentieth century, causing damage amounting to billions of US dollars. On a larger scale, evidence exists for tsunami with heights of 100 m or more formed by the collapse of large oceanic islands such as those of the Hawaiian and Canarian archipelagoes (e.g. Moore and Moore, 1984; Simon Day, personal communication). Constraints on the scale and extent of tsunami generated by variously sized impacts into ocean targets are derived primarily from modelling studies. Distinctive lithologies associated with the Chicxulub impact have, however, been interpreted as tsunami deposits which, if true, provides some constraint on wave parameters. Bourgeois *et al.* (1988) for example, report sand beds at the K/T boundary at Brazos River (Texas, USA), which they interpret to be the result of a major disruption of the local depositional environment due to a tsunami between 50 and 100 m high. Similar sequences of coarse clastics found in north-east Mexico and containing impact ejecta (shocked quartz and spherules) are also interpreted as having been rapidly (within a few days) deposited by tsunami (e.g. Smit *et al.*, 1996).

As for 'normal' tsunami, those produced by ocean impacts move at velocities in excess of 500 km/h, and are characterised by deep-water waves that are many times smaller than those that eventually hit land. The *run-up factor* (the vertical height above sea level of the tsunami at its furthest point inland divided by the deep-water wave amplitude) varies considerably and is particularly dependent on local topography and the direction of travel of the waves relative to the coastline. Similarly, estimates also vary, and while Hills and Goda (1998) reported seismogenic tsunami run-up factors as high as 40, Crawford and Mader (1998) suggested that the typical value is nearer 2 or 3. The choice of value clearly has major implications for the destructive capabilities of impactorgenic tsunami. When an asteroid or comet hits an ocean impact point, both the impactor and the water vaporise, forming a transient crater in the ocean that is approximately 20 times the diameter of the impacting object. The ocean rushes back in, forming a 'mountain' of water in the middle, which spreads rapidly and radially outwards as a giant tsunami. The centre of the transient crater continues to oscillate up and down several times as it fills, generating a series of further tsunami.

Considerable disagreement exists between researchers on the deep-water wave height, and therefore the destructive potential, of impactorgenic tsunami (Table 7.6). For example, while Crawford and Mader (1998) proposed that a 200-m stony asteroid will generate a negligible deep-water wave at a point 1000 km from the impact site, Ward and Asphaug (2000) forecast a wave 5 m high. Even assuming a conservative run-up factor of 10 (which is not unusual for seismogenic

TABLE 7.6 Estimated deep-water wave height (above sea level) at a point 1000 km from the site of an asteroid ocean impact

Stony asteroid diameter (m)	Hills and Goda (1998)	Ward and Asphaug (2000)	Crawford and Mader (1998)
200	1 m	5 m	Negligible
500	11 m	15 m	<2 m
1000	35 m	50 m	6 m

tsunami), this is the difference between a land-falling tsunami a few metres high and one with a run-up height of 50 m – the difference between minor damage and total devastation. The dramatic variation in tsunami height estimates arises from different ideas about wave dissipation. Crawford and Mader (1998) proposed that to produce a coherently propagating wave (one that does not lose most of its energy as it travels over ocean-basin-scale distances), the transient crater must be between 3 and 5 times wider than the depth of the ocean. For a typical 4-km ocean depth, and a crater diameter 20 times that of the impactor, this means that only objects 1 km or more across can produce tsunami that are persistent over great distances. The corollary of this is that impactors smaller than this would not pose a severe threat in terms of tsunami that are hugely destructive on an ocean basin scale. It should be noted, first, however, that no direct evidence exists for the size and destructive capacity of tsunami generated by small impactors, and second, that the wave trains generated by ocean impacts up to a kilometre across would still be large enough to cause local to regional devastation along adjacent coastlines. In addition, other forecasts exist that predict considerably larger waves from even small impacts (Fig. 7.8). Hills and Mader (1997), for example, proposed that an impact anywhere in the Atlantic by an asteroid 400 m across would devastate the coastlines on both sides of the ocean by generating tsunami over 100 m high. Even smaller impacts could generate destructive tsunami on short timescales, and Ward and Asphaug (2000) quote 1-in-24 and 1-in-35 probabilities that, respectively, New York and Tokyo will be hit by impact tsunami in excess of 5 m in the next millennium. On a considerably larger scale, they also reported that an asteroid 5 km across striking in mid-Atlantic would inundate the entire upper east coast of the United States to the base of the Appalachian Mountains.

7.5.2 Long-term effects

Local to regional damage to the environment resulting from small to medium-size impactors (with explosive yields in the range 10–10^3 Mt) is liable, at the very least, to impose severe stress on ecosystems and likely to result in their complete destruction over areas determined by the energy of the impact. In marine environments, effects may be more widespread as a result of tsunami formation and the disturbance and redeposition of sea-floor sediments, and coastal and reef ecosystems are likely to be particularly badly affected. Again, depending on the scale of devastation, recolonisation and a return to pre-impact conditions are likely to take decades to centuries. For medium-sized objects that are still below the threshold for triggering global catastrophe (explosive yields in the range 10^4–10^5 Mt), strong regional climatic effects are predicted. Toon *et al.* (1994) identify 10^5 Mt as the energy for impacts in land targets at which sub-micrometre dust yields an atmospheric opacity approaching 1, similar to that produced by very large (7 on the Volcanic Explosivity Index) volcanic eruptions (see Chapter 5). By comparison with the 1815 eruption of Tambora (Indonesia) (Stothers, 1984), Toon and his colleagues forecast significantly lower temperatures on at least a regional scale, which could

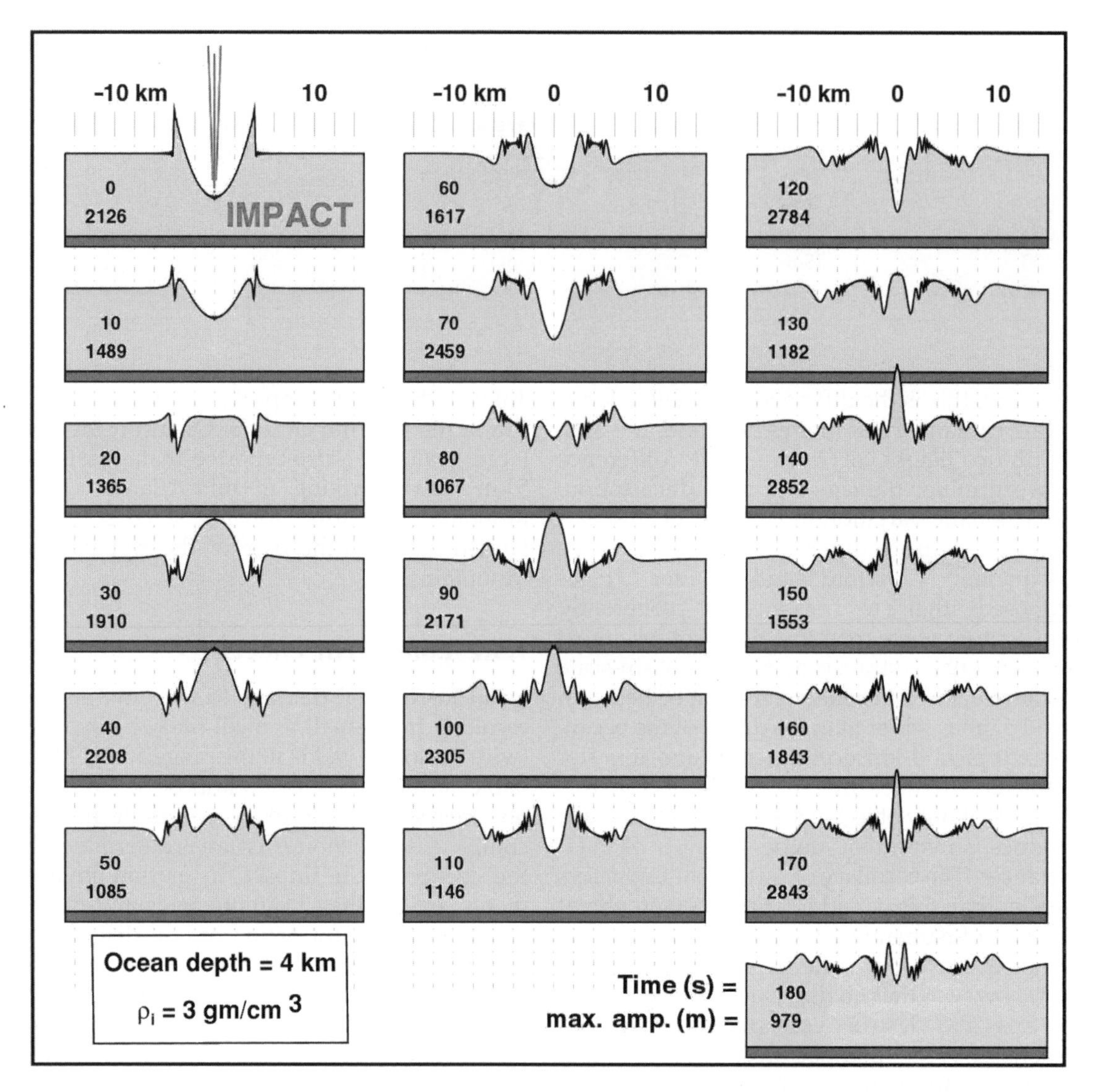

FIGURE 7.8 Tsunami induced by the impact of a 200-m-diameter asteroid striking deep ocean at 20 km/s. After 3 minutes the maximum wave amplitude remains almost 1km high. After 5 minutes (not shown) the tsunami would have covered an area of ocean of around 2500 km^2 and the leading wave would be about 325 m high. (From Ward and Asphaug, 2000)

lead to serious crop loss. For larger impacts, with energies of between 10^6 and 10^7 (corresponding to comet and asteroid diameters of up to 4 and 6.5 km respectively), Toon *et al.* (1994) propose that dust levels would be sufficiently high as to reduce light levels below those required for photosynthesis, leading to devastating implications for both plant and animal life on a global scale. The predicted long-term effects for the environment arising from large impacts are manifold (Fig. 7.9), and range from severely reduced temperatures leading to a so-called *impact winter*, to greenhouse warming and destruction of the ozone layer.

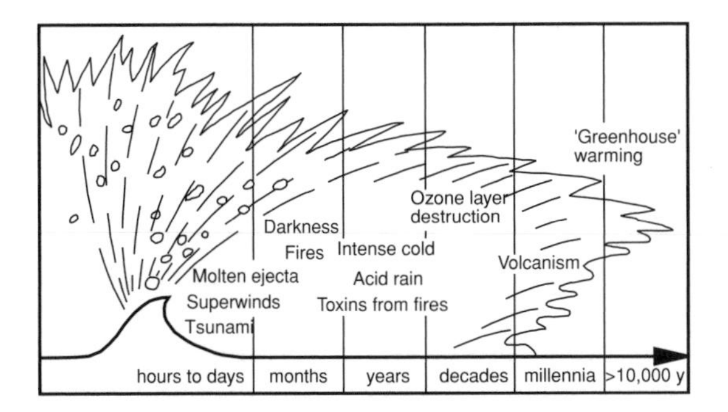

FIGURE 7.9 Environmental effects attributed to the K/T impact and charged with triggering widespread extinctions are manifold and operate on timescales ranging from hours to millennia. (From McGuire, 1999)

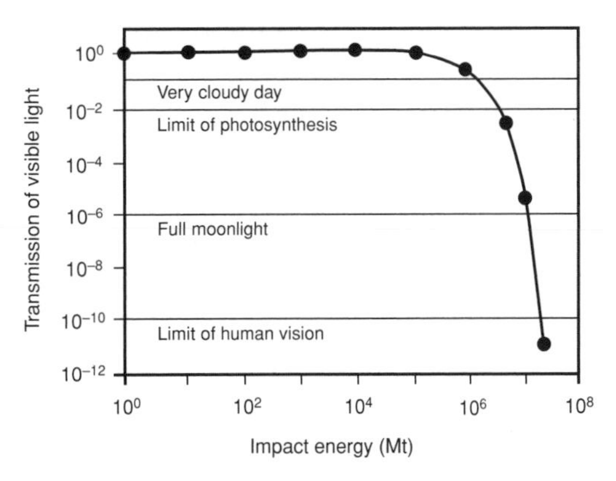

FIGURE 7.10 Owing to the injection of sub-micrometre dust into the atmosphere, light transmission following an impact with an energy of 5 × 10^6 Mt would probably drop below that required for photosynthesis. For larger impacts of the order of 10^7 Mt, light levels may fall below the limit of human vision. Model assumes that 10 per cent of the mass of the impactor is injected into the atmosphere as sub-micrometre dust. (After Toon *et al.*, 1994)

7.5.2.1 Impact winter

In their seminal 1980 paper, Alvarez *et al.* proposed that the impact of an asteroid would inject 60 times the object's mass into the atmosphere as pulverised rock, where it would contribute to global darkness. In the context of the K/T impact, the authors suggested that this would have been sufficient to cause the cessation of photosynthesis, the death of land plants and photosynthesising marine organisms, and the consequent demise, through starvation, of first herbivores, then their carnivorous predators. In 1983, Pollack *et al.* focused in more detail on the environmental effects of an impact-generated dust cloud at the K/T boundary. Through modelling the evolution and radiative effects of a debris cloud from a hypothesised impact, the authors forecast a fall in light levels at the Earth's surface to below those required for photosynthesis for several months and, for a shorter interval, to levels too low for animals to see. They also predicted a rapid cooling, over a period of six months to a year, during which time ocean temperatures would cool by a few degrees Celsius and land temperatures fall by as much as 40 degrees C. More recently, Toon *et al.* (1994) determined that a mass of material approaching 10 per cent of the impactor mass would enter the stratosphere in the form of sub-micrometre dust. They showed (Fig. 7.10) that for objects with impact energies of between 10^7 and 10^8 Mt, smaller than the K/T object, the quantity of sub-micrometre dust would be sufficient to reduce the transmission of light in the atmosphere to below the level of human vision and far below the level required for photosynthesis. Such impacts would produce severe and prolonged cooling.

A number of researchers have drawn attention to the fact that the composition of a large-impact site may also be critical in determining the effect on the global environment. Following the identification of the Chicxulub site as the likely source of the K/T boundary event, the role of anhydrite and gypsum – the target lithologies – have been considered by a number of authors. Sigurdsson *et al.* (1992) proposed, for example, that the impact into a thick evaporite succession must have led to the formation of between 3.8 × 10^{18} and 1.3 × 10^{19} g of sulphate aerosol, leading to a global atmospheric mass loading of at least 1–2.6 g/cm^2. The authors go on to suggest that in combination with dust loading, this would have contributed to a decline in surface temperatures to near freezing within a week, together with prolonged cooling for several years following. Similarly, Pope *et al.* (1994) proposed that the large-scale vaporisation of sulphur-rich rocks at Chicxulub would have led to a decade of freezing and near-freezing temperatures worldwide. These numbers have been disputed, however, and Chen *et al.* (1994) advocated, on the basis of laboratory experiments, that the amount of sulphate aerosol released as a result of Chicxulub-related shock-induced devolatilisation might have been up to 100 times less. Consequently, the authors suggested that environmental stress due to post-impact global cooling may have been much less than that forecast by Sigurdsson and his co-workers. Most recently, however, shock vaporisation experiments undertaken by Yang and Ahrens (1998) suggest that sulphur aerosol release at Chicxulub was sufficient to cause a global temperature fall of at least 10 degrees C, large enough to cause serious disruption to the existing food chains.

The environmental effects of dust clouds generated by smaller impactors appear to be significantly less severe than for K/T-size objects. Covey *et al.* (1990), for example, used an atmospheric general circulation model (GCM) to investigate the impact of a dust cloud that might be generated by an impactor with energy of around 6×10^5 Mt (broadly corresponding to an asteroid around 2 km across). They found that although within two weeks the global average land temperature fell by around 8 degrees C (with temperatures in many regions falling from summer values to between 0 and 10°C), within 30 days temperatures recovered to pre-impact values. At a critical time of year, for example shortly before harvest, such a dramatic temperature fall could severely affect the world's annual crop yield, but long-term environmental effects are likely to be minimal. In a second study, Covey *et al.* (1994) used a GCM to simulate the effects of a dust cloud generated by a K/T-size event, this time showing that global average land temperatures would rapidly fall to around 0°C and stay there for a year, despite most of the dust settling out within 6 months. Low temperatures were accompanied by a reduction in global precipitation levels of 95 per cent for several months due to suppressed convection, a situation likely to lead to severe drought conditions.

7.5.2.2 Acid rain

A number of authors have implicated acid rain in extinctions due to large impacts (e.g. Prinn and Fegley, 1987). There are a number of ways in which acid rain can be produced during such collisions: first, as a result of nitric oxide produced by strong shock waves in the air, and second, from nitrogen oxides and other compounds produced by wildfires. In addition, should the target be rich in sulphate or carbonate, then sulphur dioxide and carbon dioxide may also contribute to the acid rain budget. Notwithstanding the Chicxulub case, however, the relative paucity of these lithologies means that the former mechanisms are the most likely, leading to nitric acid-dominated acid rain. Nitric oxide (NO) is produced when shock waves caused by the impact or the passage of the impactor through the atmosphere cause nitrogen and oxygen to combine (Zahnle, 1990). Typical yields of NO from shocked air are of the order of 10^9 moles per megaton of energy released in the atmosphere. Toon *et al.* (1994) proposed that even large impacts on the scale of Chicxulub will not generate sufficient nitric acid to acidify the ocean surface layers, but will produce a significant amount of acid rain. The authors suggest, however, that levels are unlikely to be much above those already experienced in heavily industrialised areas of Europe and North America.

7.5.2.3 Destruction of the ozone layer

Nitric oxide production during impacts is also incriminated in depletion of the ozone layer. Even for the small Tunguska event, Turco *et al.* (1981) reported that NO could have been responsible for a 45 per cent depletion of the ozone layer in the Northern Hemisphere. For larger impactors, ozone depletion would be global and severe, with serious implications for surface life. After a major impact the ozone layer would be subject to depletion due to a number of sources: most importantly from the injection of large quantities of nitrogen oxides, but also by reaction with smoke and dust particles, and heating of the ozone layer due to re-entering ejecta. Toon *et al.* (1994) forecast that impacts with energies in excess of 10^5 Mt would significantly deplete the ozone layer, with the effects becoming most serious following a clearing of the atmosphere of dust and NO (through acid rain), when the Earth's surface would be bombarded by increased doses of ultraviolet radiation, perhaps for several years. More recently, Cockell and Blaustein (2000) have proposed that ozone depletion would result in a doubling in the levels of ultraviolet radiation reaching the surface. They envisage an *ultraviolet spring* following the impact winter, which would cause severe damage to surviving animal and plant life. To some extent the nature of the impact target may be important in determining potential damage to the ozone layer, and for the K/T impact it is likely that sulphur aerosols liberated from the target evaporites may have acted to block some of the increased ultraviolet radiation from the surface, thereby counteracting – to some extent – ozone loss.

Marine impacts may also deplete the ozone layer through the ejection into the upper atmosphere of large quantities of salt. For an Eltanin-size (~1 km) impactor, Klumov (1999) proposes a consequent 50 per cent fall in ozone concentrations within a few thousand kilometres of the impact site. The depletion is transient, however, and persists for only a few days in the upper stratosphere and for a few years in the lower stratosphere.

7.5.2.4 Greenhouse warming

Following impact winter and ultraviolet spring, the longer-term legacy of a large impact may be a period of planetary warming, resulting from the massive injection of water vapour (with or without carbon dioxide) into the upper atmosphere. For marine impacts, Toon *et al.* (1994) calculated that an object substantially smaller than the water depth would vaporise water equivalent to 11 times the mass of the impactor. Even impacts with energies as low as 10^4 Mt (an asteroid ~650 m across) could produce a global water vapour cloud with a mass equal to or larger than all the water currently in the upper atmosphere, and even for a 10^6-Mt impact – still a hundred times smaller than the K/T event – this could rise to 250 times. The details of how much warming is likely to result from such an injection of water vapour remain unresolved, however, and to some extent at least, ice clouds forming from the elevated water vapour levels may actually contribute to a counteractive cooling effect. Toon *et al.* (1994) suggested that residence time in the atmosphere for an impact-generated water cloud is short – perhaps only a few years, with horizontal transport processes carrying water-rich air to colder parts of the atmosphere, such as the winter polar regions. Given such circumstances, it is unlikely that additional impact-related water vapour is likely to result in a prolonged period of greenhouse warming.

For impacts into carbonate-rich targets, dramatically enhanced levels of atmospheric carbon dioxide might also be expected to contribute towards greenhouse warming. For the Chicxulub impact, however, the degree of warming is disputed, and although O'Keefe and Ahrens (1989) proposed carbon dioxide levels several times those of current values, recent estimates are lower. As Cretaceous carbon dioxide levels were substantially higher than those today, the addition of an impact-related component may not have been significant, and in fact Pierazzo *et al.* (1998) suggested that the end-Cretaceous atmospheric carbon dioxide inventory may have been enhanced only by 40 per cent at the most. Any resulting warming arising from an increase on this scale is likely to have been easily counteracted by the cooling effect of sulphur aerosols generated at the same time.

Using a different approach, Brinkhuis *et al.* (1998) studied marine dinoflagellate cyst (dinocyst) assemblages from samples straddling the K/T boundary around the world, in order to get a proxy measure of changes in contemporary sea surface temperatures. They report a cooling across the K/T boundary immediately followed by an interval of pronounced global warming, which may have been long-lasting enough to ensure climate instability for ~100,000 years after the K/T impact. Clearly, then, constraints on both the degree and duration of post-impact global warming are currently poor, but a better picture may emerge with further research targeted at determining proxy temperature records of the period.

7.5.3 Impacts and mass extinctions

Although the role of asteroids and comets in triggering mass extinctions has found serious consideration only following the Alvarez *et al.* (1980) paper, if we look back in the literature we find – as often happens – that the hypothesis had already found some favour. Trailing the ideas of Alvarez and his colleagues, De Laubenfels (1956) suggested in a paper in the *Journal of Paleontology* that

> the survivals and extinctions at the close of the Cretaceous are such as might be expected to result from intensely hot winds such as would be generated by extra large meteorite or planetesimal impacts. It is suggested that, when the various hypotheses as to dinosaur extinction are being considered, this one be added to the others.

Although the precise mechanism(s) now put forward to account for extinctions at the K/T boundary differ from the 'hot wind' model of De Laubenfels, his ideas must be applauded as being far ahead of their time.

In considering impact events and mass extinctions, three questions need to be addressed: (a) Did an impact cause the K/T mass extinction? (b) Did impacts cause other mass extinctions? (c) Are all mass extinctions the result of impacts? All are difficult questions to answer. Although many aspects of a purely impact-triggered K/T extinction are challenged, few, if any, now deny the existence of a major impact event at the K/T boundary. Some do, however, argue that it had little or nothing to do with the end-Cretaceous extinction. McLean (1981) and Officer and Drake (1985) among others, for example, prefer a volcanic cause, while Stanley (1987) and Hallam and Wignall (1999) implicate the dramatic (~150 m) late-Cretaceous sea-level drawdown. Still others (e.g. MacLeod, 1996) concluded, on the basis of cladistic surveys, that there was no catastrophic extinction at the K/T boundary, instead preferring a gradual decline in species, perhaps related to major climate changes and the coldest seawater temperatures of any period within the Cretaceous. This view is disputed, however, and Smit *et al.* (1996) recognised a mass mortality in the oceans immediately following the K/T impact, which continued at a slower rate for several thousand years due to post-impact raised sea surface temperatures.

In a summary of the characteristics of catastrophic extinctions, Rampino and co-workers (1997) listed five pieces of evidence that they believe support categorically the existence of a mass extinction at the K/T boundary. These are:

1. a globally synchronous or near-synchronous mass mortality level or levels, marked by a negative shift in carbon isotope ratios, $\delta^{13}C$, in marine carbonates, indicating a biomass loss and drop in productivity;
2. a proliferation of opportunistic species, followed by recovery and radiation of surviving species;
3. a marked negative shift in oxygen isotope ratios, $\delta^{18}O$, suggesting a brief global warming;
4. a positive shift in sulphur isotope ratios, $\delta^{34}S$, suggesting a trend towards anoxia in ocean waters;
5. a sharp reduction in biogenic $CaCO_3$.

The voluminous Deccan Trap continental flood basalt volcanism in India occurred, like the Chicxulub impact, around 65 million years ago, and has been held responsible by some authors for playing at least some role in the K/T extinction (see Chapter 5). Courtillot *et al.* (1996) pointed out, for example, that when the impact occurred, the global environmental effects of Deccan volcanism meant that the extinction was already under way, the implication being that the ramifications of the impact merely enhanced the environmentally detrimental effects of the volcanism. Diverging from the general consensus, Courtillot and his colleagues also proposed that the Chicxulub impactor was a one in a billion year event, implying that similar impacts could not be held responsible for other Phanerozoic mass extinctions. In support of an alternative mechanism, they note that 7 out of 10 mass extinction events in the geological record also coincide with major episodes of flood basalt magmatism, suggesting that this – rather than asteroid or comet impacts – is the major candidate for most mass extinctions (see Chapter 5). Interestingly, Hallam and Wignall (1999) purport to recognise a correlation between large eustatic inflections (sea-level changes) and the biggest five mass extinctions, including those at the K/T boundary, the end of the Permian and the end of the Triassic, all of which Courtillot and his colleagues attribute to flood basalt volcanism.

In contrast to both Hallam and Wignall (1999) and Courtillot *et al.* (1996), Rampino *et al.* (1997) prefer an impact cause for mass extinctions in general, observing that seven 'extinction pulses' – including the K/T and end-Triassic events – coincide with stratigraphic markers that suggest a major impact event, for example elevated iridium levels, shocked minerals or microtektites (Fig. 7.11). Rampino and his colleagues address the important issue of the rapidity of extinction, the most common source of argument between 'catastrophists' and 'gradualists' in the mass extinction debate. A central issue in this debate focuses on whether or not a stratigraphic hiatus

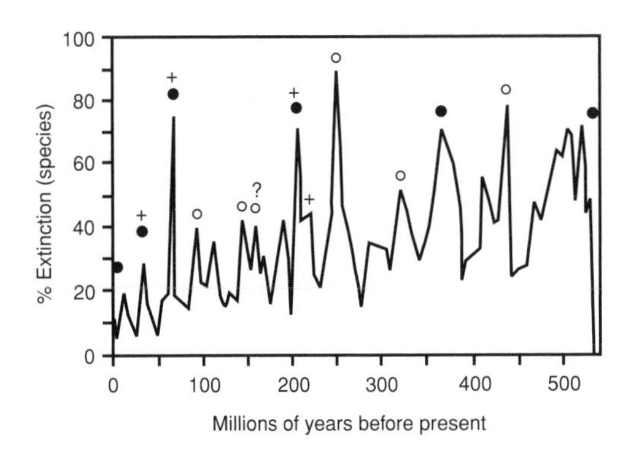

FIGURE 7.11 Correlation of impact events with mass extinctions in the geological record. The curve shows extinctions of marine species per geological stage (or sub-stage) during the Phanerozoic. Key: diagnostic stratigraphical evidence of impact (closed circles), possible stratigraphical evidence of impact (open circles), large dated impact craters (crosses). (After Rampino *et al.*, 1997)

between the last occurrence of a group of fossils and a major stratigraphic boundary actually indicates that the organisms died out before the boundary event. In this context, Rampino and his co-authors point out that the problem with using biostratigraphic last occurrences to infer patterns of extinction is that the method almost always underestimates the rapidity of extinction. This is because, as shown by Signor and Lipps (1982), sampling errors and the vagaries of preservation can cause apparent truncations of species ranges below the stratigraphic level of extinction, thereby making sudden extinctions appear gradual. This effect is invoked by Rampino and colleagues to explain apparently gradual faunal extinctions at the K/T boundary and other major stratigraphic horizons coincident with mass extinctions; their interpretation being rapid extinction due to impact. In the same vein, and focusing specifically on the vertebrate population, Cutler and Behrensmeyer (1996) show that even 100 per cent mortality at the K/T boundary would not provide enough bones to form widespread bone beds and their absence does not therefore rule out sudden mortality. As shown in Fig. 7.12, only in a relatively small region around the impact site would animals (except those in herds) have the opportunity to congregate in severely damaged ecosystems before dying through injury or starvation.

In an attempt to explain the K/T extinctions in terms of a combination of events, Glasby and Kunzendorf (1996) proposed that extinctions at the boundary resulted from already falling sea levels compounded by the environmental effects of the Deccan eruptions. These, the authors suggested, ejected sufficient gas and dust into the atmosphere to cool the climate, lower sea levels further, and reduce ocean temperatures, thereby causing a major reduction in ocean productivity. Within the Glasby and Kunzendorf model, the Chicxulub impact plays only a supporting, regional role in the coincident mass extinction.

Considerable controversy still exists over whether mass extinctions in the stratigraphic record are characterised by periodic components and – if they exist – what they mean (e.g. Raup and Sepkoski, 1984). Reported periodicities lie in

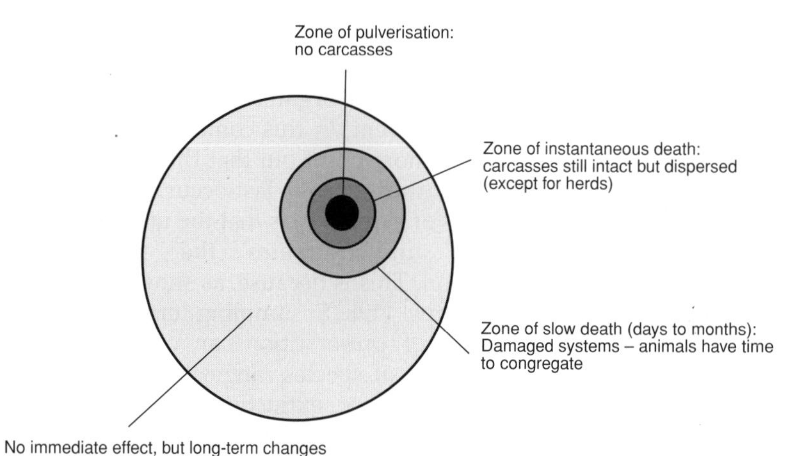

FIGURE 7.12 The hypothetical effect of an impact on land vertebrate taphonomy. Even 100 per cent mortality at the K/T boundary would not result in widespread bone beds, and their absence does not rule out sudden mortality. (After Cutler and Behrensmeyer, 1996)

the range of 26–30 million years (e.g. Raup and Sepkoski, 1986; Rampino and Stothers, 1987), and have been linked by Rampino and Haggerty (1996), among many others, with an extra-terrestrial origin. Pointing to the coincidence of around half (13 out of 25) of the recognised mass extinction events with either major impact craters or possible second-order evidence for impact, they proposed a mechanism to explain a periodic increase in the flux of solar system debris and therefore collisions with Earth. In their *Shiva* hypothesis, Rampino and Haggerty suggest that the passage of the solar system through the galactic plane – which occurs every 30 ± 3 million years – dislodges comets from the deep space comet source known as the Oort Cloud, sending them into the inner solar system, where they increase the flux of Earth impactors. Other researchers have also pursued this line of thinking and their evidence has been published in a series of papers of which Napier and Clube (1979) is the earliest and Clube and Napier (1996) the latest. Although heated debate on the subject is certain to continue for many years before being resolved, there does now appear to be reasonable evidence for comparable periodicities in mass extinctions (Raup and Sepkoski, 1984), the terrestrial cratering flux (Napier, 1998), and the passage of the solar system through the galactic plane. This has led to some authors linking periodic bombardment during so-called 'impact epochs' (Seyfert and Simkin, 1979) with the ~30-million-year geological cycles proposed by Arthur Holmes (1927) and characterised by a range of traumatic phenomena including climate and sea-level changes, mass extinctions, geomagnetic reversals and global volcanism. The corollary of this is that global environmental disturbances may well be forced in a quasi-periodic fashion through a combination of impacts and consequent prolonged climate stress.

8

Environmental change and natural hazards: prospects for the future

8.0 Chapter summary

In this short concluding chapter we look to the future and address the issue of how contemporary environmental change driven by anthropogenic global warming might lead to an increase in hazard. We examine one recent forecast for climate change over the next 80 years and explore its hazard implications. We also look at the human dimension and evaluate the critical role increasing vulnerability plays in exacerbating the impact of natural hazards on human society. In conclusion, we present a none-too-rosy hazard forecast for the future, while drawing attention to the fact that this and all forecasts of an increase in natural hazards due to global warming remain – at present – poorly constrained.

8.1 Environmental change and natural hazards: the impact in the twenty-first century

From the perspective of the first year of a new millennium, the impact of rapid-onset geophysical hazards over the course of the next century is difficult to forecast, particularly in light of the major uncertainties attached to predictions of environmental change over this period. Perhaps some constraints can be imposed, however, on the basis of climate change impact forecasts made in a recent report by the UK Meteorological Office (1999) (Box 8.1), and based upon the second Hadley Centre climate model. Assuming unmitigated greenhouse emissions, the report predicts a global temperature rise of 3 degrees Celsius by 2080, accompanied by a 41-cm rise in global mean sea level. Substantial diebacks of tropical forests and grasslands are forecast, while forest growth is promoted at higher latitudes. Water availability is predicted to fall in some parts of the world while rising in others, and patterns of cereal yields are expected to undergo similar dramatic changes. In natural hazard terms, an increase in coastal flooding is the most obvious consequence of the forecast changes, and the UK Met Office itself predicts a rise in the number of people affected annually from 13 to 94 million. Population growth in coastal zones is estimated using a projection of existing trends and assumes that the frequency of storms remains constant. Changes in either of these parameters would result in an even greater rise in coastal flood impact. In addition to an increase in major discrete flood events, elevated and rising sea levels will ensure progressive impacts on many coastlines, including inundation and increased erosion with the potential to cause instability, slumping and landslide formation. A consequence of the dieback of tropical forests and grasslands – forecast to affect 3 million km^2 by the end of the century – is an increase in desertification, which is likely to lead to a rise in the numbers and severity of dust storms in the affected regions. Desertification is likely also to be exacerbated in some regions, such as Australia and southern Europe, by reduced river run-off, while

Box 8.1 Forecast environmental change: the world in 2080 assuming unmitigated greenhouse gas emissions

Climate change

Global average temperature increase 3 degrees C

Natural ecosystems

Substantial dieback of tropical forests and tropical grasslands (particularly in northern South America and central South Africa)

Considerable growth of forests in North America, northern Asia and China

Water resources

Substantial decrease in the availability of water from rivers in Australia, India, Southern Africa, most of South America, and Europe

Increase in the availability of water from rivers in North America, Asia (particularly Central Asia), and central eastern Africa

Three billion people will suffer increased water resource stress

Agriculture

Increase in cereal yields at high and mid-latitudes such as North America, China, Argentina and much of Europe

Decrease in cereal yields in Africa, the Middle East and, particularly, India

Coastal effects

Sea level will be about 41 cm higher than today

The average annual number of people flooded will increase from 13 million to 94 million. The majority of the increase will occur in South Asia (from Pakistan, through India, Sri Lanka and Bangladesh, to Burma) and South-East Asia (from Thailand to Vietnam, and including Indonesia and the Philippines)

Sea-level rise will compound the decline of coastal wetlands

Sea-level rise will produce a range of progressive impacts on coastal lowlands and on low-lying coastal islands

Human health

An estimated 290 million additional people will be at risk of *falciparum* malaria (clinically more dangerous than the more widespread *vivax* malaria), mainly in China and Central Asia

Source: UK Meteorological Office (1999)

increased run-off in North America and parts of Asia and Africa may contribute to more flooding.

Prospects for windstorm frequency and severity in the late twenty-first century are unclear and at present difficult to model, owing to inadequate resolution in current climate models. Nevertheless, some recent modelling and theoretical studies predict a modest increase in the intensity of both tropical and extratropical cyclones, which, coupled to possible changes in circulation patterns, could result in significant increases in the windstorm hazard for some regions.

The UK Met Office report (1999) also includes alternative scenarios assuming stabilisation of atmospheric carbon dioxide at 550 ppm and 750 ppm. Stabilisation at the lower figure would delay a rise of 2 degrees C (expected by the 2050s with unmitigated emissions) for 100 years, while stabilisation at the higher figure would result in a 50-year delay. The coastal flood impacts of sea-level rise would also be reduced under the stabilisation scenarios. However, the warming already incurred will ensure that global sea-level rise continues for many centuries.

8.2 Natural hazards: the human dimension

The impact of natural hazards on society is clearly on the rise, although it still falls far below that due to environmental degradation and, in particular, civil strife (Fig. 8.1). Figures for the period 1900–90 indicate that almost 90 per cent of disaster-related deaths over the period can be attributed to war and famine, with all the natural hazards together making up the remainder. Notwithstanding this, the numbers of people affected by natural hazards during the 1970s and 1980s fell little short of a billion – somewhere between a fifth and a quarter of the Earth's population. With over 250 million people being affected by the 1996 and 1998 Chinese floods alone, similar figures for the last decade of the millennium are likely easily to top the billion mark. The increasing impact of natural hazards over the past half-century is without doubt linked to rapidly rising populations in particularly vulnerable regions. At greatest risk are the poorest inhabitants of developing countries, forced to live on marginal land in coastal zones and around cities. Around 50 per cent of the Earth's 6 billion inhabitants are now urban dwellers. For historical and geographical reasons, large and growing numbers of these people are concentrated in coastal environments, where flat land permits the rapid expansion of urban centres, and an estimated 75 per cent of the population of coastal zones now reside in towns and cities. Unfortunately, coastal environments are also the most vulnerable to natural hazards, and particularly the flooding associated with windstorms and tsunami caused by offshore earthquakes. Because of the common coincidence of coastlines and plate margins, volcanic eruptions and earthquakes often pose an additional threat. The Earth's population is currently growing at around 90 million people a year, with the largest increases concentrated in developing countries. Currently, 96 per cent of all deaths due to natural hazards and environmental degradation occur in developing countries, and this situation is unlikely to change as the rate of urbanisation increases in the most vulnerable regions. In 2007, for the first time more people will live in cities and towns than in the countryside. A little over 15 per cent of the planet's population currently live in *megacities* (defined as having populations in excess of 8 million people) and by 2020 this figure is expected to have climbed to over 30 per cent. As shown in Fig. 8.2 and Table 8.1, most of this increase will be accommodated by the growth of cities in developing countries characterised by low income and high sensitivity to natural hazards. Assuming a projected population by this time of between 7 and 8 billion, this places over 2

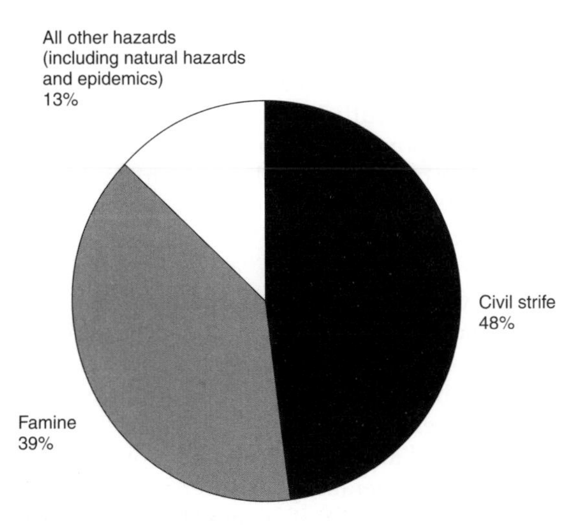

FIGURE 8.1 Breakdown of disaster-related deaths 1900–90. (After Blaikie *et al.*, 1994)

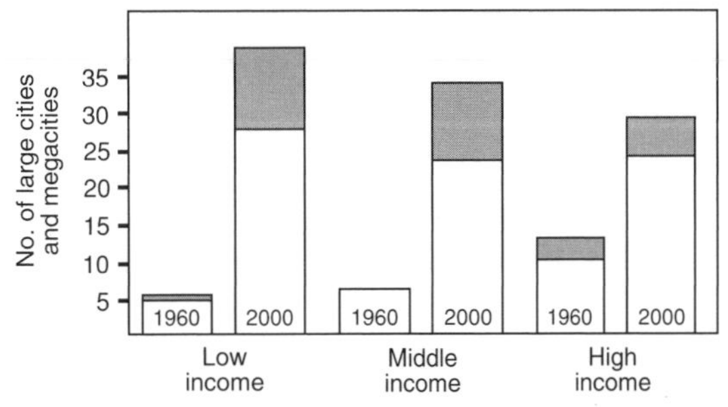

FIGURE 8.2 Predicted growth of large (3–8 million) cities and megacities (>8 million). Megacities shaded. Many of the most rapidly growing population concentrations are in developing countries that are most vulnerable to natural hazards.

TABLE 8.1 The ten most populous cities in the world in 1950 and 2015 (predicted). Reproduced with permission from Munich Reinsurance Group, major natural catastrophes from the 11th to the 19th century. MRNat Catservice, December 1999.

City	Population
1950	
1. New York	12,300,000
2. London	8,700,000
3. Tokyo	6,900,000
4. Moscow	5,400,000
5. Paris	5,400,000
6. Rhine-Ruhr (Essen)	5,300,000
7. Shanghai	5,300,000
8. Buenos Aires	5,000,000
9. Chicago	4,900,000
10. Calcutta	4,400,000
2015	
1. Tokyo	28,900,000
2. Mumbai	26,900,000
3. Lagos	24,600,000
4. São Paulo	20,300,000
5. Dhaka	19,500,000
6. Karachi	19,400,000
7. Mexico City	19,200,000
8. Shanghai	18,000,000
9. New York	17,600,000
10. Calcutta	17,300,000

Note (1) the huge increase in absolute numbers over the period, and (2) the dominance of the 2015 table by low per-capita income cities in developing countries.

billion people in particularly vulnerable circumstances.

8.3 Forecast for the future

Broadly speaking, the forecast for the twenty-first century and beyond is not, from a natural hazard perspective, a particularly promising one. Although increased losses from weather and climate extremes have been explained away in terms of human factors such as greater vulnerability and wealth concentration (e.g. Changnon *et al.*, 2000), it is likely that continued anthropogenic global warming will result in more extreme, and therefore more hazardous, meteorological phenomena. Observations of small numbers of extreme weather events are difficult to relate to climate change, as natural variability in the climate system – including spatial variations – is great, and statistical variations are large. Sometimes the scale and extent of these variations are not appreciated, because records are either too short or too inconsistent. Simplistic cause-and-effect linkage of natural hazard intensity and frequency with changing climatic parameters can be misleading, and can be supported only by the results of careful observation and modelling. Nevertheless, observations using homogeneous climate data sets and improved modelling of the ocean–atmosphere system at higher resolutions are starting now to agree with one another and bear fruit in terms of forecasts of future climate that are reasonably certain. Easterling *et al* (2000b) summarise the probability of various meteorological hazards changing over the twenty-first century, as determined by modelling and/or theory, together with the current consensus on observed hazard trends over the twentieth century (Table 8.2). In addition, rising sea level will certainly make coastal flooding worse and will exacerbate the impact of storm surges and tsunami.

In terms of geological hazards, we have no reason to believe yet that either the forecast climate change or accompanying sea-level change will be sufficient to trigger a volcanic or seismic response. Similarly, we know of no asteroid or comet on a collision course with the planet, although this could change at short notice. Equally, while the next volcanic super-eruption is statistically overdue, there is currently no evidence of an imminent eruption on this scale. It is most likely, therefore, that the increasing impact of natural hazards – at least for the coming century – will result from a rise in the aforementioned meteorological hazards accompanied and exacerbated by a progressive increase in vulnerability. In recent decades this has been a product of a number of different factors, most of which were brought about by rapid population growth, poor education and insufficient monetary resources. In regions threatened by windstorms, earthquakes, floods and landslides, widespread, uncontrolled construction is now a major problem, particularly in the shantytown sprawls that encompass megacities in South America and Asia. Planning and building regulations rarely exist, and if they do, they are unlikely to be enforced. This situation was highlighted particularly well by the poor response of thousands of apartment blocks to the earthquake of Richter magnitude 7.4 that took almost 20,000 lives in the Izmit region of north-west Turkey in 1999. Add to this limited public awareness of the hazard threat, minimal education and training in how to cope with hazard impacts, insufficient financial means and technical resources, exacerbation of the hazard problem by mismanagement of the environment, and the failure of countries in the developed world to address the issue seriously,

TABLE 8.2 Summary of analyses of different types of climate extremes (Reproduced with permission from Easterling, D.R., Meehl, G.A., Parmesan, C., Changnon, S.A., Karl, T.R. and Mearns, L.O. 2000. Climate extremes, observations, modeling and impacts. *Science* **289**, 2068–2074. Copyright (2000) American Association for the Advancement of Science. Easterling *et al.*, 2000b, who also provide detailed definitions of the uncertainty estimates)

	Observed (twentieth century)	**Modelling (end of twenty-first century)**
Simple extremes based on climate statistics		
Higher maximum temperatures	Very likely	Very likely
More hot summer days	Likely	Very likely
Increase in heat index	Likely	Very likely
Higher minimum temperatures	Virtually certain	Very likely
Fewer frost days (higher minimum temperatures)	Virtually certain	Likely*
More heavy 1-day precipitation events (increased intensity of precipitation events)	Likely	Very likely
More heavy multiday precipitation events (increased intensity of precipitation events)	Likely	Very likely
Complex event-driven climate extremes		
More heatwaves	Possible	Very likely* (higher maximum temperatures)
Fewer cold waves	Very likely	Very likely* (higher minimum temperatures)
More drought	Unlikely	Very likely (reduced mid-latitude summer soil moisture)
More wet spells	Likely	Likely (increased precipitation at mid- and high latitudes in winter)
More tropical storms	Unlikely	Possible
More intense tropical storms	Unlikely	Possible
More intense mid-latitude storms	Possible	Possible
More intense El Niño events	Possible	Possible
More common El Niño-like conditions	Likely	Likely*

*No direct model analyses, but these changes are physically plausible on the basis of other simulated model changes; comparable changes simulated by the models are noted in parentheses.
The assessment in this table is a general one where observed and model changes appear to be representative and physically consistent with a majority of changes globally. Where the observed changes agree with the models, they are qualitatively consistent with climate changes expected from increasing greenhouse gases.

and the rapid rise of natural catastrophes in vulnerable developing countries is easily explained. Despite attention being drawn to the problem in the 1990s through the UN International Decade for Natural Disaster Reduction (IDNDR) initiative, prospects for reducing the impact of natural hazards on the populations of developing countries in the short to medium term look bleak. The follow-up UN programme – the International Strategy for Natural Disaster Reduction (ISDR) – is admirable in both its aims and objectives, but it is highly unlikely to offer workable and rapid solutions on a global scale.

It might be hoped that increasing vulnerability to natural hazards could be at least partially mitigated by a reduction in greenhouse gas emissions and a consequent slowdown in the rate of anthropogenic environmental change. From the perspective of late summer 2000, however, the prospects do not look bright. Under the 1997 Kyoto Protocol, industrialised countries signed up to legally binding targets to reduce their greenhouse gas emissions, which for the period 2008–12 should result in an overall reduction to at least 5 per cent below 1990 levels. Sadly, continuing procrastination and political manoeuvring since the Kyoto meeting have ensured that the protocol still has not come into force. Some countries have made significant progress. In Germany, for example, CO_2 emissions in 1999 were over 15 per cent lower than in 1990, while in the UK, greenhouse gas emissions in 1998 were 8.5 per cent below 1990 values. These moderate 'success stories' contrast sharply with the spectacular failures of other industrialised countries. Japan, the United States and Australia, for example, have seen significant rises in their CO_2 in the past few years, of between 9 and 12 per cent. Given such a lack of political will, it seems unlikely that we will be able to avoid the consequences of significant environmental change in the coming century or two, unless and until the more hazardous aspects of such a dramatic climatic shift make themselves felt.

References

Adams, J., Maslin, M. and Thomas, E. 1999. Sudden climate transitions during the Quaternary. *Progress in Physical Geography* **23**, 1–36.

Adushkin, V.V. and Nemchimov, I.V. 1994. Consequences of impacts of cosmic bodies on the surface of the Earth. In Gehrels, T. (ed.) *Hazards due to asteroids and comets*. University of Arizona Press, Tucson, 721–778.

Agee, E.M. 1991. Trends in cyclone and anticyclone frequency and comparison with periods of warming and cooling over the Northern Hemisphere. *Journal of Climate* **4**, 263–267.

Alexandersson, H., Schmith, T., Iden, K. and Tuomenvirta, H. 1998. Long-term variations of the storm climate over NW Europe. *The Global Atmosphere and Ocean System* **6**, 97–120.

Alexandersson, H., Tuomenvirta, H., Schmith, T. and Iden, K. 2000. Trends of storms in NW Europe derived from an updated pressure data set. *Climate Research* **14**, 71–73.

Alibés, B., Rothwell, R.G., Canals, M., Weaver, P.P.E. and Alonso, B. 1999. Determination of sediment volumes, accumulation rates and turbidite emplacement frequencies on the Madeira Abyssal Plain (NE Atlantic): a correlation between seismic and borehole data. *Marine Geology* **160**, 225–250.

Alley, R.B., Meese, D.A., Schuman, C.A. *et al.* 1993. Abrupt increase in Greenland snow accumulation at the end of the Younger Dryas event. *Nature* **362**, 527–529.

Alvarez, W. and Muller, R. 1984. Evidence for crater ages for periodic impact on Earth. *Nature* **308**, 718–720.

Alvarez, L.W., Alvarez, W., Asaro, F. and Michel, H.V. 1980. Extraterrestrial cause for the Cretaceous–Tertiary extinction. *Science* **208**, 1095–1108.

Ambrose, S.H. 1998. Late Pleistocene human population bottlenecks, volcanic winter, and differentiation of modern humans. *Journal of Human Evolution* **34**, 623–651.

Anderson, D.C. 1974. Earthquakes and the rotation of the Earth. *Science* **186**, 49–50.

Angel, J.R. and Isard, S.A. 1998. The frequency and intensity of Great Lake cyclones. *Journal of Climate* **11**, 61–71.

Angell, J.K. 1988. Impact of El Niño on the delineation of tropospheric cooling due to volcanic eruptions. *Journal of Geophysical Research* **93**, 3697–3704.

Angell, J.K. and Koshover, J. 1988. Impact of El Niño on the delineation of tropospheric cooling due to volcanic eruptions. *Journal of Geophysical Research* **93**, 3697–3704.

Argyle, E. 1989. The global fallout signature of the K–T bolide impact. *Icarus* **77**, 220–222.

Artemieva, N. and Shuvalov, V. 1999. Shock zones on the ocean floor: numerical simulations. *Reports on Polar Research* **343**, 16–18.

Atkinson, B.K. 1984. Subcritical crack growth in geological materials. *Journal of Geophysical Research* **89**, 4077–4114.

Bagnold, R.A. 1954. Experiments on a gravity free dispersion of large solid particles in a Newtonian fluid under shear. *Proceedings of the Royal Society of London A* **225**, 49–63.

Baillie, M.G.L. 1991. Suck in and smear, two related chronological problems for the 90s. *Journal of Theoretical Archaeology* **2**, 12–16.

Bedient, P.B. and Huber, W.C. 1992. *Hydrology and floodplain analysis*, 2nd edition. Addison-Wesley, Reading, MA.

Beer, T. and Williams, A. 1995. Estimating Australian forest fire danger under conditions of doubled carbon dioxide concentrations. *Climatic Change* **29**, 169–188.

Begét, J.E. and Kienle, J. 1992. Cyclic formation of debris avalanches at Mount St. Augustine volcano. *Nature* **356**, 701–704.

Bekki, S. 1995. Oxidation of volcanic SO_2: a sink for stratospheric OH and H_2O. *Geophysical Research Letters* **22**, 913–916.

Bekki, S., Pyle, J.A. and Pyle, D.M. 1996. The role of microphysical and chemical processes in prolonging climate forcing of the Toba eruption. *Geophysical Research Letters* **23**, 2669–2672.

Bell, B. 1999. The liquid Earth. *The Atlantic Monthly*, January. Online version at www3.theatlantic.com/issues/99jan/mudslide.htm

Belousov, A.B. 1994. Large-scale sector collapses at Kurile–Kamchatka volcanoes in the 20th century. In *Abstracts of the International Conference on Volcano Instability on the Earth and Other Planets*. Geological Society, London.

Bender, M.A. and Ginis, I. 2000. Real-case simulations of hurricane-ocean interaction using a high resolution coupled model: effects on hurricane intensity. *Monthly Weather Reiew.* **128**, 917–946.

Bengtsson, L., Botzet, M. and Esch, M. 1996. Will greenhouse gas-induced warming over the next 50 years lead to higher frequency and greater intensity of hurricanes? *Tellus* **48A**, 57–73.

Bentley, M.L. and Mote, T.L. 1998. A climatology of derecho-producing mesoscale convective systems in the

Central and Eastern United States, 1986–1995: I. Temporal and spatial distribution. *Bulletin of the American Meteorological Society* **79**, 2527–2540.

Berger, A. and Loutre, M.F. 1991. Insolation values for the climate of the last 10 million years. *Quaternary Science Reviews* **10**, 297–317.

Bernardet, L.R., Grasso, L.D., Nachamkin, J.E., Finley, C.A. and Cotton, W.R. 2000. Simulating convective events using a high-resolution mesoscale model. *Journal of Geophysical Research* **105**, 14963–14982.

Blackford, J.J., Edwards, K.J., Dugmore, A.J., Cook, J.T. and Buckland, P.C. 1992. Hekla-4 Icelandic volcanic ash and the mid-Holocene Scots Pine decline in northern Scotland. *The Holocene* **2**, 260–265.

Blaikie, P., Cannon, T., Davis, I. and Wisner, B. 1994. *At Risk*. Routledge, London.

Blanchon, P. and Shaw, J. 1995. Reef drowning during the last deglaciation: evidence for catastrophic sea-level rise and ice-sheet collapse. *Geology* **23**, 4–8.

Blodgett, T.A., Blizard, C. and Isacks, B.L. 1998. Andean landslide hazards. In Kalvoda, J. and Rosenfeld, C.L. (eds) *Geomorphological hazards in high mountain areas*, Kluwer, Dordrecht, 211–227.

Bloom, A.L. 1971. Glacial-eustatic and isostatic controls of sea level since the last glaciation. In Turekian, K.K. (ed.) *The late Cenozoic glacial ages*. Yale University Press, New Haven, CT, 355–379.

Bloom, A.L., Broecker, W.S., Chappell, J.M.A., Matthews, R.K. and Mesolella, K.J. 1974. Quaternary sea-level fluctuations on a tectonic coast: New $^{230}Th/^{234}Ur$ dates from the Huon Peninsula, New Guinea. *Quaternary Research* **4**, 185–205.

Bluestein, H.B. 1999a. *Tornado alley*. Oxford University Press, Oxford.

Bluestein, H.B. 1999b. A history of severe-storm-intercept field programs. *Weather and Forecasting* **14**, 558–577.

Bluestein, H.B., LaDue, J.G., Stein, H. and Speheger, D. 1993. Doppler radar wind spectra of supercell tornadoes. *Monthly Weather Review* **121**, 2200–2221.

Bluth, G.J.S., Doiron, S.D., Krueger, A.J., Walter, L.S. and Schnetzler, C.C. 1992. Global tracking of the SO_2 clouds from the June 1991 Mount Pinatubo eruptions. *Geophysical Research Letters* **19**, 151–154.

Bluth, G.J.S., Rose, W.I., Sprod, I.E. and Krueger, A.J. 1997. Stratospheric loading of sulphur from explosive volcanic eruptions. *Journal of Geology* **105**, 671–683.

Bluth, G.J.S., Schnetzler, C.C., Krueger, A.J. and Walter, L.S. 1993. The contribution of explosive volcanism to global atmospheric sulphur dioxide concentrations. *Nature* **366**, 327–329.

Bohor, B.F. 1990. Shocked quartz and more; impact signatures in Cretaceous/Tertiary boundary clays. In Sharpton, V.L. and Ward, P.D. (eds) *Global catastrophes in Earth history: an interdisciplinary conference on impacts, volcanism, and mass mortality*. Geological Society of America Special Paper **247**, 335–342.

Bohor, B.F., Foord, E.E., Modreski, P.J. and Triplehorn, D.M. 1984. Mineralogical evidence for an impact event at the Cretaceous/Tertiary boundary. *Science* **224**, 867–869.

Bondevik, S. and Svendson, S. 1995. Palaeotsunamis in the Norwegian and North Seas. University of Bergen. Unpublished.

Borgia, A. 1994. Dynamic basis of volcano spreading. *Journal of Geophysical Research* **99**, 17791–17804.

Boslough, M.B., Chael, E.P., Trucano, T.G., Crawford, D.A. and Campbell, D.L. 1996. Axial focusing of impact energy in the Earth's interior: a possible link to flood basalts and hotspots. In Ryder, G., Fastovsky, D. and Gartner, S. (eds) *The Cretaceous-Tertiary Event and other Catastrophes in Earth History*. Geological Society of America Special Paper 307.

Boslough, M.B.E. and Crawford, D.A. 1997. Shoemaker-Levy 9 and plume-forming collisions on Earth. In Remo, J.L. (ed.) Near Earth objects. *Annals of the New York Academy of Sciences* **822**, 236–282.

Bourgeois, J., Hansen, T.A., Wiberg, P.L., and Kaufman, E.G. 1988. A tsunami deposit at the Cretaceous–Tertiary boundary in Texas. *Science* **241**, 567–570.

Bouws, E., Jannink, D. and Komen, G.J. 1996. The increasing wave height in the North Atlantic Ocean. *Bulletin of the American Meteorological Society* **77**, 2275–2277.

Brasseur, G. and Granier, C. 1992. Mount Pinatubo aerosols, chlorofluorocarbons, and ozone depletion. *Science* **257**, 1239–1242.

Bray, J.R. 1976. Volcanic triggering of glaciation. *Nature* **260**, 414–415.

Bray, J.R. 1977. Pleistocene volcanism and glacial initiation. *Science* **197**, 251–253.

Brinkhuis, H., Bujak, J.P., Smit, J., Versteegh, G.J.M. and Visscher, H. 1998. Dinoflagellate-based sea surface temperature reconstructions across the Cretaceous-Tertiary boundary. *Palaeogeography and Palaeobiology* **141**, 67–83.

Broccoli, A.J., Manabe, S., Mitchell, J.F.B. and Bengtsson, L. 1995. Comments on 'Global climate change and tropical cyclones': II. *Bulletin of the American Meteorological Society* **76**, 2243–2245.

Brooks, C.E.P. and Carruthers, N. 1953. *Handbook of statistical methods in meteorology*. HMSO, London.

Brown, P. 2001. Melting permafrost threatens the Alps. *Guardian* (London), 4 January, 3.

Brunsden, D. 1995. Learning to live with landslides: some British examples. In Horlick-Jones, T., Amendola, A. and Casale, R. (eds) *Natural risk and civil protection*. E. & F.N. Spon, London, 268–282.

Brunsden, D., Ibsen, M.-L. 1996. Mudslide. In Dikau, R., Brunsden, D., Schrott, L. and Ibsen, M.-L. (eds) *Landslide recognition, identification, movement and causes*. John Wiley, Chichester, 103–119.

Bryan, K. 1996. The steric component of sea-level rise associated with enhanced greenhouse warming. *Climate Dynamics* **12**, 545–555.

Bryant, E.A. 1991. *Natural hazards*. Cambridge University Press, Cambridge.

Bugge, T. 1983. Submarine slides on the Norwegian continental margin, with special emphasis on the Storegga area. *Continental Shelf Institute Publication* 110.

Bugge, T., Befring, S., Belderson, R.H. *et al.* 1987. A giant three-stage submarine slide off Norway. *Geo-Marine Letters* **7**, 191–198.

Bugge, T., Belderson, R.H. and Kenyon, N.H. 1988. The Storegga Slide. *Philosophical Transactions of the Royal Society A* **325**, 357–388.

Bühring, C., Sarnthein, M. and Leg 184 Shipboard Scientific Party. 2000. Toba ash layers in the South China Sea: evidence of contrasting wind directions during eruption *ca.* 74 ka. *Geology* **28**, 275–278.

Burgess, C. 1989. Volcanoes, catastrophe and the global crisis of the late second millennium BC. *Current Archaeology* **117**, 325–329.

Caine, N. 1980. Rainfall intensity–duration control of shallow landslides and debris flows. *Geografiska Annaler* **62A**, 23–27.

Camargo, Z.A. and Suárez, R.G. 1994. Evidencia sísmica del crater de impacto de Chicxulub (seismic evidence of the Chicxulub impact crater). *Boletín de la Asociación Mexicanade Geofísico de Exploración* **34**, 1–28 (in Spanish).

Campagnoni, F., De Antonis, L., Magosso, P., Perrone, R., Salsotto, L. and Tonanzi, P. 1998. Regione Campania. Emergenza idrogeologica del 5 maggio 1998. Rilievo geologico e geomorfologico dei versanti sovrastanti gli abitati di Sarno, Quindici, Siano e Bracigliano. *Regione Piemonte, Direzione Regionale Servizi Tecnici di Prevenzione, Quaderno* **11**, 52.

Campbell, C.S. 1989. Self lubrication for long runout landslides. *Journal of Geology* **97**, 653–665.

Campbell, C.S., Cleary, P.W. and Hopkins, M. 1995. Large-scale landslide simulations, global deformation, velocities and basal friction. *Journal of Geophysical Research* **100**, 8267–8283.

Cannon, S.H. and Ellen, S.D. 1985. Rainfall conditions for abundant debris avalanches, San Francisco Bay Region, California. *California Geology* **38**, 267–272.

Carnell, R.E. and Senior, C.A. 1998. Changes in mid-latitude variability due to increasing greenhouse gases and sulphate aerosols. *Climate Dynamics* **14**, 369–383.

Carnell, R.E., Senior, C.A. and Mitchell, J.F.B. 1996. An assessment of measures of storminess: simulated changes in northern hemisphere winter due to increasing CO_2. *Climate Dynamics* **12**, 467–476.

Carracedo, J.C., Day, S.J., Guillou, H. and Pérez Torrado, F.J. 1999. Quaternary collapse structures and the evolution of the western Canaries (La Palma and El Hierro). *Journal of Volcanology and Geothermal Research* **94**, 169–190.

Casale, R. and Margottini, C. (eds) 1996. *Meteorological events and natural disasters*. European Commission, Brussels.

Chadwick, W., Moore, J.G., Fox, C.G. and Christie, D.M. 1992. Morphologic similarities of submarine slope failures: south flank of Kilauea, Hawaii, and the southern Galápagos platform. *Eos* **73**, 507 (abstract).

Chan, J.C.L. and Shi, J. 1996. Long-term trends and inter-annual variability in tropical cyclone activity over the western North Pacific. *Geophysical Research Letters* **23**, 2765–2767.

Changnon, S.A. and Changnon, D. 2000. Long-term fluctuations in hail incidences in the United States. *Journal of Climate* **13**, 658–664.

Changnon, D., Noel, J.J. and Maze, L.H. 1995. Determining cyclone frequencies using equal-area circles. *Monthly Weather Review* **123**, 2285–2294.

Changnon, S.A., Pielke, R.A., Changnon, D., Sylves, R.T. and Pulwarty, R. 2000. Human factors explain the increased losses from weather and climate extremes. *Bulletin of the American Meteorological Society* **81**, 437–442.

Chapman, C.R. and Morrison, D. 1994. Impacts on the Earth by asteroids and comets: assessing the hazard. *Nature* **367**, 33–39.

Chappell, J. 1974. Late Quaternary glacio- and hydro-isostasy on a layered Earth. *Quaternary Research* **4**, 405–428.

Chappell, J. 1975. On possible relationships between Upper Quaternary glaciations, geomagnetism, and volcanism. *Earth and Planetary Science Letters* **26**, 370–376.

Chappell, J. and Shackleton, N.J. 1986. Oxygen and isotopes and sea level. *Nature* **349**, 137–140.

Chen, G.Q., Tyburczy, J.A. and Ahrens, T.J. 1994. Shock-induced devolatilization of calcium sulfate and implications for K–T extinctions. *Earth and Planetary Science Letters* **128**, 615–628.

Chesner, C.A., Rose, W.I., Deino, A., Drake, R. and Westgate, J.A. 1991. Eruptive history of Earth's largest Quaternary caldera (Toba, Indonesia) clarified. *Geology* **19**, 200–203.

Chu, P.-S. and Clark, J.D. 1999. Decadal variations of tropical cyclone activity over the central North Pacific. *Bulletin of the American Meteorological Society* **80**, 1875–1881.

Chu, P-S. and Wang, J. 1997. Tropical cyclone occurrences in the vicinity of Hawaii: are the differences between El Niño and non-El Niño years significant? *Journal of Climate* **10**, 2683–2689.

Church, M. and Miles, M.J. 1987. Meteorological antecedents to debris flow in southwestern British Columbia; some case studies. *Geological Society of America Reviews in Engineering Geology* **7**, 63–79.

Chyba, C.F., Thomas, P.J. and Zahnle, K.J. 1993. The 1908 Tunguska explosion: atmospheric disruption of a stony asteroid. *Nature* **361**, 40–44.

Clark, J.A. and Primus, J.A. 1987. Sea-level changes resulting from future retreat of ice sheets: an effect of CO_2 warming of the climate. In Tooley, M.J. and Shennan, I. (eds) *Sea-level changes*. Blackwell, Oxford, 365–370.

Clube, S.V.M. and Napier, W.M. 1996. Galactic dark matter and terrestrial periodicities. *Quarterly Journal of the Royal Astronomical Society* **37**, 617–642.

Cockell, C.S. and Blaustein, A.R. 2000. 'Ultraviolet spring' and the ecological consequences of catastrophic impacts. *Ecology Letters* **3**, 77–81.

Collins, J.M. and Mason, I.M. 2000. Local environmental conditions related to seasonal tropical cyclone activity in the Northeast Pacific basin. *Geophysical Research Letters* **27**, 3881–3884.

Coltorti, M., Dramis, F., Gentili, B., Cambianchi, G., Crescenti, U. and Sorriso-Valvo, M. 1985. The December 1982 Ancona landslide, a case of deep-seated gravitational slope deformation evolving at unsteady rate. *Zeitschrift für Geomorphologie* NF **29**, 335–345.

Corominas, J. 1996. The angle of reach as a mobility index for small and large landslides. *Canadian Geotechnical Journal* **33**, 260–271.

Courtillot, V., Jaeger, J.-J., Yang, Z., Féraud, G. and Hofmann, C. 1996. The influence of continental flood basalts on mass extinctions: where do we stand? In Ryder, G., Fastovsky, D. and Gartner, S. (eds) *The Cretaceous–Tertiary event and other catastrophes in Earth history*. Geological Society of America Special Publication Paper 307, 513–525.

Covey, C., Ghan, S.J., Walton, J.J. and Weissman, P.R. 1990. Global environmental effects of impact-generated aerosols: results from a general circulation model. In Sharpton, W.L. and Ward, P.D. (eds) *Global catastrophes in Earth history*. Geological Society of America Special Paper 247, 263–270.

Covey, C., Thompson, S.L., MacCracken, M.C. and Weissman, P.R. 1994. Global climatic effects of atmospheric dust from an asteroid or comet impact on Earth. *Global and Planetary Change* **9**, 263–273.

Crawford, D.A. 1997. Comet Shoemaker-Levy 9 fragment size estimates: how big was the parent body? In Remo, J.L. (ed.) Near Earth objects. *Annals of the New York Academy of Sciences* **822**, 155–173.

Crawford, D.A. and Mader, C.L. 1998. Modelling asteroid impact and tsunami. *Science of Tsunami Hazards* **16**, 21–30.

Croft, S.K. 1985. The scaling of complex craters. *Proceedings of the 15th Lunar and Planetary Science Conference*, 828–842.

Crowley, T.J. and Baum, S.K. 1995. Is the Greenland ice sheet bistable? *Palaeoceanography* **10**, 357–363.

Crowley, T.J. and Kim, K.-Y. 1999. Modelling the temperature response to forced climate change over the last six centuries. *Geophysical Research Letters* **26**, 1901–1904.

Cruden, D.M. and Varnes, D.J. 1996. Landslide types and processes. In Turner, A.K. and Schuster, R.L. (eds) *Landslides: investigation and mitigation*. Transportation Research Board Special Report 247. National Academy Press, Washington, DC, 36–75.

Cutler, A.H. and Behrensmeyer, A.K. 1996. *Models of vertebrate mass mortality events at the K/T boundary*. Geological Society of America Special Paper **307**, 375–379.

Dai, A., Fung, I.Y. and Del Genio, A.D. 1997. Surface observed global land precipitation variations during 1900–88. *Journal of Climate* **10**, 2943–2962.

Dansgaard, W., White, J.W.C. and Johnsen, S.J. 1989. The abrupt termination of the Younger Dryas climate event. *Nature* **339**, 532–534.

Darwin, C. 1859. *The Origin of Species*. John Murray, London.

Davis, R.E. and Dolan, R. 1993. Nor'easters. *American Scientist* **81**, 428–439.

Davis, M., Hut, P. and Muller, R.A. 1984. Extinction of species by periodic comet showers. *Nature* **308**, 715–717.

Dawson, A.G. 1992. *Ice Age Earth: late Quaternary geology and climate*. Routledge, London.

Dawson, A.G., Long, D. and Smith, D.E. 1988. The Storegga Slides: evidence from eastern Scotland for a possible tsunami. *Marine Geology* **82**, 271–276.

Day, S. J., Elsworth, D. and Maslin, M. 2000. *A possible connection between sea surface temperature variations, orographic rainfall patterns, water table fluctuations and giant lateral collapses of oceanic island volcanoes*. Abstract volume. Western Pacific Geophysics Meeting. Tokyo. WP251.

Day, S.J., Heleno da Silva, S.I.N. and Fonseca, J.F.B.D. 1999. A past giant lateral collapse and present-day flank instability of Fogo, Cape Verde Islands. *Journal of Volcanology and Geothermal Research* **94**, 191–218.

Day, S.J., McGuire, W.J., Elsworth, D., Carracedo, J.C. and Guillou, H. 1999. Do giant collapses of volcanoes tend to occur in warm, wet, interglacial periods and if so, why? *Abstract volume, week B, IUGG XXII General Assembly*. Birmingham, UK, 174.

De Laubenfels, M.W. 1956. Dinosaur extinction: one more hypothesis. *Journal of Paleontology* **30**, 207–218.

de Wolde, J.R., Bintanja, R. and Oerlemans, J. 1995. On thermal expansion over the last one hundred years. *Journal of Climate* **8**, 2881–2891.

Decker, R.W. 1990. How often does a Minoan eruption occur? In Hardy, D.A. (ed.) *Thera and the Aegean World III 2*. Thera Foundation, London, 444–452.

Dent, J.D. 1986. Flow properties of granular materials with large overburden loads. *Acta Mechanica* **64**, 111–122.

Dikau, R., Brunsden, D., Schrott, L. and Ibsen, M.-L. (eds) 1997. *Landslide recognition: identification, movement and causes*. John Wiley, Chichester, 251.

Dodge, R.E., Fairbanks, R.G., Benninger, L.K. and Maurrasse, F. 1983. Pleistocene sea levels from raised coral reefs of Haiti. *Science* **219**, 1423–1425.

Dodgshon, R.A., Gilbertson, D.D. and Grattan, J.P. 2000. Endemic stress, farming communities and the influence of Icelandic volcanic eruptions in the Scottish Highlands. In McGuire, W, J., Griffiths, D.R., Hancock, P.L. and Stewart, I.S. (eds) *The Archaeology of Geological Catastrophes*. Special Publication of the Geological Society 171. The Geological Society, London, 267–280.

Dorland, C., Tol, R.S.J., Olsthoorn, A.A. and Palutikof, J.P. 1999. Impacts of windstorms in the Netherlands: present risk and prospects for climate change. In Downing, T.E., Olsthoorn, A.J. and Tol, R.S.J. (eds) *Climate, change and risk*. Routledge, London.

Douglas, B.C. 1992. Global sea-level acceleration. *Journal of Geophysical Research* **97**, 12699–12706.

Douglas, B.C. 1995. Global sea-level change: determination and interpretation. *Reviews of Geophysics* (supplement), 1425–1432.

Douglas, B.C. 1997. Global sea-level rise: a redetermination. *Surveys in Geophysics* **18**, 279–292.

Downing, T.E., Gawith, M.J., Olsthoorn, A.J., Tol, R.S.J. and Vellinga, P. 1999a. Introduction. In Downing, T.E., Olsthoorn, A.J. and Tol, R.S.J. (eds) *Climate, change and risk*. Routledge, London.

Downing, T.E., Olsthoorn, A.J. and Tol, R.S.J. 1999b. *Climate, change and risk*. Routledge, London.

Driscoll, N.W., Weissel, J.K. and Goff, J.A. 2000. Potential for large-scale submarine slope failure and tsunami generation along the U.S. mid-Atlantic coast. *Geology* **28**, 407–410.

Druyan, L.M., Lonergan, P. and Eichler, T. 1999. A GCM investigation of global warming impacts relevant to tropical cyclone genesis. *International Journal of Climatology* **19**, 607–617.

Easterling, D.R., Evans, J.L., Groisman, P.Ya., Karl, T.R., Kunkel, K.E. and Ambenje, P. 2000a. Observed variability and trends in extreme climate events: a brief review. *Bulletin of the American Meteorological Society* **81**, 417–425.

Easterling, D.R., Meehl, G.A., Parmesan, C., Changnon, S.A., Karl, T.R. and Mearns, L.O. 2000b. Climate extremes: observations, modeling and impacts. *Science* **289**, 2068–2074.

Easterling, W.E. and Kates, R.W. 1995. Indexes of leading climate indicators for impact assessment. *Climatic Change* **31**, 623–648.

Ellen, S.D., Wieczorek, G.F., Brown, W.M. III and Herd, D.G. 1988. Introduction. In Ellen, S.D. and Wieczorek, G.F. (eds) *Landslides, floods, and marine effects of the storm of January 3–5, 1982, in the San Francisco Bay Region, California*. US Geological Survey Professional Paper **1434**, 1–5.

Elsner, J.B. and Kara, A.B. 1999. *Hurricanes of the North Atlantic: climate and society*. Oxford University Press, Oxford.

Elsner, J.B. and Kocher, B. 2000. Global tropical cyclone activity: a link to the North Atlantic Oscillation. *Geophysical Research Letters* **27**, 129–132.

Elsner, J.B., Jagger, T. and Niu, X.-F. 2000. Changes in the rates of North Atlantic major hurricane activity during the 20th century. *Geophysical Research Letters* **27**, 1743–1746.

Elsworth, D. and Day, S.J. 1999. Flank collapse triggered by intrusion: the Canarian and Cape Verde archipelagoes. *Journal of Volcanology and Geothermal Research* **94**, 323–340.

Elsworth, D. and Voight, B. 1995. Dyke intrusion as a trigger for large earthquakes and the failure of volcano flanks. *Journal of Geophysical Research* **100**, 6005–6024.

Emanuel, K.A. 1986. An air–sea interaction theory for tropical cyclones: I. Steady-state maintenance. *Journal of Atmospheric Science* **43**, 585–604.

Emanuel, K.A. 1987. The dependence of hurricane intensity on climate. *Nature* **326**, 483–485.

Emanuel, K.A. 1988. Towards a general theory of hurricanes. *American Scientist* **76**, 371–379.

Emanuel, K.A. 1991. The theory of hurricanes. *Annual Review of Fluid Mechanics* **23**, 179–196.

Emanuel, K.A. 1995. Comments on 'Global climate change and tropical cyclones': I. *Bulletin of the American Meteorological Society* **76**, 2241–2243.

Emanuel, K.A. 1997. Climate variations and hurricane activity: some theoretical issues. In Diaz, H.F. and Pulwarty, R.S. (eds) *Hurricanes: climate and socioeconomic impacts*. Springer, New York.

Emanuel, K.A. 1999. Thermodynamic control of hurricane intensity. *Nature* **401**, 665–669.

Emanuel, K.A. 2000. A statistical analysis of tropical cyclone intensity. *Monthly Weather Review* **128**, 1139–1152.

Engel, H. 1997. The flood events of 1993/4 and 1995 in the Rhine River basin. In *Destructive water: water-caused natural disasters, their abatement and control*. IAHS Publication **239**, 21–32.

Engelhardt, W. von., Luft, E., Arndt, J., Schock, H. and Weiskirchner, W. 1987. Origin of moldavites. *Geochimica et Cosmochimica Acta* **51**, 1425–1443.

Erismann, T.H. 1979. Mechanisms of large landslides. *Rock Mechanics* **12**, 15–46.

Essenwanger, O. 1976. *Applied statistics in atmospheric science*. Elsevier, Amsterdam.

Etkin, D.A. 1995. Beyond the year 2000, more tornadoes in western Canada? Implications from the historical record. *Natural Hazards* **12**, 19–27.

Fairbanks, R.G. 1989. A 17,000 year glacio-eustatic sea-level record: influence of glacial melting rates on the Younger Dryas event and deep ocean circulation. *Nature* **342**, 637–642.

Filmer, P.E., McNutt, M.K., Webb, H.F. and Dixon, D.J. 1994. Volcanism and archipelagic aprons in the Marquesas and Hawaiian Islands. *Marine Geophysics Research* **16**, 385–406.

Flageollet, J.-C. 1989. Landslides in France: a risk reduced by recent legal provisions. In Brabb, E.E. and Harrod, B.L. (eds) *Landslides: extent and economic significance*. Balkema, Rotterdam, 157–167.

Flannigan, M.D., Bergeron, Y., Engelmark, O. and Wotton, B.M. 1998. Future wildfire in circumboreal forests in relation to global warming. *Journal of Vegetation Science* **9**, 469–476

Folland, C.K., Miller, C., Bader, D. *et al.* 1999. Workshop on indices and indicators for climate extremes, Asheville, NC, USA, 3–6 June 1997. Breakout group C: temperature indices for climate extremes. *Climatic Change* **42**, 31–43.

Francis, P.W. 1994. Large volcanic debris avalanches in the central Andes. In *Abstracts of the International Conference on Volcano Instability on the Earth and Other Planets*. Geological Society, London.

Franklin, B. 1784. Meteorological imaginations and conjectures. *Memoir of the Manchester Literary and Philosophical Society* **2**, 373–377.

Free, M. and Robock, A. 1999. Global warming in the context of the Little Ice Age. *Journal of Geophysical Research* **104**, 19057–19070.

Genin, A., Lazar, B. and Brenner, S. 1995. Vertical mixing and coral death in the Red Sea following the eruption of Mount Pinatubo. *Nature* **377**, 507–510.

Gersonde, R. and Deutsch, A. 2000. Oceanic impacts: mechanisms and environmental perturbations – insight in a new field of impact research. *Eos* **81** (20), 221, 223, 228.

Gersonde, R., Kyte, F.T., Bleil, U. *et al.* 1997. Geological record and reconstruction of the late Pliocene impact of the Eltanin asteroid in the Southern Ocean. *Nature* **390**, 357–363.

Glasby, G.P. and Kunzendorf, H. 1996. Multiple factors in the origin of the Cretaceous-Tertiary boundary. The role of environmental stress and Deccan Trap volcanism. *Geologische Rundschau* **85**, 191–210.

Glass, B.P., Burns, C.A., Crosbie, J.R. and DuBois, D.L. 1985. Late Eocene North American microtektites and clinopyroxene-bearing spherules. *Proceedings of the 16th Lunar and Planetary Science Conference. Journal of Geophysical Research* (supplement) **90**, D175–D196.

Glasstone, S. and Dolan, P.J. 1977. *The effects of nuclear weapons*, 3rd edition. US Government Printing Office, Washington, DC.

Glazner, A.F., Manley, C.R., Marron, J.S. and Rojstaczer, S. 1999. Fire or ice: anticorrelation of volcanism and glaciation in California over the past 800,000 years. *Geophysical Research Letters* **26**, 1759–1762.

Goguel, J. 1978. Scale-dependent rockslide mechanisms, with emphasis on the role of pore fluid vaporization. In Voight, B. (ed.) *Rockslides and avalanches.1. Natural phenomena*. Elsevier, New York, 693–708.

Goudie, A.S. and Middleton, N.J. 1992. The changing frequency of dust storms through time. *Climatic Change* **20**, 197–225.

Grattan, J.P. and Brayshay, M.B. 1995. An amazing and portentous summer: environmental and social responses in Britain to the 1783 eruption of an Icelandic volcano. *Geographical Journal* **161**, 125–134.

Grattan, J.P. and Charman, D.J. 1994. Non-climatic factors and the environmental impact of volcanic volatiles: implications of the Laki fissure eruption of AD 1783. *Holocene* **4**, 101–106.

Grattan, J.P. and Pyatt, F.B. 1999. Volcanic dry fogs and the European palaeoenvironmental record: localised phenomena or hemispheric impacts? *Global and Planetary Change* **21**, 173–179.

Gray, W. 1979. Hurricanes: their formation, structure and likely role in the tropical circulation. In Shaw, D.B. (ed.) *Meteorology over the tropical oceans*. Royal Meteorological Society, James Glaisher House, 155–218.

Gray, W.M., Landsea, C.W., Mielke, P.W. and Berry, K.J. 1993. Predicting Atlantic Basin seasonal tropical cyclone activity by 1 August. *Weather and Forecasting* **8**, 73–86.

Gray, W.M., Sheaffer, J.D. and Landsea, C.W. 1997. Climate trends associated with multidecadal variability of Atlantic hurricane activity. In Diaz, H.F. and Pulwarty, R.S. (eds) *Hurricanes: climate and socioeconomic impacts*. Springer, New York, 15–53.

Grazulis, T.P. 1993. *Significant tornadoes, 1680–1991: a chronology and analysis of events*. St Johnsonbury, V: The Tornado Project of Environmental Films.

Grieve, R.A.F. 1997. Target Earth: evidence for large-scale impact events. In Remo, J.L. (ed.) Near Earth objects. *Annals of the New York Academy of Sciences* **822**, 319–352.

Grieve, R.A.F. 1998. Extraterrestrial impacts on Earth: the evidence and the consequences. In Grady, M.M., Hutchinson, R., McCall, G.J.H. and Rothery, D.A. (eds) *Meteorites: Flux with Time and Impact Effects*. Geological Society of London Special Publication 140, 105–131.

Grieve, R.A.F. and Pesonen, L.J. 1996. Terrestrial impact craters: their spatial and temporal distribution and impacting bodies. *Earth, Moon, and Planets* **72**, 357–376.

Grieve, R.A.F. and Shoemaker, E.M. 1994. The record of past impacts on Earth. In Gehrels, T. (ed.) *Hazards Due to Asteroids and Comets*. University of Arizona Press, Tucson, 417–462.

Griffiths, D.J., Colquhoun, J.R., Blatt, K.L. and Casinader, T.R. 1993. *Climatic Change* **25**, 369–388.

Groisman, P.Ya. 1985. Regional climate consequences of volcanic eruptions (in Russian). *Meteorology and Hydrology* **4**, 39–45.

Groisman, P.Ya. 1992. Possible regional climate consequences of the Pinatubo eruption: an empirical approach. *Geophysical Research Letters* **19**, 1603–1606.

Groisman, P.Ya., Karl, T.R., Easterling, D.R. *et al.* 1999. Changes in the probability of heavy precipitation: important indicators of climatic change. *Climatic Change* **42**, 243–283.

Grudd, K., Briffa, K.R., Gunnarson, B.E. and Linderholm, H.W. 2000. Swedish tree rings provide new evidence in support of a major, widespread environmental disruption in 1628 BC. *Geophysical Research Letters* **27**, 2957–2960.

Hadley, J.B. 1978. Madison Canyon rockslide, Montana, U.S.A. In Voight, B. (ed.) *Rockslides and avalanches.1. Natural phenomena*. Elsevier, Amsterdam, 167–180.

Hálfdánarson, G. 1984. Loss of human lives following the Laki eruption. In Gunnlaugsson, G.A. and Rafnsson, S. (eds) *Skaftáreldar 1783–1784: Ritgerdir og heimildir*. Mál og Menning, Reykjavik, 139–162 (in Icelandic).

Hall, K. 1982. Rapid deglaciation as an initiator of volcanic activity: an hypothesis. *Earth Surface Processes and Landforms* **7**, 45–51.

Hall, N.M.J., Hoskins, B.J., Valdes, P.J. and Senior, C.A. 1994. Storm tracks in a high-resolution GCM with doubled carbon dioxide. *Quarterly Journal of the Meteorological Society* **120**, 1209–1230.

Hallam, A. and Wignall, P.B. 1999. Mass extinctions and sea-level changes. *Earth Science Reviews* **48**, 217–250.

Hallbauer, D.K,. Wager, H. and Cook, N.G.W. 1973. Some observations concerning the microscopic and mechanical behaviour of quarzite specimens in stiff, triaxial compression tests. *International Journal of Rock Mechanics and Mining Science* **10**, 713–726.

Hammer, C.U., Clausen, H.B. and Dansgaard, W. 1980. Greenland ice sheet evidence of post-glacial volcanism and its climatic impact. *Nature* **288**, 230–235.

Handler, P. 1986. Possible association between the climatic effects of stratospheric aerosols and sea-surface temperatures in the eastern tropical Pacific Ocean. *Journal of Climatology* **6**, 31–41.

Hanks, T.C. 1977. Earthquake stress-drops, ambient tectonic stresses, and stresses that drive plates. *Pure and Applied Geophysics* **115**, 441–458.

Haq, B.U., Hardenbol, J. and Vail, P.R. 1987. Chronology of fluctuating sea levels since the Triassic. *Science* **235**, 1156–1166.

Harbitz, C.B. 1992. Model simulations of tsunamis generated by the Storegga Slides. *Marine Geology* **105**, 1–21.

Harrison, D.E. and Larkin, N.K. 1997. Darwin sea level pressure, 1876–1996: evidence for climate change? *Geophysical Research Letters* **24**, 1779–1782.

Harvey, L.D.D. 1994. Transient temperature and sea-level response of a two-dimensional ocean-climate model to greenhouse gas emissions. *Journal of Geophysical Research* **99**, 18447–18466.

Harvey, L.D.D. 2000. *Global warming: the hard science*. Prentice Hall, Harlow.

Hayashi, J.N. and Self, S. 1992. A comparison of pyroclastic flows and debris avalanche mobility. *Journal of Geophysical Research* **97**, 9063–9071.

Hearty, P.J. 1997. Boulder deposits from large waves during the last interglaciation on north Eleuthera Island, Bahamas. *Quaternary Research* **48**, 326–338.

Hearty, P.J., Conrad Neumann, A. and Kaufman, D.S. 1998. Chevron ridges and runup deposits in the Bahamas from storms late in oxygen-isotope substage 5e. *Quaternary Research* **50**, 309–322.

Heino, R., Brázdil, R., Førland, E. *et al.* 1999. Progress in the study of climatic extremes in Northern and Central Europe. *Climatic Change* **42**, 151–181.

Henderson, K.G. and Robinson, P.J. 1994. Relationships between Pacific/North American teleconnection patterns and precipitation events in the south-eastern USA. *International Journal of Climatology* **14**, 307–323.

Henderson-Sellers, A., Zhang, H., Berz, G. *et al.* 1998. Tropical cyclones and global climate change: a post-IPCC assessment. *Bulletin of the American Meteorological Society* **79**, 19–38.

Hendron, A.J. and Patton, F.D. 1985. *The Vaiont slide, a geotechnical analysis based on new geologic observations of the failure surface*. US Army Corps of Engineers Technical Report GL-85-5 (2 volumes).

Herbert, D.W.M. and Merkens, J.C. 1961. The effect of suspended mineral solids on the survival of trout. *International Journal of Air and Water Pollution* **5**, 46–55.

Hewitt, K. 1988. Catastrophic landslide deposits in the Karakoram Himalaya. *Science* **242**, 64–67.

Hildebrand, A.R., Penfield, G.T., Kring, D.A. *et al.* 1991. Chicxulub crater: a possible Cretaceous/Tertiary boundary impact crater on the Yucatán peninsula, Mexico. *Geology* **19**, 867–871.

Hillaire-Marcel, C., Occhietti, S. and Vincent, J.-S. 1981. Sakami moraine, Quebec: a 500 km long moraine without climate control. *Geology* **9**, 210–214.

Hills, J.G. and Goda, M.P. 1993. The fragmentation of small asteroids in the atmosphere. *Astronomical Journal* **105**, 1114–1144.

Hills, J.G. and Goda, M.P. 1998. Tsunami from asteroid and comet impacts: the vulnerability of Europe. *Science of Tsunami Hazards* **16**, 3–10.

Hills, J.G. and Mader, C.L. 1997. Tsunami produced by the impacts of small asteroids. In Remo, J.L. (ed.) Near Earth objects. *Annals of the New York Academy of Sciences* **822**, 381–393.

Hills, J.G., Nemchinov, I.V., Popov, S.P. and Teterev, A.V. 1994. Tsunami generated by small asteroid impacts. In Gehrels, T. (ed.) *Hazards Due to Asteroids and Comets*. University of Arizona Press, Tucson, 779–789.

Hirono, M. 1988. On the trigger of El Niño–Southern Oscillation by the forcing of early El Chichón volcanic aerosols. *Journal of Geophysical Research* **93**, 5365–5384.

Hodell, D.A., Curtis, J.H. and Brenner, M. 1995. Possible role of climate in the collapse of Classic Maya civilisation. *Nature* **375**, 391–394.

Hoek, E. and Bray, J.W. 1981. *Rock slope engineering*, 3rd edition (rev.). IMM and E. & F.N. Spon, London, 358.

Hofman, D.J. and Rosen, J.M. 1987. On the prolonged lifetime of the El Chichón sulfuric acid aerosol cloud. *Journal of Geophysical Research* **92**, 9825–9830.

Holcomb, R.T. and Searle, R.C. 1991. Large landslides from oceanic volcanoes. *Marine Geotechnology* **10**, 19–32.

Holland, G.J. 1997. The maximum potential intensity of tropical cyclones. *Journal of Atmospheric Science* **54**, 2519–2541.

Holmes, A. 1927. *The age of the Earth – an introduction to geological ideas*. Benn, London.

Hopkins, L.C. and Holland, G.J. 1997. Australian heavy-rain days and asociated east coast cyclones: 1958–92. *Journal of Climate* **10**, 621–635.

Houghton, J.T. 1997. *Global warming: the complete briefing*, 2nd edition. Cambridge University Press, Cambridge.

Howard, K.E. 1973. Avalanche mode of motion, implication from lunar examples. *Science* **180**, 1052–1055.

Hsü, K.J. 1969. Role of cohesive strength in the mechanics of overthrust faulting and of landsliding. *Geological Society of America Bulletin* **80**, 927–952.

Hsü, K.J. 1975. On sturzstroms – catastrophic debris streams generated by rockfalls. *Geological Society of America Bulletin* **86**, 129–140.

Hsü, K.J. 1978. Albert Heim: observations on landslides and relevance to modern interpretations. In Voight, B. (ed.) *Rockslides and avalanches. 1. Natural phenomena*. Elsevier, New York, 71–93.

Hubbert, M.K. and Rubey, W.W. 1959. Role of fluid pressure in mechanics of overthrust faulting. *Geological Society of America Bulletin* **70**, 115–160.

Huder, J. 1976. *Creep in Bundner schist*, Norwegian Geotechnical Institute (Laurits Bjerrum Memorial Volume), Oslo, 125–153.

Humphreys, W.J. 1913. Volcanic dust and other factors in the production of climatic changes, and their possible relation to ice gases. *Journal of the Franklin Institute*, August, 131–172.

Humphreys, W.J. 1940. *Physics of the Air*. Dover, New York.

Hurrell, J.W. 1995. Decadal trends in the North Atlantic oscillation: regional temperatures and precipitation. *Science* **269**, 676–679.

Hutchinson, J.N. 1988. General report: morphological and geotechnical parameters of landslides in relation to geology and hydrogeology. *Proceedings of the 5th International Symposium on Landslides* (Lausanne) **1**, 353–358.

Hutchinson, J.N. and Bhandari, R.K. 1971. Undrained loading, a fundamental mechanism of mudflows and other mass movements. *Geotechnique* **21**, 353–358.

Hutton, J. 1788. Theory of the Earth; or an investigation of the laws observable in the composition, dissolution, and restoration of land upon the globe. *Transactions of the Royal Society of Edinburgh* **3** (2), 209–304.

Huybrechts, P. and de Wolde, J. 1999. The dynamic response of the Greenland and Antarctic ice sheets to multiple-century climatic warming. *Journal of Climate* **12**, 2169–2187.

Hyde, W.T. and Crowley, T.J. 2000. Probability of future climatically significant volcanic eruptions. *Journal of Climate* **13**, 1445–1450.

Inokuchi, T. 1988. Gigantic landslides and debris avalanches on volcanoes in Japan. In *Proceedings of the Kagoshima International Conference on Volcanoes*. National Institute for Research Administration, Japan, 456–459.

International Federation of Red Cross and Red Crescent Societies. 2000. *World Disaster Report 2000*. Geneva.

IPCC 1994. Summary for policy makers. In Houghton, J.T., Meira Filho, L.G., Bruce, J. *et al.* (eds) *Climate Change 1994: radiative forcing of climate change*. The 1994 Report of the Scientific Assessment Group of the Intergovernmental Panel on Climate Change. Cambridge University Press, Cambridge.

IPCC 1996. Summary for policy makers. In Houghton, J.T., Meira Filho, L.G., Callander, B.A., Harris, N., Kattenberg, A. and Maskell, K. (eds) *Climate change 1995: the science of climate change*. The Second Assessment Report of the Intergovernmental Panel on Climate Change: Contribution of Working Group 1. Cambridge University Press, Cambridge.

Ivanov, B.A. 1999. Comet impacts to the ocean: numerical analysis of Eltanin scale events. *Reports on Polar Research* **343**, 47.

Iverson, R.M. 1997. The physics of debris flows. *Reviews in Geophysics* **35**, 245–296.

Iverson, R.M., Schilling, S.P. and Vallance, J.W. 1998. Objective delineation of lahar-inundation hazard zones. *Geological Society of America Bulletin* **110**, 972–984.

Izett, G.A. 1991. Tektites in the Cretaceous–Tertiary boundary rocks on Haiti and their bearing on the Alvarez extinction hypothesis. *Journal of Geophysical Research* **96**, 20879–20905.

Jacobi, R.D. and Hayes, D.E. 1982. Bathymetry, microphysiography, and reflectivity characteristics of the West African margin between Sierra Leone and Mauritania. In Von Rad, U., Hinz, K., Sarnthein, M. and Seibold, E. (eds) *Geology of the northwest African continental margin*. Springer-Verlag, Berlin, 182–212.

Jaeger, J.C. 1969. *Elasticity, fracture and flow*, 3rd edition. Chapman and Hall, London.

Jäger, H. and Carnuth, W. 1987. The decay of the El Chichón stratospheric perturbation, observed by lidar at northern midlatitudes. *Geophysical Research Letters* **14**, 696–699.

Jahn, A. 1964. Slopes morphological features resulting from gravitation. *Zeitschrift für Geomorphologie SB* **5**, 59–72.

Jakosky, B.M. 1986. Volcanoes, the stratosphere, and climate. *Journal of Volcanology and Geothermal Research* **28**, 247–255.

Jaupart, C. and Allégre, C.J. 1991. Gas content, eruption rate, and instabilities of eruption regime in silicic volcanoes. *Earth and Planetary Research Letters* **102**, 413–429.

Johnson, R.W. 1987. Large-scale volcanic cone collapse: the 1888 slope failure of Ritter volcano, and other examples from Papua New Guinea. *Bulletin of Volcanology* **49**, 669–679.

Jones, A.T. 1992. Comment on 'Catastrophe wave erosion on the southeastern coast of Australia: impact of the Lanai tsunami *ca.* 105 ka?'. *Geology* **20**, 1150–1151.

Jones, A.T. and Mader, C.L. 1996. Wave erosion on the southeastern coast of Australia: tsunami propagation and modelling. *Australian Journal of Earth Science* **43**, 479–483.

Jones, P.D. and Kelly, P.M. 1996. The effect of tropical explosive eruptions on surface air temperature. In Fiocci, G., Fu'a, D. and Visconti, G. (eds) *The Pinatubo eruption: effects on the atmosphere and climate.* NATO ASI Series **42**, 95–112. Springer-Verlag, Berlin and Heidelberg.

Jones, P.D., Briffa, K.R. and Schweingruber, F.H. 1995. Tree-ring evidence of the widespread effects of explosive volcanic eruptions. *Geophysical Research Letters* **22**, 1333–1336.

Jones, P.D., Briffa, K.R. and Tett, S.F.B. 1998. High resolution palaeoclimatic records for the last millennium. *The Holocene* **8**, 455–471.

Jones, P.D., Horton, E.B., Folland, C.K., Hulme, M., Parker, D.E. and Basnett, T.A. 1999. The use of indices to identify changes in climatic extremes. *Climatic Change* **42**, 131–149.

Karl, T.R. and Easterling, D.R. 1999. Climate extremes: selected review and future research directions. *Climatic Change* **42**, 309–325.

Karl, T.R. and Haeberli, W. 1997. Climate extremes and natural disasters: trends and loss reduction prospects. In *Proceedings of the Conference on the World Climate Research Programme: Achievements, Benefits, and Challenges*, 26–28 August, Geneva.

Karl, T.R. and Knight, R.W. 1998. Secular trends of precipitation amount, frequency, and intensity in the United States. *Bulletin of the American Meteorological Society* **79**, 231–241.

Karl, T.R., Knight, R.W., Easterling, D.R. and Quayle, R.G. 1995a. Indices of climate change for the United States. *Bulletin of the American Meteorological Society* **77**, 279–292.

Karl, T.R., Knight, R.W. and Plummer, N. 1995b. Trends in high-frequency climate variability in the twentieth century. *Nature* **377**, 217–220.

Karl, T.R., Nicholls, N. and Gregory, J. 1997. The coming climate. *Scientific American*, May, 78–83.

Kattenberg, A., Giorgi, F., Grassl, H. *et al.* 1996. Climate models: projections of future climate. In Houghton, J.T., Meira Filho, L.G., Callander, B.A., Harris, N., Kattenberg, A. and Maskell, K. (eds) *Climate change 1995: the science of climate change.* Cambridge University Press, Cambridge.

Katz, R.W. and Brown, B.G. 1992. Extreme events in a changing climate: variability is more important than averages. *Climatic Change* **21**, 289–302.

Katzfey, J.J. and McInnes, K.L. 1996. GCM simulations of Eastern Australian cutoff lows. *Journal of Climate* **9**, 2337–2355.

Keating, B.H. and McGuire, W.J. 2000. Island edifice failures and associated tsunami hazards. *Pure and Applied Geophysics* **157**, 899–955.

Keefer, D.K. 1984. Landslides caused by earthquakes. *Geological Society of America Bulletin* **95**, 406–421.

Keefer, D.K. and Johnson, A.M. 1983. *Earthflows: morphology, mobilization and movement.* US Geological Survey Professional Paper, **1264**: 1–56.

Keller, J., Ryan, W.B., Ninkovich, D. and Altherr, R. 1978. Explosive volcanic activity in the Mediterranean over the past 200,000 years as recorded in deep-sea sediments. *Bulletin of the Geological Society of America* **89**, 591–604.

Kelly, P.M. and Sear, C.B. 1984. Climatic impact of explosive volcanic eruptions. *Nature* **311**, 740–743.

Kelly, P.M. and Wigley, T.M.L. 1992. Solar cycle length, greenhouse forcing and global climate. *Nature* **360**, 328–330.

Kennett, J.P. and Thunell, R.C. 1975. Global increase in Quaternary explosive volcanism. *Science* **187**, 497–503.

Kent, P.E. 1966. The transport mechanism in catastrophic rockfalls. *Journal of Geology* **74**, 79–83.

Kiem, B.D. and Cruise, J.F. 1998. A technique to measure trends in the frequency of discrete random events. *Journal of Climate* **11**, 848–855.

Kilburn, C.R.J. and Petley, D.N. In press. Forecasting giant, catastrophic slope collapse: lessons from Vajont, northern Italy. *Geomorphology*.

Kilburn, C.R.J. and Sørensen, S.A. 1998. Runout lengths of sturzstroms: the control of initial conditions and of fragment dynamics. *Journal of Geophysical Research* **103**, 17877–17884.

Kilburn, C.R.J. and Voight, B. 1998. Slow rock fracture as eruption precursor at Soufrière Hills volcano, Montserrat. *Geophysical Research Letters* **25**, 3665–3668.

Kimberlain, T.B. and Elsner, J.B. 1998. The 1995 and 1996 North Atlantic hurricane seasons: a return of the tropical-only hurricane. *Journal of Climate* **11**, 2062–2069.

Klumov, B.A. 1999. Destruction of the ozone layer as a result of a meteoroid falling into the ocean. *JETP Letters* **70**, 363–370.

Knaff, J.A. 1997. Implications of summertime sea level pressure anomalies in the tropical Atlantic region. *Journal of Climate* **10**, 789–804.

Knutson, T.R., Tuleya, R.E. and Kurihara, Y. 1998. Simulated increases of hurricane intensities in a CO_2-warmed world. *Science* **279**, 1018–1020.

Kokelaar, P. and Romagnoli, C. 1995. Sector collapse, sedimentation, and clast population evolution at an active island-arc volcano: Stromboli, Italy. *Bulletin of Volcanology* **57**, 240–262.

Kondratyev, K.Y. and Galindo, I. 1997. *Volcanic activity and climate.* Deepack, Hampton, VA.

Kring, D.A. and Boynton, W.V. 1991. Altered spherules of impact melt and associated relic glass from K/T boundary sediments in Haiti. *Geochimica et Cosmochimica Acta* **55**, 1737–1742.

Krishnamurti, T.N., Correa-Torres, R., Latif, M. and Daughenbaugh, G. 1998. The impact of current and possibly future sea surface temperature anomalies on the frequency of Atlantic hurricanes. *Tellus* **50A**, 186–210.

Kunkel, K.E., Pielke, R.A. and Changnon, S.A. 1999. Temporal fluctuations in weather and climate extremes that cause economic and human health impacts: a review. *Bulletin of the American Meteorological Society* **80**, 1077–1098.

Kwadijk, J. 1993. *The impact of climate change on the discharge of the River Rhine.* Netherland Geographical Studies, University of Utrecht.

Kyte, F.T. and Smit, J. 1986. Regional variations in spinel compositions: an important key to the Cretaceous/ Tertiary event. *Geology* **14**, 485–487.

Kyte, F.T., Zhou, L. and Wasson, J.T. 1981. High noble metal concentrations in a late Pliocene sediment. *Nature* **292**, 417–420.

Labazuy, P. 1996. Recurrent landsliding events on the submarine flank of Piton de la Fournaise volcano (Réunion Island). In McGuire, W.J., Jones, A.P. and Neuberg, J. (eds) *Volcano instability on the Earth and other planets.* Geological Society of London Special Publication **110**, 293–306.

Lamb, H.H. 1970. Volcanic dust in the atmosphere, with a chronology and assessment of its meteorological significance. *Philosophical Transactions of the Royal Society A* **266**, 425–533.

Lamb, H.H. 1977. *Climate present, past and future*, vol. 2: *Climatic history and the future.* Methuen, London.

Lamb, H.H. 1979. Climatic variation and changes in the wind and ocean circulation: the Little Ice Age in the Northeast Atlantic. *Quaternary Research* **11**, 1–20.

Lamb, H.H. 1988. *Weather, climate and human affairs.* Routledge, London.

Lamb, H.H. 1991. *Historic storms of the North Sea, British Isles and Northwest Europe.* Cambridge University Press, Cambridge.

Lambert, S.J. 1995. The effect of enhanced greenhouse warming on winter cyclone frequencies and strengths. *Journal of Climate* **8**, 1447–1452.

Lambert, S.J. 1996. Intense extratropical northern hemisphere winter cyclone events: 1899–1991. *Journal of Geophysical Research* **101**, 21319–21325.

Landsea, C.W., Nicholls, N., Gray, W.M. and Avila, L.A. 1996. Downward trends in the frequency of intense Atlantic hurricanes during the past five decades. *Geophysical Research Letters* **23**, 1697–1700.

Landsea, C.W., Pielke, R.A., Mestas-Nuñez, A.M. and Knaff, J.A. 1999. Atlantic basin hurricanes: indices of climate changes. *Climatic Change* **42**, 89–129.

Lawn, B. 1993. *Fracture of brittle solids*, 2nd edition. Cambridge University Press, Cambridge.

Lean, J. 1997. The sun's radiation and its relevance for Earth. *Annual Review of Astronomy and Astrophysics* **35**, 33–67.

Leathers, D.J., Yarnal, B. and Palecki, M.A. 1991. The Pacific North-American teleconnection pattern and United States climate: 1. Regional temperature and precipitation associations. *Journal of Climate* **4**, 517–528.

Ledbetter, M. and Sparks, R.S.J. 1979. Duration of large magnitude explosive eruptions deduced from graded bedding in deep-sea ash layers. *Geology* **7**, 240–244.

Letréguilly, A., Huybrechts, P. and Reeh, N. 1991. Steady-state characteristics of the Greenland ice sheet under different climates. *Journal of Glaciology* **37**, 149–157.

Levy, D.H. 1998. The collision of Comet Shoemaker-Levy 9 with Jupiter. *Space Science Reviews* **85**, 523–545.

Lighthill, J., Holland, G., Gray, W. *et al.* 1994. Global climate change and tropical cyclones. *Bulletin of the American Meteorological Society* **75**, 2147–2157.

Lipman, P.W. and Mullineaux, D. (eds) 1981. *The 1980 Eruptions of Mount St. Helens.* US Geological Survey Professional Paper **1250**.

Lockner, D.A. 1995. Rock failure. In Ahrens, T.J. (ed.) *Rock physics and phase relations: a handbook of physical constants.* AGU Reference Shelf **3**, 127–147.

Lomoschitz, A. and Corominas, J. 1996. Los depositos de desliziamentos gravitacionales del centro-sur de Gran Canaria. *Geogaceta* **20**, 1346–1348.

Long, C. 1999. Survivors of Violent Nicaragua Landslide Rebuild, Remember. Disaster Relief website, www.DisasterRelief.org.

Lunkeit, F., Ponater, M., Sausen, R., Sogalla, M., Ulbrich, U. and Windelband, M. 1996. Cyclonic activity in a warmer climate. *Beiträge zur Physik der Atmosphäre* **69**, 393–407.

Lyell, C. 1830–33. *Principles of Geology.* John Murray, London. 3 volumes.

Macleod, N. 1996. k/T redux. *Palaeobiology* **22**, 311–317.

Mader, C.L. 1998. Modelling the Eltanin asteroid tsunami. *Science of Tsunami Hazards* **16**, 17–20.

Main, I.G. and Meredith, P.G. 1991. Stress corrosion constitutive laws as a possible mechanism of intermediate-term and short-term seismic event rates and b-values. *Geophysical Journal International* **107**, 363–372.

Main, I.G., Sammonds, P.R. and Meredith, P.G. 1993. Application of a modified Griffith criterion to the evolution of fractal damage during compressional rock failure. *Geophysical Journal International* **115**: 367–380.

Maloney, E.D. and Hartmann, D.L. 2000a. Modulation of hurricane activity by the Madden–Julian Oscillation. *Science* **287**, 2002–2004.

Maloney, E.D. and Hartmann, D.L. 2000b. Modulation of Eastern North Pacific hurricanes by the Madden–Julian Oscillation. *Journal of Climate* **13**, 1451–1460.

Mantua, N.J., Hare, S.R., Zhang, Y., Wallace, J.M. and Francis, R.C. 1997. A Pacific interdecadal climate oscillation with impacts on salmon production. *Bulletin of the American Meteorological Society* **78**, 1069–1079.

Marcus, S.L., Chao, Y., Dickey, J.O. and Gegout, P. 1998. Detection and modelling of non-tidal oceanic effects on Earth's rotation rate. *Science* **281**, 1656–1659.

Maslin, M. 1998. Why study past climate in marine sediments? *Globe* **41**, 11–12.

Maslin, M., Mikkelsen, N., Vilela, C. and Haq, B. 1998. Sea-level- and gas-hydrate-controlled catastrophic sediment failures of the Amazon Fan. *Geology* **26**, 1107–1110.

Mass, C. and Robock, A. 1982. The short-term influence of the Mount St. Helens volcanic eruption on surface temperature in the northwest United States. *Monthly Weather Review* **110**, 614–622.

Mass, C.F. and Portman, D.A. 1989. Major volcanic eruptions and climate: a critical evaluation. *Journal of Climate* **2**, 566–593.

Masson, D.G. and Weaver, P.P.E. 1992. Changing sea levels, erupting volcanoes, and submarine landslides – a wider perspective. *Geology Today* **8**, 201–202.

Matthews, R.K. 1969. Tectonic implication of glacio-eustatic sea level fluctuations. *Earth and Planetary Science Letters* **5**, 459–462.

Mayewski, P.A., Meeker, L.D., Whitlow, S. *et al.* 1993. The atmosphere during the Younger Dryas. *Science* **261**, 195–197.

McGuire, W.J. 1996. Volcano instability: a review of contemporary themes. In McGuire, W.J., Jones, A.P. and Neuberg, J. (eds) *Volcano instability on the Earth and other planets.* Geological Society of London Special Publication **110**, 1–23.

McGuire, W.J. 1999. *Apocalypse: a natural history of global disasters*. Cassell, London.

McGuire, W.J., Howarth, R.J., Firth, C.R. *et al.* 1997. Correlation between rate of sea-level change and frequency of explosive volcanism in the Mediterranean. *Nature* **389**, 473–476.

McGuire, W.J. and Kilburn, C.R.J. 1997. Forecasting volcanic events: some contemporary issues. *Geologische Rundschau* **86**: 439–445.

McLean, D.M. 1981. Deccan volcanism and the Cretaceous-Tertiary transition scenario: a unifying causal mechanism. In Russell, D.A. and Rice, G. (eds) *K-Tec II, Cretaceous-Tertiary extinctions and possible terrestrial and extraterrestrial causes*. National Museum of Canada, Ottawa, p143–144.

McLean, D.M. 1982. Flood basalt volcanism and global extinctions at the Cretaceous–Tertiary transition. *American Association for the Advancement of Science Meeting, Washington, DC*, 47.

McLean, D.M. 1985. Deccan traps mantle degassing in the terminal Cretaceous marine extinctions. *Cretaceous Research* **6**, 235–259.

McSaveney, M.J. 1978. Sherman glacier rock avalanche, Alaska, U.S.A. In Voight, B. (ed.) *Rockslides and avalanches.1. Natural phenomena*. Elsevier, Amsterdam, 197–258.

McTainsh, G.H. and Lynch, A.W. 1996. Quantitative estimates of the efect of climate change on dust storm activity in Australia during the last glacial maximum. *Geomorphology* **17**, 263–271.

Mearns, L.O., Katz, R.W. and Schneider, S.H. 1984. Extreme high-temperature events: changes in their probabilities with changes in mean temperature. *Journal of Climate and Applied Meteorology* **23**, 1601–1613.

Meehan, W.R. 1974. Fish habitat and timber harvest in southeast Alaska. *Naturalist* **25**: 28–31.

Meehl, G.A., Karl, T., Easterling, D.R. *et al.* 2000a. An introduction to trends in extreme weather and climate events: observations, socioeconomic impacts, terrestrial ecological impacts, and model projections. *Bulletin of the American Meteorological Society* **81**, 413–416.

Meehl, G.A., Zwiers, F., Evans, J., Knutson, T., Mearns, L. and Whetton, P. 2000b. Trends in extreme weather and climate events: issues related to modeling extremes in projections of future climate change. *Bulletin of the American Meteorological Society* **81**, 427–436.

Melosh, H.J. 1979. Acoustic fluidization: a new geologic process? *Journal of Geophysical Research* **87**, 7513–7520.

Melosh, H.J. 1981. Atmospheric breakup of terrestrial impactors. In Schultz, P.H. and Merrill, P.B. (eds) *Multi-ring basins*. Pergamon Press, New York, 29–35.

Melosh, H.J. 1986. The physics of very large landslides. *Acta Mechanica* **64**, 89–99.

Melosh, H.J. 1987. The mechanics of large rock avalanches. *Reviews in Engineering Geology* **7**, 41–49.

Melosh, H.J. 1997. Atmospheric screening of comet and asteroid impacts. In Remo, J.L. (ed.) Near Earth objects. *Annals of the New York Academy of Sciences* **822**, 283 (abstract).

Melosh, H.J., Schneider, N.M., Zahnle, K. and Latham, D. 1990. Ignition of global wildfires at the Cretaceous/Tertiary boundary. *Nature* **343**, 251–254.

Mercer, J.H. 1978. West Antarctic ice sheet and CO_2 greenhouse effect: a threat of disaster. *Nature* **271**, 321–325.

Mitchell, J.M. 1961. Recent secular changes of the global temperature. *Annals of the New York Academy of Sciences* **95**, 235–250.

Mitchell, J.M. 1970. A preliminary evaluation of atmospheric pollution as a cause of the global temperature fluctuation of the past century. In Singer, S.F. (ed.) *Global effects of environmental pollution*. Reidel, Norwell, MA, 139–155.

Mitchell, J.F.B., Manabe, S., Meleshko, V. and Tokioka, T. 1990. Equilibrium climate change – and its implications for the future. In Houghton, J.T., Jenkins, G.J. and Ephraums, J.J. (eds) *Climate change: the IPCC scientific assessment*. Cambridge University Press, Cambridge.

Monfredo, W. 1999. Relationship between phases of the El Niño–Southern Oscillation and character of the tornado season in the south-central United States. *Physical Geography* **20**, 413–421.

Moore, J.G. and Moore, G.W. 1984. Deposit from a giant wave on the island of Lanai, Hawaii. *Science* **226**, 1312–1315.

Moore, J.G., Normark, W.R. and Holcomb, R.T. 1994. Giant Hawaiian landslides. *Annual Review of Earth and Planetary Sciences* **22**, 119–144.

Morgan, W.J. 1986. Flood basalts and mass extinctions. *Eos* **67**, 391.

Mörner, N.-A. 1980. The Fennoscandian uplift: geological data and their geodynamical implication. In Mörner, N.-A. (ed.) *Earth Rheology, Isostasy, and Eustasy*. Wiley, Chichester, 251–284.

Morrison, D., Chapman, C.R. and Slovic, P. 1994. The impact hazard. In Gehrels, T. (ed.) *Hazards Due to Comets and Asteroids*. University of Arizona Press, Tucson, 59–91.

Müller, L. 1964. The rock slide in the Vaiont valley. *Felsmechanik und Ingenieur-geologie* **2**, 148–212.

Munich Re 1999. Major natural catastrophes from the 11th to the 19th century. *MRNatCatservice*. December 1999. Munich Reinsurance Group, Munich.

Murty, T.S. and Wigen, S.O. 1976. Tsunami behaviour on the Atlantic coast of Canada and some similarities to the Peru coast. *Royal Society of New Zealand Bulletin* **15**, 51–60.

Nakada, M. and Yokose, H. 1992. Ice age as a trigger of active Quaternary volcanism and tectonism. *Tectonophysics* **212**, 321–329.

Nakamura, K. 1980. Why do long rift zones develop in Hawaiian volcanoes? A possible role of thick ocean sediments. *Bulletin of the Volcanological Society of Japan* **25**, 255–269.

Napier, W.M. 1998. Galactic periodicity and the geological record. In Grady, M.M., Hutchinson, R., McCall, G.J.H. and Rothery, D.A. (eds) *Meteorites: Flux with Time and Impact Effects*. Geological Society of London Special Publication 140, 19–29.

Napier, W.M. and Clube, S.V.M. 1979. A theory of terrestrial catastrophe. *Nature* **282**, 455–459.

Nash, D., Brunsden, D.K., Hughes, R.E., Jones, D.K.C. and Whalley, B.F. 1985. A catastrophic debris flow near Gupis, Northern areas, Pakistan. *Proceedings of the 11th International Conference on Soil Mechanics and Foundation Engineering* **3**: 1163–1166.

Nesje, A. and Johannessen, T. 1992. What were the primary forcing mechanisms of high-frequency Holocene climate and glacier variations? *The Holocene* **2**, 79–84.

Newhall, C.G. and Self, S. 1982. The volcanic explosivity index (VEI): an estimate of explosive magnitude for historical volcanism. *Journal of Geophysical Research* **87**, 1231–1238.

Nicholls, N. 1995. Long term climate monitoring and extreme events. *Climatic Change* **31**, 231–245.

Nicholls, N. and Murray, W. 1999. Workshop on indices and indicators for climate extremes, Asheville, NC, USA, 3–6 June 1997. Breakout group B: precipitation. *Climatic Change* **42**, 23–29.

Nicholls, N., Gruza, G.V., Jouzel, J., Karl, T.R., Ogallo, L.A. and Parker, D.E. 1996. Observed climate variability and change. In Houghton, J.T., Meira Filho, L.G., Callander, B.A., Harris, N., Kattenberg, A. and Maskell, K. (eds) *Climate change 1995: the science of climate change*. Cambridge University Press, Cambridge.

Nicholls, N., Landsea, C. and Gill, J. 1998. Recent trends in Australian region tropical cyclone activity. *Meteorology and Atmospheric Physics* **65**, 197–205.

Niino, H., Fujitani, T. and Watanabe, N. 1997. A statistical study of tornadoes and waterspouts in Japan from 1961–1993. *Journal of Climate* **10**, 1730–1752.

NOAA 1999. Mitch: the deadliest Atlantic hurricane since 1780. NOAA website, http://www.ncdc.noaa.gov/ol/reports/mitch/mitch.html

Officer, C.B. and Drake, C.L. 1985. Terminal Cretaceous environmental events. *Science* **227**, 1161–1167.

O'Keefe, J.D. and Ahrens, T.J. 1982. The interaction of the Cretaceous/Tertiary extinction bolide with the atmosphere, ocean, and the solid Earth. In Silver, L.T. and Schultz, P.H. (eds) *Geological implications of the impacts of large asteroids and comets on the Earth*. Geological Society of America Special Paper 190, 103–120.

O'Keefe, J.D. and Ahrens, T.J. 1989. Impact production of CO_2 by the Cretaceous/Tertiary extinction bolide and the resultant heating of the Earth. *Nature* **338**, 247–249.

Ormo, J. and Lindstrom, M. 2000. When a cosmic impact strikes the sea bed. *Geological Magazine* **137**, 67–80.

PACE 2001. Permafrost and climate in Europe. Website at: www.cf.ac.uk/earth/pace

Palais, J.M. and Sigurdsson, H. 1989. Petrologic evidence of volatile emissions from major historic and prehistoric volcanic eruptions. In Berger, A., Dickinson, R.E. and Kidson, J.W. (eds) *Understanding Climate Change*. American Geophysical Union Monograph **52**, 31–53.

Palutikof, J.P. and Downing, T.E. 1994. European windstorms. In Downing, T.E., Favis-Mortlock, D.T. and Gawith, M.J. (eds) *Climate Change and Extreme Events: Scenarios of Altered Hazards for Future Research*. Environmental Change Unit, University of Oxford, 61–78.

Parry, M.L. 1978. *Climatic change and agricultural settlement*. Dawson/Archon, Folkestone.

Parry, M.L. and Carter, T.R. 1985. The effect of climatic variations on agricultural risk. *Climatic Change* **7**, 95–110.

Paterne, M., Guichard, F. and Labeyrie, J. 1988. Explosive activity of the south Italian volcanoes during the past 80,000 years as determined by marine tephrochronology. *Journal of Volcanology and Geothermal Research* **34**, 153–172.

Peltier, W.R. and Tushingham, A.M. 1989. Global sea-level rise and the greenhouse effect: might they be connected? *Science* **244**, 806–810.

Penfield, G.T. and Camargo, Z.A. 1981. Definition of a major igneous zone in the central Yucatan platform with aeromagnetics and gravity. *Society of Exploration Geophysicists Technical Program, Abstracts and Biographies* **51**, 37 (abstract).

Peterson, R.E. and Gregory, J.M. 1993. Blowing dust and climate change. In Bras, R. (ed.) The world at risk: natural hazards and climate change. *AIP Conference Proceedings* **277**, 125–130.

Petley, D.N. 1999. Failure envelopes of mudrocks at high effective stresses. In Aplin, A.C., Fleet, A.J., Macquaker, J.H.S. (eds) *Physical properties of muds and mudstones*. Geological Society of London Special Publication **158**: 61–71.

Phillips, R.W. 1971. Effects of sediment on the gravel environment and fish production. In: *Forest land use and stream environment*. Oregon State University, Corvallis, 64–74.

Pielke, R.A. and Pielke, R.A. 1997. Hurricanes: their nature and impact on society. Wiley, Chichester.

Pierazzo, F., Kring, D.A. and Melosh, H.J. 1998. Hydrocode simulation of the Chicxulub impact event and the production of climatically active gases. *Journal of Geophysical Research* **103**, 2860–28625.

Pierson, T.C. and Costa, J.E. 1987. A rheological classification of subaerial sediment-water flows. *Geological Society of America, Reviews in Engineering Geology* **7**: 1–12.

Plafker, G. and Ericksen, G.E. 1978. Nevados Husacarán avalanches, Peru. In Voight, B. (ed.) *Rockslides and avalanches.1. Natural phenomena*. Elsevier, Amsterdam, 277–314.

Pollack, J.B., Toon, O.B., Ackerman, T.P., McKay, C.P. and Turco, R.P. 1983. Environmental effects of an impact-generated dust cloud: implications for the Cretaceous-Tertiary extinctions. *Science* **219**, 247-249.

Pollack, J.B., Toon, O.B., Sagan, C., Summers, A., Baldwin, B. and Van Camp, W. 1976. Volcanic explosions and climate change: a theoretical assessment. *Journal of Geophysical Research* **81**, 1071–1083.

Pope, K., Baines, K.H., Ocampo, A.C. and Ivanov, B.A. 1994. Impact winter and the Cretaceous/Tertiary extinctions: results of a Chicxulub asteroid impact model. *Earth and Planetary Science Letters* **128**, 719–725.

Porter, S.C. 1981. Recent glacier variations and volcanic eruptions. *Nature* **291**, 139–141.

Porter, S.C. 1986. Pattern and forcing of Northern Hemisphere glacier variation during the last millennium. *Quaternary Research* **26**, 27–48.

Prather, M. 1992. Catastrophic loss of stratospheric ozone in dense volcanic clouds. *Journal of Geophysical Research* **97**, 10187–10191.

Price, C. and Rind, D. 1994. The impact of a $2 \times CO_2$ climate on lightning-caused fires. *Journal of Climate* **7**, 1484–1494.

Prinn, R.G. and Fegley, B. 1987. Bolide impacts, acid rain and biospheric traumas at the Cretaceous–Tertiary boundary. *Earth and Planetary Science Letters* **83**, 1–15.

Prior, D.B., Doyle, E.H. and Neurauter, T. 1986. The Currituck slide, mid-Atlantic continental slope revisited. *Marine Geology* **73**, 25–45.

Pyle, D.M., Beattie, P.D. and Bluth, G.J.S. 1996. Sulphur emissions to the stratosphere from explosive volcanic eruptions. *Bulletin of Volcanology* **57**, 663–671.

Rabinowitz, D., Helin, E., Lawrence, K. and Pravdo, S. 2000. A reduced estimate of the number of kilometre-sized near-Earth asteroids. *Nature* **403**, 165–166.

Rampino, M.R. and Haggerty, B.M. 1996. The 'Shiva hypothesis': impact crises, mass extinctions, and the galaxy. *Earth, Moon and Planets* **72**, 441–460.

Rampino, M.R. and Self, S. 1984. Sulphur-rich volcanic eruptions and stratospheric aerosols. *Nature* **310**, 677–679.

Rampino, M.R. and Self, S. 1992. Volcanic winter and accelerated glaciation following the Toba super-eruption. *Nature* **359**, 50–52.

Rampino, M.R. and Self, S. 1993a. Climate–volcanism feedback and the Toba eruption of ~74,000 years ago. *Quaternary Research* **40**, 269–280.

Rampino, M.R. and Self, S. 1993b. Bottleneck in human evolution and the Toba eruption (correspondence). *Nature* **262**, 1954.

Rampino, M.R. and Stothers, R.B. 1984. Terrestrial mass extinction, cometary impacts and the Sun's motion perpendicular to the galactic plane. *Nature* **308**, 709–712.

Rampino, M.R. and Stothers, R.B. 1987. Episodic nature of the Cenozoic marine record. *Paleoceanography* **2**, 255–258.

Rampino, M.R. and Stothers, R.B. 1988. Flood basalt volcanism during the past 250 million years. *Science* **241**, 663–668.

Rampino, M.R., Haggerty, B.M. and Pagano, T.C. 1997. A unified theory of impact crises and mass extinctions: quantitative tests. In Remo, J.L. (ed.) Near Earth objects. *Annals of the New York Academy of Sciences* **822**, 403–431.

Rampino, M.R., Self, S. and Fairbridge, R.W. 1979. Can rapid climatic change cause volcanic eruptions? *Science* **206**, 826–829.

Raup, D.M. and Sepkoski, J.J. 1984. Periodicity of extinctions in the geological past. *Proceedings of the National Academy of Science* **81**, 801–805.

Ringrose, P.S. 1989. Palaeoseismic (?) liquefaction event in late Quaternary lake sediment at Glen Roy, Scotland. *Terra Nova* **1**, 57–62.

Roberts, J.A. and Cramp, A. 1996. Sediment stability on the western flanks of the Canary Islands. *Marine Geology* **134**, 13–30.

Robock, A. 1989. Volcanoes and climate. In *Climate and geosciences: a challenge for science and society in the 21st century*. NATO ASI Ser. C **285**, 309–314.

Robock, A. 2000. Volcanic eruptions and climate. *Reviews of Geophysics* **38**, 191–219.

Robock, A. and Free, M.P. 1995. Ice cores as an index of global volcanism from 1850 to the present. *Journal of Geophysical Research* **100**, 11549–11567.

Robock, A. and Free, M.P. 1996. The volcanic record in ice cores for the past 2000 years. In Jones, P. D., Bradley, R. S. and Jouzel, J. (eds) *Climatic Variations and Forcing Mechanisms of the Last 2000 Years*. Springer Verlag, New York, 533–546.

Robock, A. and Liu, Y. 1994. The volcanic signal in Goddard Institute for Space Studies three-dimensional model simulations. *Journal of Climate* **7**, 44–55.

Robock, A. and Mao, J. 1992. Winter warming from large volcanic eruptions. *Geophysical Research Letters* **19**, 2405–2408.

Robock, A. and Mao, J. 1995. The volcanic signal in surface temperature observations. *Journal of Climate* **8**, 1086–1103.

Robock, A. and Matson, M. 1983. Circumglobal transport of the El Chichón volcanic dust cloud. *Science* **221**, 195–197.

Robock, A., Taylor, K.E., Stenchikov, G.L. and Liu, Y. 1995. GCM evaluation of a mechanism for El Niño triggering by the El Chichón ash cloud. *Geophysical Research Letters* **22**, 2369–2372.

Rodbell, D.T., Seltzer, G.O., Anderson, D.M., Abbott, M.B., Enfield, D.B. and Newman, J.H. 1999. An ~15,000-year record of El Niño-driven alluviation in southwestern Equador. *Science* **283**, 516–520.

Rogers, J.C. 1990. Patterns of low frequency monthly sea level pressure variability (1899–1986) and associated wave cyclone frequencies. *Journal of Climate* **3**, 1364–1379.

Rogers, J.C. 1997. North Atlantic storm track variability and its association to the North Atlantic Oscillation and climate variability of Northern Europe. *Journal of Climate* **10**, 1635–1647.

Rogers, J.C. and van Loon, H. 1979. The seesaw in winter temperatures between Greenland and Northern Europe: I. Some oceanic and atmospheric effects in middle and high latitudes. *Monthly Weather Review* **107**, 509–519.

Rose, W.I. and Chesner, C.A. 1990. Worldwide dispersal of ash and gases from earth's largest known eruption. *Palaeontology* **89**, 269–275.

Rothwell, R.G., Thomson, J. and Kähler, G. 1998. Low sea-level emplacement of a very large Late Pleistocene 'megaturbidite' in the western Mediterranean Sea. *Nature* **392**, 377–380.

Royer, J.F., Chauvin, F., Timbal, B., Araspin, P. and Grimal, D. 1998. *Climatic Change* **38**, 307–343.

Ruddiman, W.F. and McIntyre, A. 1984. Ice-age thermal response and climatic role of the surface Atlantic Ocean, 40º N to 63º N. *Geological Society of America Bulletin* **95**, 381–396.

Ryan, B.F., Watterson, I.G. and Evans, J.L. 1992. Tropical cyclone frequencies inferred from Gray's yearly genesis parameter: validation of GCM tropical climates. *Geophysical Research Letters* **19**, 1831–1834.

Sadler, J.P. and Grattan, J.P. 1999. Volcanoes as agents of past environmental change. *Global and Planetary Change* **21**, 181–196.

Sato, M., Hansen, J.E., McCormick, M.P. and Pollack, J.B. 1993. Stratospheric aerosol optical depth 1850–1990. *Journal of Geophysical Research* **98**, 22987–22994.

Saunders, M.A. 1999. Earth's future climate. *Philosophical Transactions of the Royal Society of London A* **357**, 3459–3480.

Saunders, M.A. 2001. Earth's future climate. In Thompson, J.M.T. (ed.) *Visions of the future: astronomy and earth science*. Cambridge University Press, 203–220.

Saunders, M.A. and Harris, A.R. 1997. Statistical evidence links exceptional 1995 Atlantic hurricane season to record sea warming. *Geophysical Research Letters* **24**, 1255–1258.

Schaefer, J.T., Kelly, D.L. and Abbey, R.F. 1986. A minimum assumption tornado-hazard probability model. *Journal of Climate and Applied Meteorology* **25**, 1934–1945.

Schaefer, J.T., Kelly, D.L., Doswell, C.A. *et al.* 1980. Tornadoes: when, where, how often. *Weatherwise*, April, 53–59.

Schmidt, H. and von Storch, H. 1993. German Bight storms analysed. *Nature* **365**, 791.

Schmith, T., Kaas, E. and Li, T.-S. 1998. Northeast Atlantic winter storminess 1875–1995 re-analysed. *Climate Dynamics* **14**, 529–536.

Schöenwiese, C.-D., Ullrich, R., Beck, F. and Rapp, J. 1994. Solar signals in global climatic change. *Climatic Change* **27**, 259–281.

Schubert, M., Perlwitz, R., Blender, R., Fraedrich, K. and Lunkeit, F. 1998. North Atlantic cyclones in CO_2-induced warm climate simulations: frequency, intensity, tracks. *Climate Dynamics* **14**, 827–837.

Schuster, R.L. 1996. Socioeconomic significance of landslides. In Turner, A.K. and Schuster, R.L. (eds) *Landslides: investigation and mitigation*. TRB Special Report **247**, 12–35. National Research Council, Washington, DC.

Schweingruber, F.H. 1988. *Tree rings, basics and applications of dendrochronology*. Reidel, Dordrecht.

Scuderi, L.A. 1990. Tree-ring evidence for climatically effective volcanic eruptions. *Quaternary Research* **34**, 67–85.

Sear, C.B., Kelly, P.M., Jones, P.D. and Goodess, C.M. 1987. Global surface temperature responses to major volcanic eruptions. *Nature* **330**, 365–367.

SEAVOLC 1995. Sea-level change and the stability and activity of coastal and island volcanoes. Environment Programme Contract EV5V-CT92–0170 Final Report. Commission of the European Communities, Brussels.

Selby, M.J. 1993. *Hillslope materials and processes*. Oxford University Press, Oxford.

Self, S., Rampino, M.R. and Barbera, J.J. 1981. The possible effects of large 19th and 20th century volcanic eruptions on zonal and hemispherical surface temperatures. *Journal of Volcanology and Geothermal Research* **11**, 41–60.

Self, S., Rampino, M.R., Zhao, J. and Katz, M.G. 1997. Volcanic aerosol perturbations and strong El Niño events: no general correlation. *Geophysical Research Letters* **24**, 1247–1250.

Sereze, M.C., Carse, F. and Barry, R.C. 1997. Icelandic low cyclone activity: climatological features, linkages with the NAO, and relationships with recent changes in the Northern Hemisphere circulation. *Journal of Climate* **10**, 453–464.

Seyfert, C.R. and Simkin, L.A. 1979. *Earth history and plate tectonics*. Harper Row, New York.

Shapiro, L.J. and Goldenberg, S.B. 1998. Atlantic sea surface temperatures and tropical cyclone formation. *Journal of Climate* **11**, 578–590.

Shackleton, N.J. 1987. Oxygen isotopes, ice volume, and sea level. *Quaternary Science Reviews* **6**, 183–190.

Shackleton, N.K. and Opdyke, N.D. 1973. Oxygen isotopes and palaeomagnetic stratigraphy of equatorial Pacific core V28–238: oxygen isotope temperatures and ice volumes on a 100,000 year and 1,000,000 year scale. *Quaternary Research* **3**, 39–55.

Sharpton, V.L., Dalrymple, G.B., Marin, L.E., Ryder, G., Schuraytz, B.C. and Urrutia-Fucugauchi, J. 1992. New links between the Chicxulub impact structure and the Cretaceous/Tertiary boundary. *Nature* **359**, 819–821.

Sharpton, V.L. and Marín, L.E. 1997. The Cretaceous–Tertiary impact crater and the cosmic projectile that produced it. In Remo, J.L. (ed.) Near Earth objects. *Annals of the New York Academy of Sciences* **822**, 353–380.

Sharpton, V.L., Marín, L.E., Carney, J.L. *et al.* 1996. A model of the Chicxulub impact basin based on evaluation of geophysical data, well logs, and drill core samples. *Geological Society of America Special Paper* **307**, 55–74.

Shen, W.X., Tuleya, R.E. and Ginis, I. 2000. A sensitivity study of the thermodynamic environment on GFDL model hurricane intensity: implications for global warming. *Journal of Climate* **13**, 109–121.

Shoemaker, E.M. 1983. Asteroid and comet bombardment of the Earth. *Annual Review of Earth and Planetary Sciences* **11**, 461–494.

Shoemaker, E.M., Wolfe, R.F. and Shoemaker, C.S. 1990. Asteroid and comet flux in the neighborhood of Earth. *Geological Society of America Special Paper* **247**, 155–170.

Shreve, R.L. 1968. The Blackhawk landslide. *Special Paper of the Geological Society of America*, **108**, 1–47.

Sickmöller, M., Blender, R. and Fraedrich, K. 2000. Observed winter cyclone tracks in the northern hemisphere in re-analysed ECMWF data. *Quarterly Journal of the Royal Meteorological Society* **126**, 591–620.

Sidle, R.C., Pearce, A.J. and O'Loughlin, C.L. 1985. *Hillslope stability and land use*. Water Resources Monograph, American Geophysical Union, Washington, DC.

Siebert, L. 1984. Large volcanic debris avalanches: characteristics of source areas, deposits, and associated eruptions. *Journal of Volcanology and Geothermal Research* **22**, 163–197.

Siebert, L. 1992. Threats from debris avalanches. *Nature* **356**, 658–659.

Siebert, L. 1996. Hazards of large volcanic debris avalanches and associated eruptive phenomena. In Scarpa, R. and Tilling, R.I. (eds) *Monitoring and mitigation of volcano hazards*. Springer-Verlag, Berlin, 541–572.

Signor, P.W. and Lipps, J.H. 1982. Sampling bias, gradual extinction patterns and catastrophes in the fossil record. In Silver, L.T. and Schultz, P.H. (eds) *Geological implications of the impacts of large asteroids and comets on the Earth*. Geological Society of America Special Paper 190, 291–296.

Sigurdsson, H. and Carey, S. 1989. Plinian and co-ignimbrite tephra fall from the 1815 eruption of Tambora volcano. *Bulletin of Volcanology* **51**, 243–270.

Sigurdsson, H., D'Hont, S., Arthur, M.A. *et al.* 1991. Glass from the Cretaceous/Tertiary boundary in Haiti. *Nature* **349**, 482–487.

Sigurdsson, H., D'Hondt, S. and Carey, S. 1992. The impact of the Cretaceous/Tertiary bolide on evaporite terrain and generation of major sulphuric acid aerosol. *Earth and Planetary Science Letters* **109**, 543–559.

Sigvaldason, G.E., Annertz, K. and Nilsson, M. 1992. Effect of glacier loading/deloading on volcanism: post-glacial volcanic production rate of the Dyngjufjöll area, central Iceland. *Bulletin of Volcanology* **54**, 385–392.

Simkin, T. and Siebert, L. 1994. *Volcanoes of the World*. Geoscience Press, Tucson, AZ.

Simpson, D.W., Leith, W.S. and Scholz, C.H. 1988. Two types of reservoir induced seismicity. *Bulletin of the Seismological Society of America* **78**, 2025–2040.

Sissons, J.B. and Cornish, R. 1982. Differential glacio-isostatic uplift of crustal blocks at Glen Roy, Scotland. *Quaternary Research* **18**, 268–288.

Smit, J. 1996. 'Strangelove', a tragic episode. *Recherche* **293**, 62–64.

Smit, J. and Hertogen, J. 1980. An extraterrestrial event at the Cretaceous–Tertiary boundary. *Nature* **285**, 198–200.

Smit, J. and Klaver, G. 1981. Sanidine spherules at the Cretaceous-Tertiary boundary indicate a large impact event. *Nature* **292**, 47–49.

Smit, J., Montanari, A., Swinburne, N.H.M. *et al.* 1992. Tektite-bearing, deep-water clastic unit at the Cretaceous–Tertiary boundary in northeastern Mexico. *Geology* **20**, 99–103.

Smit, J., Roep, Th. B., Alvarez, W. *et al.* 1996. Coarse-grained clastic sandstone complex at the K/T boundary around the Gulf of Mexico: deposition by tsunami waves induced by the Chicxulub impact? *Geological Society of America Special Paper* **307**, 151–182.

Smith, K. 1996. *Environmental hazards,* 2nd edn. Routledge, London.

Smith, K. and Ward, R. 1998. *Floods: physical processes and human impacts.* Wiley, Chichester.

Smith, M.S. and Shepherd, J.B. 1996. Tsunamigenic landslides at Kick 'em Jenny. In McGuire, W.J., Jones, A.P. and Neuberg, J. (eds) *Volcano instability on the Earth and other planets.* Geological Society of London Special Publication **110**, 293–306.

Smith, R.B. and Braile, L.W. 1994. The Yellowstone hotspot. *Journal of Volcanology and Geothermal Research* **61**, 121–187.

Sobel, A.H. and Maloney, E.D. 2000. Effect of ENSO and the MJO on Western North Pacific tropical cyclones. *Geophysical Research Letters* **27**, 1739–1742.

Solana, M.C. and Kilburn, C.R.J. 2001. Reducing the vulnerability of people to landslides in the Barranco de Tirajana, Gran Canaria, Spain. *Geomorphology* (in press).

Solomon, S. 1999. Stratospheric ozone depletion: a review of concepts and and history. *Reviews in Geophysics* **37**, 275–316.

Stanley, S.M. 1987. *Extinctions.* Freeman, New York.

Steel, D. 1995. *Rogue asteroids and doomsday comets.* Wiley. New York.

Stoopes, G.R. and Sheridan, M.F. 1992. Giant debris avalanches from the Colima Volcanic Complex, Mexico: implications for long-runout landslides (>100 km) and hazard assessment. *Geology* **20**: 299–302.

Stothers, R.B. 1984. The great Tambora eruption of 1815 and its aftermath. *Science* **224**, 1191–1198.

Stothers, R.B. 1993. Flood basalts and extinction events. *Geophysical Research Letters* **20**, 1399–1402.

Stothers, R.B. and Rampino, M.R. 1983. Volcanic eruptions in the Mediterranean before AD 630 from written and archaeological sources. *Journal of Geophysical Research* **88**, 6357–6371.

Swisher, C.C., Grajales-Nishimura, J.M., Montanari, A. *et al.* 1992. Coeval $^{40}Ar/^{39}Ar$ ages of 65.0 million years ago from the Chicxulub crater melt rock and Cretaceous–Tertiary boundary tektites. *Science* **257**, 954–958.

Swiss Re 2000. Natural catastrophes and man-made disasters in 1999: storms and earthquakes lead to the second-highest losses in insurance history. *Sigma 2/2000.* Swiss Reinsurance Company, Zurich.

Talandier, J. and Bourrouilh-le-Jan, F. 1988. High energy sedimentation in French Polynesia: cyclone or tsunami? In El-Sabh, M.I. and Murty, T.S. (eds) *Natural and man-made hazards.* Department of Oceanography, University of Quebec, Rimiuski PQ, Canada, 193–199.

Tappin, D.R., Matsumoto, T., Watts, P. *et al.* 1999. Sediment slump likely caused 1998 Papua New Guinea tsunami. *EOS* **80**, 329–340.

Taylor, K.C., Alley, R.B., Doyle, G.A. *et al.* 1993. The 'flickering switch' of late Pleistocene climate change. *Nature* **361**, 432–436.

Taylor, K.C., Mayewski, P.A., Alley, R.B. *et al.* 1997. The Holocene–Younger Dryas transition recorded at Summit, Greenland. *Science* **278**, 825–827.

Tett, S.F.B., Stott, P.A., Allen, M.R., Ingram, W.J. and Mitchell, J.F.B. 1999. Causes of twentieth-century temperature change near the Earth's surface. *Nature* **399**, 569–572.

Thórarinsson, S. 1953. Some new aspects of the Grímsvötn problem. *Journal of Glaciology* **14**, 267–276.

Thórarinsson, S. 1971. Damage caused by tephra fall in some big Icelandic eruptions and its relation to the thickness of the tephra layers. In *Acta of the First International Scientific Congress on the Volcano Thera,* September 1969, 213–236.

Thordarson, Th. and Self, S. 1993. The Laki (Skaftár Fires) and Grímsvötn eruptions in 1783–1785. *Bulletin of Volcanology* **55**, 233–263.

Thordarson, Th., Self, S., Óskarsson, N. and Hulsebosch, T. 1996. Sulfur, chlorine, and fluorine degassing and atmospheric loading by the 1783–1784 AD Laki (Skaftár Fires) eruption in Iceland. *Bulletin of Volcanology* **58**, 205–225.

Tobin, G.A. and Montz, B.E. 1997. *Natural hazards: explanation and integration.* Guilford Press, New York.

Tonkin, H., Holland, G.J., Holbrook, N. and Henderson-Sellers, A. 2000. An evaluation of thermodynamic estimates of climatological maximum potential tropical cyclone intensity. *Monthly Weather Review* **128**, 746–762.

Tooley, M.J. and Turner, K. 1995. The effects of sea level rise. In Parry, M. and Duncan, R. (eds) *The economic implications of climate change in Britain.* Earthscan, London, 8–27.

Toon, O.B. and Pollack, J.B. 1980. Atmospheric aerosols and climate. *American Scientist* **68**, 268–278.

Toon, O.B., Zahnle, K., Morrison, D., Turco, R.P. and Covey, C. 1997. Environmental perturbations caused by the impacts of asteroids and comets. In Remo, J.L. (ed.) Near Earth objects. *Annals of the New York Academy of Sciences* **822**, 403–431.

Toon, O.B., Zahnle, K., Turco, R.P. and Covey, C. 1994. Environmental perturbations caused by asteroid impacts. In Gehrels, T. (ed.) *Hazards due to asteroids and comets.* University of Arizona Press, Tucson, 791–826.

Trenberth, K.E. 1990. Recent observed interdecadal climate changes in the Northern Hemisphere. *Bulletin of the American Meteorological Society* **71**, 988–993.

Trenberth, K.E. 1998. Atmospheric moisture residence times and cycling: implications for rainfall rates and climate change. *Climatic Change* **39**, 667–694.

Trenberth, K.E. 1999. Conceptual framework for changes of extremes of the hydrological cycle with climate change. *Climatic Change* **42**, 327–339.

Trenberth, K.E. and Hoar, T.J. 1997. El Niño and climate change. *Geophysical Research Letters* **24**, 3057–3060.

Trenberth, K.E. and Owen, T.W. 1999. Workshop on indices and indicators for climate extremes, Asheville, NC, 3–6 June 1997. Breakout group A: storms. *Climatic Change* **42**, 9–21.

Turco, R.P., Toon, O.B., Ackerman, T.P., Pollack, J.B. and Sagan, C. 1991. Nuclear winter: physics and physical mechanisms. *Annual Review of Earth and Planetary Sciences* **19**, 383–422.

Turco, R., Toon, O.B., Park, C., Whitten, R., Pollack, J.B. and Noerdlinger, P. 1981. Tunguska meteor fall of 1908: effects on stratospheric ozone. *Science* **214**, 19–23.

Turner, A.K. and Schuster, R.L. (eds) 1996. *Landslides. Investigation and mitigation.* TRB Special Report **247**, 12–35. National Research Council, Washington, DC.

Ui, T. 1983. Volcanic dry avalanches deposits: identification and comparison with non-volcanic debris stream deposits. *Journal of Volcanology and Geothermal Research* **18**, 135–150.

UK Meteorological Office 1999. *Climate change and its impacts: stabilisation of CO_2 in the atmosphere*. Meteorological Office, London.

UK Task Force on Potentially Hazardous Near Earth Objects 2000. *Report of the Task Force*. The Stationery Office, London.

Ulbrich, U. and Christoph, M. 1999. A shift of the NAO and increasing storm track activity over Europe due to anthropogenic greenhouse gas forcing. *Climate Dynamics* **15**, 551–559.

Ungar, S. 1999. Is strange weather in the air? A study of U.S. national network news coverage of extreme weather events. *Climatic Change* **41**, 133–150.

van Loon, H. and Rogers, J.C. 1978. The seesaw in winter temperatures between Greenland and Northern Europe: I. General description. *Monthly Weather Review* **106**, 296–310.

Varnes, D.J. 1958. Landslide types and processes. In Eckel, E.B. (ed.) *Landslides and engineering practice*. HRB Special Report **29**, 20–47. National Research Council, Washington, DC.

Varnes, D.J. 1978. Slope movement types and processes. In Schuster, R.L. and Krizek, R.J. (eds) *Landslides: analysis and control*. TRB Special Report **176**, 11–33. National Research Council, Washington, DC.

Vinnikov, K.Y., Robock, A., Stouffer, R.J. and Manabe, S. 1996. Vertical patterns of free and forced climate variations. *Geophysical Research Letters* **23**, 1801–1804.

Vogt, P.R. 1972. Evidence for global synchronism in mantle plume convection and possible significance for geology. *Nature* **240**, 338–342.

Voight, B. (ed) 1978. *Rockslides and avalanches.1. Natural phenomena*. Elsevier, Amsterdam.

Voight, B. 1988. A method for prediction of volcanic eruptions. *Nature* **332**, 125–130.

Voight, B., and Faust, C. 1982. Frictional heat and strength loss in some rapid landslides. *Géotechnique* **32**, 43–54.

von Storch, H. and Zwiers, F.W. 1999. *Statistical analysis in climate research*. Cambridge University Press, Cambridge.

Vorobiev, Y.L. (ed) 1998. *Disasters and man*. Book 1: *The Russian experience of emergency response*. AST-LTD Publishers, Moscow.

Wagner, D. 1996. Scenarios of extreme temperature events. *Climatic Change* **33**, 385–407.

Walker, D.A. 1988. Seismicity of the East Pacific Rise: correlations with the Southern Oscillation Index? *Eos* **69**, 857.

Walker, D.A. 1995. More evidence indicates link between El Niños and seismicity. *Eos* **76**, 33–36.

Wallace, J.M. and Gutzler, D.S. 1981. Teleconnections in the geopotential height field during the Northern Hemisphere winter. *Monthly Weather Review* **109**, 785–812.

Wallmann, P.C., Mahood, G.A. and Pollard, D.D. 1988. Mechanical models for correlation of ring-fracture eruptions at Pantelleria, Straits of Sicily, with glacial sea-level drawdown. *Bulletin of Volcanology* **50**, 327–339.

Walsh, K.J.E. and Katzfey, J.J. 2000. The impact of climate change on the poleward movement of tropical-cyclone-like vortices in a regional climate model. *Journal of Climate* **13**, 1116–1132.

Walsh, K. and Pittock, A.B. 1998. Potential changes in tropical storms, hurricanes, and extreme rainfall events as a result of climate change. *Climatic Change* **39**, 199–213.

Walsh, K.J.E. and Ryan, B.F. 2000. Tropical cyclone intensity increase near Australia as a result of climate change. *Journal of Climate* **13**, 3029–3036.

Walsh, K. and Watterson, I.G. 1997. Tropical cyclone-like vortices in a limited area model: comparison with observed climatology. *Journal of Climate* **10**, 2240–2259.

Ward, S.N. and Asphaug, E. 2000. Asteroid impact tsunami: a probabilistic hazard assessment. *Icarus* **145**, 64–78.

Warrick, R.A., Le Provost, C., Meier, M.F., Oerlemans, J. and Woodworth, P.L. 1996. Changes in sea level. In Houghton, J.T., Meira Filho, L.G., Callander, B.A., Harris, N., Kattenberg, A. and Maskell, K. (eds) *Climate Change 1995: The Science of Climate Change*. Cambridge University Press, Cambridge, 359–405.

WASA Group 1998. Changing waves and storms in the Northeast Atlantic? *Bulletin of the American Meteorological Society* **79**, 741–760.

Watterson, I.G., Evans, J.L. and Ryan, B.F. 1995. Seasonal and Interannual variability of tropical cyclogenesis: diagnostics from large-scale fields. *Journal of Climate* **8**, 3052–3066.

Watts, A.B. and Masson, D.G. 1995. A giant landslide on the north flank of Tenerife, Canary Islands. *Journal of Geophysical Research* **100**, 24487–24498.

Weaver, P.P.E. and Kuijpers, A. 1983. Climate control of turbidite deposition on the Madeira Abyssal Plain. *Nature* **306**, 360–366.

Weaver, P.P.E., Masson, D.G. and Kidd, R.B. 1994. Slumps, slides, and turbidity currents: sea level change and sedimentation in the Canary Basin. *Geoscientist* **4**, 14–16.

Weissman, P.R. 1990. The cometary impactor flux at the Earth. In Sharpton, V.L. and Ward, P.D. (eds) *Global catastrophes in Earth history*. Geological Society of America Special Paper **247**, 171–180.

White, R. and Etkin, D. 1997. Climate change, extreme events and the Canadian insurance industry. *Natural Hazards* **16**, 135–163.

Wigley, T.M.L. 1985. Climatology: impact of extreme events. *Nature* **316**, 106–107.

Wigley, T.M.L. 1988. The effect of changing climate on the frequency of absolute extreme events. *Climate Monitor* **17**, 44–55.

Wigley, T.M.L. 1991. Climate variability on the 10–100 year time scale: observations and possible causes. In Bradley, R.S. (ed.) *Global Changes in the Past*. University Corporation for Atmospheric Research, Boulder, CO, 81–101.

Wolbach, W.S., Gilmour, I., Anders, E., Orth, C.J. and Brooks, R.R. 1988. Global fire at the Cretaceous–Tertiary boundary. *Nature* **334**, 665–669.

Wolter, K., Dole, R.M. and Smith, C.A. 1999. Short-term climate extremes over the continental United States and ENSO: I. Seasonal temperatures. *Journal of Climate* **12**, 3255–3272.

Wright, H.E. Jr 1993. Environmental determinism in near eastern prehistory. *Current Anthropology* **34**, 458–469.

Wu, P. and Johnston, P. 2000. Can deglaciation trigger earthquakes in N. America? *Geophysical Research Letters* **27**, 1323–1326.

Yang, Q., Mayewski, P.A., Zielinski, G.A. and Twickler, M. 1996. Depletion of atmospheric nitrate and chloride as a consequence of the Toba eruption. *Geophysical Research Letters* **23**, 2513–2516.

Yang, W.B. and Ahrens, T.J. 1998. Shock vaporization of anhydrite and global effects of the K/T bolide. *Earth and Planetary Science Letters* **156**, 125–140.

Young, R.W. and Bryant, E.A. 1992. Catastrophic wave erosion on the southeastern coast of Australia: impact of Lanai tsunami c 105 ka? *Geology* **20**, 199–202.

Zahnle, K. 1990. Atmospheric chemistry by large impacts. In Sharpton, V.L. and Ward, P.D. (eds) *Global catastrophes in Earth history*. Geological Society of America Special Paper 247, 271–288.

Zielinski, G.A., Mayeswki, P.A., Meeker, L.D., Whitlow, S. and Twickler, M.S. 1996. Potential atmospheric impact of the Toba mega-eruption ~71,000 years ago. *Geophysical Research Letters* **23**, 837–840.

Zielinski, G.A., Mayewski, P.A., Meeker, L.D. *et al.* 1997. Volcanic aerosol records and tephrochronology of the Summit, Greenland, ice cores. *Journal of Geophysical Research* **102**, 26625–26640.

Zuo, Z. and Oerlemans, J. 1997. Contribution of glacier melt to sea-level rise since AD 1865: a regionally differentiated calculation. *Climate Dynamics* **13**, 835–845.

Zwiers, F.W. and Kharin, V.V. 1998. Changes in the extremes of the climate simulated by CCC GCM2 under CO_2 doubling. *Journal of Climate* **11**, 2200–2222.

INDEX